THE GREAT
BOOK OF
VEGETABLES

Antonella Palazzi

THE GREAT BOOK OF VEGETABLES

SIMON & SCHUSTER

NEW YORK LONDON TORONTO SYDNEY TOKYO SINGAPORE

Photographs

Jacket: F. Pizzochero
ADNA: 294–295, 297 (A. Brusaferri)
Mondadori Archives: 41 (ADNA); 48, 159, 208
  (R. Marcialis); 42 (Visual Food); 103, 104, 117, 160, 207
L. Cretti: 99c, 100c
M. Lodi: 248, 266–267, 269, 270–271, 273, 274–275, 277, 278–279,
  281, 282–283, 285, 286–287, 289, 290–291, 293, 298–299, 301, 302–
  303, 305, 306–307, 309, 310–311, 313
F. Marcarini: 1, 2, 10, 11, 12, 13, 14, 15, 16, 17, 18, 19, 20, 21, 22, 23,
  24, 25, 26, 27, 28, 29, 30, 31, 32, 33, 34, 43, 44, 45, 46, 99, 100, 101,
  102, 155, 156, 157, 158, 203, 204, 205, 206, 243, 244, 245, 246
R. Marcialis: 45ar, 46ar, 99bl, 99c, 100ar
G. Pisacane: 153, 201, 202, 202, 207, 241, 242, 247
F. Pizzochero: 97

Home economist
A. Avallone: (jacket) and pp. 294–295, 297
C. Dumas: 248, 266–267, 269, 270–271, 273, 274–275, 277, 278–279,
  281, 282–283, 285, 286–287, 289, 290–291, 293, 298–299, 301, 302–
  303, 305, 306–307, 309, 310–311, 313

SIMON & SCHUSTER
Simon & Schuster Building
Rockefeller Center
1230 Avenue of the Americas
New York, NY 10020

Translated by Sara Harris
Printed and bound in Italy by Arnoldo Mondadori Editore, Verona,
Italy

Typeset by Tradespools Ltd, Frome, Somerset, England

10   9   8   7   6   5   4   3   2   1

Library of Congress Cataloging-in-Publication Data

Palazzi, Antonella.
  [Grande libro della verdura. English]
  The great book of vegetables/Antonella Palazzi; [translated by
Sara Harris].
    p.   ca.
  Translation of: Il grande libro delle verdure.
  Includes index.
  ISBN 0-671-79664-X
  1. Cookery (Vegetables)   I. Title.
TX801.P3213   1992
641.6'5—dc20   92-20963
                                                        CIP

·Contents·

7

INTRODUCTION

10

PREPARATION

30

COOKING METHODS

35

CLASSIC SAUCES, DRESSINGS, AND STOCKS

39

STORING AND PRESERVING

40

HINTS AND SHORT CUTS

41

LEAF AND STALK VEGETABLES

97

SHOOTS AND FRUITS

153

BULB AND ROOT VEGETABLES

201

FRESH AND DRIED LEGUMES

241

MUSHROOMS AND TRUFFLES

265

MENUS FOR ENTERTAINING

315

INDEX

To my sister Doretta

The author and the publishers extend their thanks to the following: the chefs of the Ristorante Giannino, Milan, Oliviero Fondriest and Claudio Sfiller; the pâtissier Mauro Marietta; F.lli Abbasciá of Milan, fruiterers; Ditta Alberghiera Medagliani, Milan; Dott. Luciano Cretti.
They further thank: Argenteria Dabbene; Arzberg; Beppe Morone; Ceramiche Guido De Zan; Cristallerie Saint Louis; Farnese; Florindo Besozzi; High Tech; Home Textile Emporium, Lisa Corti; Idee e Cose; Il Coccio e la Tela: India, Cina, Siam ... Qui; La Porcellana Bianca: L'Utile e il Dilettevole: Messulam: Primavera di Cristallo CALP; Ristorante Yar; Rosenthal: Shed; Taitú; Teleria SOGNA-Frette; Teleria Voglia di Casa; Telerie Sogarò; Telerie Souleiado: Villeroy & Boch.

·INTRODUCTION·

During the reign of the English king Henry VIII in the first half of the sixteenth century, vegetables were shunned by court circles and sedulously avoided by the rich and noble as being fit only for vulgar, plebeian tastes, bulk to fill the poor man's belly. Doctors in Tudor England believed that anything growing in the soil or dirt was suspect, unhealthful and probably a source of contamination leading to dread diseases such as the plague. As a result, the king and his courtiers lived on a diet of protein, fat, and carbohydrate, enjoying succulent spit-roasted cuts of meat, choice game, and poultry, followed by cloyingly sugary desserts or honeyed sweetmeats, all washed down with French wines. Henry VIII himself was one of history's most famous victims of these meals: his erratic behavior, increasingly severe fits of depression, agonizingly painful gout, and internal hemorrhaging were symptoms of scurvy, which caused a lingering and extremely unpleasant death in his mid fifties. Presumably, a number of his courtiers must have met with similar, less widely publicized ends. Centuries later, vitamin C deficiency was identified as the guilty party when modern medical knowledge was used for some historical detective work.

Leaving Tudor England for the other side of the world and traveling four centuries forward in time brings us to India where Mahatma Gandhi was campaigning for India's independence: Gandhi was a strict vegetarian, a typical meal for him consisted of a bowl of beans and some milk curds. This diet may well have helped to give him the stamina to complete the Salt March and to carry on the fight for his country's freedom. The etymology of the word vegetarian can help to explain how Gandhi retained his energy and staying power well into old age: the word is derived from the Latin vegetus, healthy, vigorous, hale, and hearty.

Nowadays, nutritionists tell us that people function better on a diet in which vegetables, not meat, predominate. The poets' idea of an earthly Paradise has always been of a garden where the fruits of the earth are there for the picking, where no one would dream of wringing a chicken's neck; but man has long been a hunter as well as a gatherer. Those of us who do not want to give up all animal protein should nevertheless take heed of the increasingly convincing evidence that a plentiful, daily intake of vegetables provides wonderful nourishment free from saturated animal fat. It also ensures that the processes of elimination are carried out efficiently and contribute to general good health. Ideally, we should start every meal with raw vegetables, followed by a selection of cooked vegetables served with legumes or cereals or with animal protein (fish being the most healthful source).

We can now evaluate the results of a particular type of diet and nutrition is fast developing into an exact science. Advances in medicine, together with many socio-economic factors, have ensured that the average life expectancy in developed countries is now far greater than it used to be; these extra years might as well be healthy and happy. Though nutritionists are emphasizing the benefits of following a mainly vegetarian diet, we don't necessarily have to deprive ourselves of all enjoyment of meat, fish or eggs when we have no moral objection to eating them. They can still be a delicious and nutritionally valuable part of a well balanced diet. If the inclusion of an anchovy makes a simple bean soup into a feast, why leave it out?

All we need to do is reverse the classic phrase "meat and potatoes" to "potatoes and (perhaps) meat." Once we learn how to make the most of the soil's bounty, how to prepare or cook the produce to its best advantage, then vegetables can provide an almost limitless variety of taste and texture in our everyday food.

Origins and distribution of vegetables

We know that cereals and vegetables have grown on earth for many millions of years. Man, or his anthropoid ape ancestors, relied on them for nourishment at a very early evolutionary stage, gathering and eating these plants wherever they grew, in their natural habitats. These wild varieties were very different, in both color and shape, to those we know today. As man developed, he became more selective and learned how to cultivate certain particularly valuable plants to help him survive. Vegetables were chosen for their nutritional value and for their taste.

The earliest known cultivated vegetable is the pea, which was grown in Turkey as far back as the sixth century B.C. Three centuries later, in fertile Mesopotamia, beet, onions, garlic, and leeks were being cultivated. The first large-scale, methodical vegetable growers lived in the Middle and Far East. Turnips were prized by the inhabitants of these regions for their high starch content and were first grown by the Chinese, as were cucumbers, radishes, and cabbages; these last were preserved by means of pickling, to be eaten with rice as the staple diet of the poor. Soon vegetable growing spread throughout the Mediterranean basin. The occupying Roman armies introduced leeks, cauliflowers, artichokes, and carrots to Britain. The carrots of ancient times were red, violet or black, rich in antocianine but devoid of carotene, nothing like those we eat today which were developed in Holland hundreds of years later. In the seventh century, Moorish invaders conquered much of Spain and introduced the eggplant and spinach; soon these were being cultivated in southern France as well. During the Middle Ages vegetables were grown extensively throughout Europe. The Dutch were particularly successful
market gardeners, thanks to their fertile alluvial soil.*

In the fifteenth century, when the Spanish conquistadores first started to explore North America, they discovered vegetables they had never seen before: maize, tomatoes, sweet peppers, pumpkins, and the staple food of the Incas, the potato, grown successfully at altitudes above 14,000 feet. A two-way traffic of vegetable plants began from the New World to the Old and vice versa, with crazes for a particular, new vegetable running their course and then falling out of favor. The potato was enthusiastically adopted in England and Ireland from the seventeenth century onward but was banned in France for a long time because it was thought it carried leprosy. Eventually, however, each and every one of the new American vegetables found its place in the cookery of various European cuisines.

The Frenchman Nicholas Appert discovered a safe canning method in the late eighteenth century and was able to put his invention into practice in the early nineteenth century; this was the first step toward the large-scale distribution and sale of vegetables out of their natural seasons. When deep-freezing was developed in the United States in the 1930s, it was only a matter of time before virtually every type of vegetable was available all the year round.

Composition and nutritional value

Most vegetables have a very high water content, a low caloric value, and contain many of the substances vital to man's health: mineral salts, vitamins, organic acids, carbohydrates, fiber, and very low quantities of fat.

The proportion of these elements varies from one vegetable to another: potatoes are very rich in starch, marrow has a very high water content, while carrots, onions, and corn contain a lot of saccharose. Vegetable plants whose "fruits" we eat are usually rich in potassium; leaf and "flower" vegetables are full of iron.

With the exception of legumes, which have a high protein content, vegetables have only traces of protein, varying from 1 per cent to 4 per cent. Vegans (strict vegetarians who eat no animal products at all, including dairy produce) have to ensure that they eat plenty of legumes (beans, peas, lentils) preferably at the same time as cereals which maximize the absorption of protein.

Not all vitamins are present in vegetables: vitamins A and C predominate and, together with mineral salts, these are found mainly in the larger, older, external leaves which are all too often discarded if these leaves are thrown away. A proportion of the vitamin content is also liable to be lost if the vegetable is not used fairly soon after it is harvested. Prolonged soaking in water and certain cooking methods further deplete their nutritional value.

All vegetables contain organic acids which are of critical importance in maintaining the alkaline balance in our bodies and in helping to eliminate toxic substances; they neutralize the acids which we produce as a result of physical or psychological stress and help to counteract the harmful side effects of a diet too high in protein and animal fats.

Fiber, in the form of cellulose and lignin, simply works its way down the digestive tract and is not absorbed into the bloodstream. This does not mean we can do without these substances: by their mere presence they help the body to eliminate potentially carcinogenic elements present in certain foods; they also help to lower the level of glucose and fatty acids in the blood and prevent chronic constipation and its consequence, painful diverticulitis, which frequently accompany an over-refined diet.

Legumes have special properties: unlike other vegetables they have a high energy value being rich in carbohydrate and protein. This is vegetable protein, often referred to as "noble protein," which is much better suited to man's digestive system than animal protein. Vegetables may be called "the poor man's meat" because they are so cheap and plentiful but an increasing number of people who can well afford to pay for the most expensive cuts of meat are opting for this healthy alternative for some, if not all, of their meals.

Note

Vegetables are usually classified according to the botanical families to which they belong: Leguminosae, Cucurbitaceae, Solanaceae, etc., but for the purposes of this cookbook it is more practical to group them according to the parts we eat: leaf and stalk vegetables; shoots and fruits; bulbs and roots, fresh and dried legumes. A separate section is devoted to those rather special products of the soil: fungi or mushrooms and truffles.

All recipes serve 4, except where otherwise indicated.

A few efficient special utensils make vegetable preparation easier. Potato peelers with fixed blade (1) and swivel blade (2); a three-tined fork (3); a potato ricer or masher (4) and the traditional vegetable mill (5); a mandoline or vegetable slicer with adjustable blade (6). A chopping board is indispensable; a half-moon cutter (7), also known as an hachoir or mezzaluna, is extremely useful. Several very sharp knives are essential: a straight-bladed peeling knife (8); a curved blade peeling knife (9); a knife with a serrated edge (10); a straight-bladed vegetable knife (11); a medium-sized filleting knife (12); a large kitchen knife (13). Scoops or melon ballers (14); a pair of good kitchen scissors (15); a brush for cleaning mushrooms (16) and, very much an optional extra, a truffle slicer (17). A fluter (18), a canelle knife (19), and a zester (20). Three bowls will be useful: one to hold the unprepared produce, one for peelings and trimmings, and one for the prepared vegetables (21). A sieve (22) and a colander (23) complete the vegetable cook's basic equipment.

ONION

Use a small peeling knife with a very sharp curved blade or serrated blade to cut off the root end of the onion. Insert the point of the knife ¼–½ in deep (1) and cut out a small cone-shaped section by turning the onion round (2) to get rid of the tough, woody part. If it is to be cooked whole, do not cut out this cone-shaped section but simply make two intersecting cuts across the base: the onion will cook evenly but will not fall apart. Peel the onion (3), removing the skin neatly. To slice: cut lengthwise in half, place cut side down on the chopping board and cut lengthwise into thin slices, holding it with your fingers away from the blade. To chop the onion, hold the long slices together and cut at right angles across them, keeping your fingertips away from the blade (4) or folded under to avoid cutting yourself.

SCALLION, LEEK

Cut off the roots, remove the outer layer and the tough, ragged ends of the green leaves (1).
Except when preparing scallions for crudités and salads, use the white bulb only and slice it thinly into rings. Take great care to wash leeks very thoroughly; they are often grown in sandy soil which lodges between their leaves, frequently far down toward the root. Slit the leek lengthwise (2) from top to middle and bottom to middle and then rinse between the layers under running cold water.

CHIVES

Trim a very small amount off both ends of the chives. Rinse briefly under running cold water, drain and dry with paper towels. Gather into a bunch, making sure all the ends are level and snip very short lengths into a bowl with kitchen scissors.

SHALLOT

Use a small peeling knife to cut off the roots and carefully pull off the outer skin, working from the roots toward the top. To chop quickly, use the same method as described for the onion.

PREPARATION·

GARLIC

To separate the cloves, place the flat of the blade of a large kitchen knife on top of the head of garlic and bring your fist sharply down on the surface of the blade (1). Use a small peeling knife with a curved or straight blade to peel the cloves one by one, pulling the skin away from the base to the top (2). Crush the cloves if necessary by placing the large knife blade flat on top of the garlic clove and giving it the same, sharp blow with your fist as before.

POTATO, CELERIAC, TURNIP

Peel with a small, sharp peeling knife with a curved or straight blade or with your usual potato peeler. Use the point of the knife or peeler to take out every trace of black spots or eyes (1). To avoid burning your hands when peeling very hot potatoes that have been boiled in their skins, spear the potato with a special three-tined fork, if you have one (2), or with a carving fork to hold the potato steady while you remove its skin.

☐ TOMATOES

Remove the stem with the tip of a sharp knife (1). The easiest way to skin tomatoes is to blanch them first: make a cross-shaped incision in the tomato (2) then plunge in boiling water for 10–20 seconds. Drain and place in a bowl of cold water to arrest the cooking process. Drain again and peel off the skin with the tip of a knife. It will come away easily. Quarter the tomatoes and extract the seeds and pulp with the knife or a spoon (3). To prepare tomatoes for stuffing, slice horizontally in half (4) then extract all the pulp. Drain well and spoon in your chosen stuffing.

CARROT, CUCUMBER, DAIKON ROOT, BLACK SALSIFY, PARSNIP

All these vegetables can be prepared in the same way. Cut off the stalk end (1) and the tip, then hold the vegetable firmly in one hand while running the potato peeler from the larger, stalk end to the tip (2) to remove a thin, outer layer all the way round.

When preparing black salsify (also called scorzonera), have a bowl of cold water mixed with freshly pressed lemon juice standing ready; then drop the pieces of vegetable into it as soon as you have peeled them to prevent discoloration.

EGGPLANT

Slice off the stalk and the tip, removing a generous slice to get rid of the tough flesh and bitter skin at each end. The rest of the skin does not have to be removed, so peeling is optional. If you intend to fry the eggplant, you can cut it lengthwise in half and then slice across as thickly or thinly as required. When small cubes or rectangles are more suitable (for casseroles, etc.), cut lengthwise into quarters (1), cut out the central seed-bearing section (2), then cut each quarter lengthwise in half and slice across these pieces.

SWEET BELL PEPPER

Rinse under running cold water, wipe dry, and cut lengthwise in half, making your first incision through the center of the stalk end (1). Remove and discard the stalk, the surrounding tough flesh, the seeds, and whiteish membrane. Slice lengthwise into strips as narrow or wide as required (2) and then cut these into shorter lengths or dice as required. To remove the thin, glossy outer layer of skin, spear the whole vegetable on a long-handled fork and place under a very hot broiler, turning frequently so the skin is evenly charred. The skin can then be easily peeled off (3) without damaging the flesh underneath; after this treatment it is usually easy to pull the stalk section out (4). Cut lengthwise

in half, remove the seeds and white membrane, and cut the flesh into strips if desired. Another easy way of removing the skin is to wrap each pepper tightly in foil and place in the oven, preheated to 400°F for about 30 minutes, turning the foil parcels over halfway through this time. Have some small sheets of newspaper ready. Take the foil parcels out of the oven one at a time, unwrap the pepper and immediately wrap it up in a piece of newspaper. Leave to cool completely. When you unwrap the peppers you will find that the skin has stuck fast to the newspaper, leaving the flesh neatly exposed.

CELERY

Trim off a thin slice from the base where it is dry and discolored, cut off the leaves, and discard. Separate the stalks. Run the potato peeler down the outside of each stalk to get rid of the stringy fibers. Wash well in cold water.

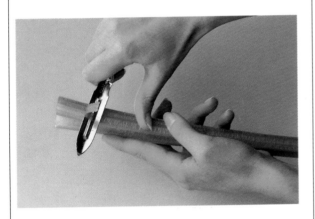

ASPARAGUS

Run the potato peeler over the surface of the paler, tough ends, stripping off the outer layers, and work toward the bottom, away from the tips.
Wash the asparagus thoroughly under running cold water; trim the lower, fatter ends to roughly the same length.

PUMPKIN

Use a large, heavy kitchen knife to cut the pumpkin in half, then into quarters; cut each of these in half. Trim off any remains of stalk; scoop out the seeds and the filaments surrounding them (1). For most recipes you will then have to peel the pumpkin with a medium-sized knife (2) and dice the flesh.

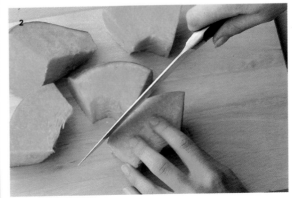

17

☐ ARTICHOKE

Pull off the lower, outermost leaves, which are tough (1). Place the artichokes immediately into a large bowl of cold water acidulated with the juice of 1 lemon or else the cut surfaces will turn black. Cut off the stem and the top of the leaves (2). If you are going to cook the whole, trimmed artichoke, you may choose to remove the hairy choke first; this grows out of the artichoke heart (3). Separate the leaves if you have merely snipped off the tips. Very young artichokes and the smaller, tender varieties can be sliced and quartered after this preparation and all the remaining sections can be eaten. But artichokes such as the Breton variety are best boiled whole and the leaves peeled off one by one as they only have tender flesh at the base of each leaf and at the heart, which is exposed when all the leaves have been pulled off. The choke, looking a little like an immature thistle flower, can then be carefully removed with a teaspoon or curved bladed peeling knife to expose the heart, a small, thick disk slightly hollowed on top. Any remaining tough parts can be pared off. Prepared artichoke hearts are available canned and frozen. Often the stalk of very young artichokes is tender enough to cook and eat but it must first be trimmed and peeled (4).

FENNEL

Slice off the tough, woody base and the top where it divides into stalks with feathery green leaves attached. Remove the outer layers if they are at all dry, tough or withered. Run a potato peeler over the surface, working from the base toward the top, even if they look fresh and tender, to remove any stringy sections on the surface (1). Part the leaves as far as you can and rinse well inside and out under running cold water. Dry and cut lengthwise into quarters. Use a small, sharp knife to trim off the woody section inside nearest the base (2). If you are going to use the fennel raw, to dip into a sauce or dressing, do not cut up into smaller pieces. For salads, cut lengthwise into thin strips.

CAULIFLOWER

Remove the outer, dark green leaves and stalks with a small, sharp knife. Cut off the stalk with a kitchen knife. Scrape away any discolored sections on the surface of the florets with the tip of the small knife. If the cauliflower head is to be cooked whole, cut a deep cross in the base (1) and rinse with cold water before boiling or steaming. If it is to be cooked in florets, simply complete the cross cuts, slicing right through the cauliflower and dividing the head in quarters. Then trim off the solid section of the base (2), releasing the separate tender stems and their florets.

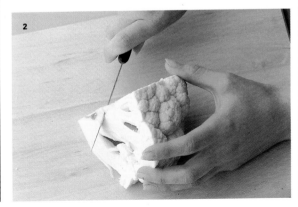

CABBAGE (GREEN, WHITE OR RED), BRUSSELS SPROUTS

Remove and discard the outer, tougher or wilted leaves and slice the cabbage from top to bottom in half, then into quarters (1) with a heavy kitchen knife. Use a small, sharp knife to cut out the hard stem (2) and the largest, hardest ribs. Place the cabbage quarters in a bowl of cold water mixed with 1 tsp vinegar, drain, and cut into pieces of suitable size for your recipe.
For Brussels sprouts it is only necessary to trim off the base; remove the outermost leaves if they are wilted or discolored. Make a cross cut in the base if you wish and rinse.

MUSHROOMS

Brush away any compost, earth or grit with a soft brush (1) if you are using field mushrooms or other wild fungi. Cultivated mushrooms are best not washed before they are cooked; wipe them if you must with a clean, damp cloth or paper towels and just trim off the ends of the stalks, cutting straight across. Wild mushrooms will probably have been twisted and pulled when picked; trim the end of the stalk into a point, just enough to remove the discolored or tough end. Rinse as little as possible or their flavor and texture will suffer. The exceptions are varieties whose undersides are spongy or have deep gills, which need very thorough rinsing in cold water.

SPINACH

If you have bought spinach that is still in little bunches, with all the stems joined at the base, remove the base by cutting the stalks about 1½–2 in above it so that you keep the tender, upper part of the stalk and the leaves (1). Rinse thoroughly. If you are serving the leaves raw as a salad, use only the tender, smaller leaves. If the spinach is to be cooked, drain briefly but do not dry and steam-boil in a saucepan with no added liquid apart from the water left clinging to the leaves (see page 32). If you want to use the base (where most of the vitamins are concentrated), leave the spinach in bunches, trim off a thin layer from the base, and cut into the remainder of the base with 3 intersecting cross-cuts (2).

BELGIAN ENDIVE

Remove the outer leaves if they are wilted or discolored. Use the tip of a small peeling knife to cut out a small cone-shaped piece from the base (1 and 2), leaving a narrow border around the hole. Doing this ensures that the leaves of the endive stay attached. The section removed is bitter and is best discarded.

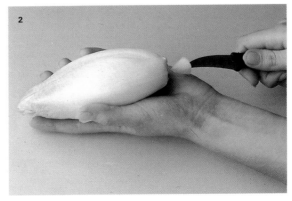

LETTUCE, CHICORY, ESCAROLE

Cut off the remains of the stalk to release all the leaves (1) and discard the outermost leaves and any that are wilted or discolored. Cut the larger leaves in half, slicing through the center rib (2), and leave the smaller leaves whole. Wash all the leaves thoroughly to remove any trace of earth or small insects, etc., then place in a salad spinner to dry thoroughly. If the salad is not to be served immediately, place in an airtight container in the refrigerator to keep it fresh and crisp.

RADICCHIO

Use a small peeling knife to trim the remains of the stalk to a neat point. If the radicchio is to be served raw, there is no need to remove a large part of the base as raw radicchio has a pleasantly bitter taste that adds contrast to salads.

FLAT GREEN BEANS, STICK BEANS, SNOW PEAS, STRING BEANS, FAVA BEANS

Cut almost through the stalk end and pull gently toward the other end. Trim neatly. Some of these vegetables may be too young and fresh to need more than trimming at both ends.

FRESH LEGUMES (PEAS, FAVA BEANS, ETC.)

Hold the full pod in your left hand with the concave side uppermost (1); press down on the seam with your thumb, making the pod split open and run your fingernail down the seam to expose all the peas or beans inside. Run your thumb along the inside of the pod (2), pushing the contents out and away from you into the waiting bowl or sieve. Fava beans can have their inner skins removed as well, either before or after cooking. The latter is easier and quicker: simply take each bean between your index finger and thumb and make a movement as if snapping your fingers gently; the skin will break easily.

HORSERADISH

Cut off a length of root and use a very sharp peeling knife or potato peeler to remove the skin, working from the top to the bottom. Grate finely; most of the flavor is contained in the thick, outer layer of the root while the core or heart is usually discarded. Wear sunglasses or, better still, goggles when grating horseradish otherwise your eyes will smart and water badly. Freshly prepared horseradish is very strong, and will dry out very quickly once grated. As it does not keep well, it is best to grate a large quantity, place it in a jar in good vinegar and store it in the refrigerator, where it will be ready whenever you need it.

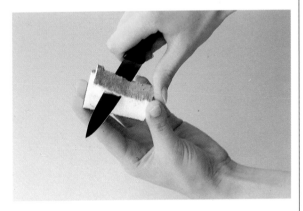

DRIED LEGUMES (DRIED BEANS, PEAS, LENTILS)

Soak dried beans and peas for 12 hours before cooking. Tiny lentils are best not pre-soaked.

SLICING

Cutting vegetables into thin round slices is best done with a narrow-bladed knife, a filleting knife or with a mandoline slicer (see page 10, fig. 6) with an adjustable blade for different thicknesses. If using a knife, hold it so that the blade is at right angles to the chopping board, hold the vegetable firmly with your free hand, keeping your fingertips well out of the way of the knife blade (1), folding them and tucking them under as you near the end of the vegetable.

To cut the vegetable into small, rectangular pieces or julienne strips, cut the vegetable lengthwise in half, place cut side down on the board and slice again lengthwise into even thicknesses. Cut these long pieces into the required lengths (2).

Various names are given to different methods of cutting vegetables to describe size and thickness. Among the most frequently used are *chiffonade* to describe very thin strips or shreds of leaf vegetables such as lettuce, and *julienne* for very thin strips of dense vegetables such as carrots.

To dice vegetables, cut into slices of the required width, then cut across these into strips and finally into small cubes or dice (3–4).

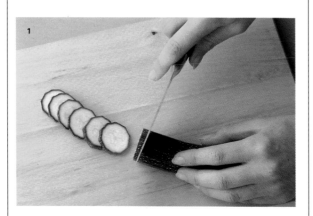

CHOPPING

A traditional utensil, the half-moon cutter (1), also known as *hachoir* or *mezzaluna*, is excellent for chopping. A double-bladed version is available for chopping small quantities of herbs. Professional cooks always use large, heavy kitchen knives for chopping and practice makes them very fast and efficient: hold the handle with your fingertips rather than with the usual cutting grasp and use your free hand to hold on to the blunt edge of the blade, toward the pointed end. Use this end as your pivot and move the handle up and down rapidly to chop, swinging it over the vegetables as you do so to ensure that they are all cut up finely (2).

"TURNING" OR SHAPING VEGETABLES

This term is a literal translation from the French *tourner*, that is trimming or cutting vegetables with a straight-bladed peeling knife so that they look elegant and are of even size.

Using carrots as an example, clean and peel medium-sized carrots as usual, then slice into sections about 2 in long (1); larger specimens can be cut lengthwise in half beforehand or into quarters if very large. Use the peeling knife to trim off the sharp edges at both ends, tapering the piece into an oval shape (2).

Potatoes look extremely attractive when prepared in the same way.

☐ BALLING OR SCOOPING

Hold the vegetable very firmly while you use a melon baller or scoop to cut into the vegetable. Hold the scoop handle tightly while pressing down hard with one edge of its bowl, rotate the scoop as you cut into the raw potato (1), and take out a ball of its flesh; practice will make the balls neater. Scoop out as many balls as you can from each potato. Scoops are available plain, fluted, and with oval bowls to cut out different decorative shapes.

☐ CANNELING OR FLUTING

You can use a cannelle knife which has one hole or a zester which has 5 smaller holes to cut channels down the outer surface of the vegetable. Run the canneling knife or zester down the vegetable from the top towards the stalk end while pressing firmly (1), making parallel channels in the skin and surface flesh. Slice the vegetable as usual: the slices will have attractive scalloped or deckled edges (2). Use for garnishes or salads.

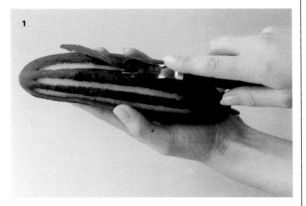

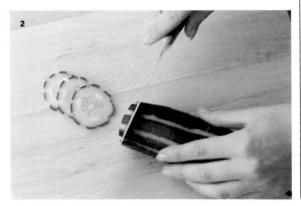

SPECIAL CUTTING METHODS FOR GARNISHES

Presentation plays an important role in serving food: decorative garnishes are pleasing to the eye and elevate any dish onto a more elegant level. All sorts of vegetables can be sculpted and cut into myriad shapes such as lattices, crabs, and butterflies, with color, shape, and texture set off to their best advantage. Japanese and Chinese chefs in particular have turned garnishing into a minor art form.

Even when serving a meal at home, it is worth taking a little extra trouble and time with garnishes that are simple yet effective. It may just be a question of cutting a thin lemon slice in quarters, arranging these so that their pointed sides meet to resemble butterfly wings, then placing two thin strips of sweet red pepper to resemble antennae. Feather the ends of scallions, cutting down into the stalk end toward the bulb in a series of cross cuts, soak the scallions in iced water and they will open out like chrysanthemums or plumes. Snow peas can be turned into small, open-mouthed fishes by simply cutting a triangle out of one end of the flat pod.
When you are really short of time, a few sprigs of fresh basil or parsley will do wonders for the appearance of any dish.

Rosebud garnish
Peel off the skin of a tomato in a thin spiral with one continuous cut (1). Roll up this spiral strip to look like a rosebud (2) and complete the effect by adding a sprig of basil.

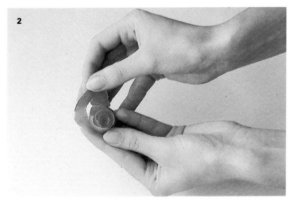

Carrot bundles

Peel the carrots, slice them lengthwise, and then cut into thin strips (1).
Cut the green leaves of a scallion into strips about ⅜ in wide and 5–7 in long.
Blanch the carrots and green strips in boiling water for 30 seconds. Drain. Tie up small bundles of carrot strips with the green scallion threads (2).

Radish flowers

To make a tulip, cut through the skin, tracing out 5 segments, then work the skin away carefully, following the shape you have traced and leaving each segment attached at the bottom (1). To make a rose, slice off the very top of the radish and then cut sideways into 5 slices, leaving these attached to the base (2).

Cucumber fans

Slice a cucumber lengthwise in half. Cut off a number of 1 – 1¼-in sections, depending on how many fan garnishes you wish to make and slice diagonally into 7 flaps (1), leaving these attached at the base (originally the center of the cucumber before it was cut in half). Fold over alternate flaps, tucking them in between the others (2).

Chili pepper flowers

Cut in half, setting aside the pointed end for other uses (1). Use scissors to cut into "petals" that widen toward the stalk end, stopping about ½ in short of the stalk base (2). Rinse under running cold water and remove the seeds. Place in a bowl of iced water and the petals will gradually open out to look like a flower.

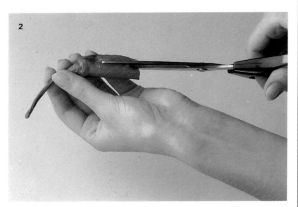

Vegetables lend themselves successfully to a wide range of cooking techniques. For boiling, use a stainless steel saucepan (1); a tall, narrow pot (2) with a removable slotted insert or basket (2) makes cooking asparagus very easy. A stainless steel or tin-lined copper fireproof casserole dish with a lid (3) is invaluable for braising vegetables. For frying you will need a nonstick and/or a cast-iron skillet (4); a wok (5) with stainless steel scoop (6) is an optional extra, ideal for stir-frying. A long-handled sieve (7) or wire mesh ladle makes it easy to remove vegetable quenelles and other preparations from water or deep oil. A steamer (8), with its slotted steaming compartments (9) is a necessity.

For deep-frying (10), use a thermostatically controlled electric deep-fryer with a frying basket for maximum safety; take great care if you use an ordinary, stove-top deep-fryer. Dry-frying is easiest on a nonstick griddle (11); broiling can be done on a separate barbecue-style broiler, with an eye-level model or underneath the broiler set in the roof of the inside of your oven. Shallow stainless steel pans (12) or fireproof gratin dishes (13) are suitable for browning and crisping toppings on finished dishes, while metal soufflé cases (14) or ceramic soufflé dishes or molds (15) will be required in one or more sizes. Choose an earthenware cooking pot for dishes that require lengthy simmering (16).

STEAMING

Place prepared vegetables in a slotted, perforated stainless steel or woven bamboo container, set them over a pan containing boiling water, and let the steam cook them. Since they do not come into contact with the water at all, the vegetables lose very little of their mineral salt and vitamin content, which makes steaming the most healthful and nourishing cooking method.

BOILING AND BLANCHING

Cook the vegetables in plenty of boiling water and then drain them. Vitamins and mineral salts leach out of the vegetables into the water, so it is best not to use this technique too often and to save the water for soups etc. when you do. To blanch vegetables, immerse them in boiling salted water for a few seconds or minutes, depending on the vegetable, then drain them.

STEAM-BOILING AND BRAISING

Place vegetables in a saucepan with a tight-fitting lid and cook them over fairly high, moderate or slow heat, with just enough water to keep them moist and steaming. Some vegetables such as Belgian endive need the addition of a little melted butter for extra flavor. These methods of cooking ensure that none of the taste or nutritional value of the vegetables is lost.

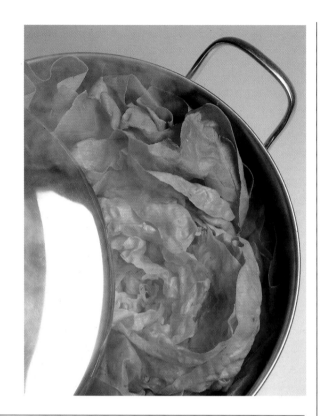

SAUTÉING, STIR-FRYING

Heat oil, clarified butter or fat in a skillet until it is very hot. Then add the vegetables, which are usually cut into small pieces and cook quickly while stirring over high heat. They will remain pleasantly crisp and retain their original color and texture. Chinese cuisine makes much use of this cooking method.

DEEP-FRYING

Heat oil or other types of fat to a temperature of 350°F then deep-fry the prepared vegetables as they are or dip them first in a frying batter. The raw vegetables may be fried in several small batches so as not to lower the temperature of the oil too much; (this should remain constant and should not rise above 360°F). As the vegetables fry, a thin, crisp, golden brown layer forms all over their surface.

BROILING

Using a brush, coat the vegetables with a thin film of oil. Then season them, if you wish, with freshly ground pepper and herbs. You may salt them before or after cooking. Place the vegetables under or over a hot broiler and cook them for a short time first on one side and then turn and cook on the other side. Nonstick dry-frying inserts, frying hot-plates or griddles can be used to dry-fry vegetables without any coating of oil. The heat must be sudden and intense for both methods to make a crisp outer coating, which will act as a temporary seal and prevent the moisture inside from escaping.

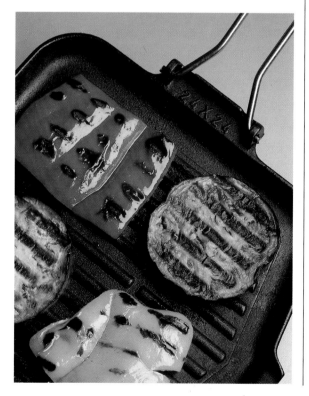

AU GRATIN

This is a finishing process for vegetables, a secondary cooking stage after they have been boiled in salted water or steamed. Place the vegetables in a single layer in an ovenproof dish greased with butter and cover them with a protective coating sauce such as béchamel, or simply dot them with small pats of butter, then sprinkle with grated cheese and/or breadcrumbs. Place in a very hot, preheated oven or under a preheated broiler and leave to cook until the surface is crisp and golden brown.

MICROWAVE

This relatively recent addition to the choice of cooking methods means that vegetables can be cooked without any added water, fat or flavoring: the vegetables cook extremely quickly in their own moisture and this means that little or none of their taste, texture or nutritional value is lost. Using a microwave is a very practical method, since the vegetables can be cooked in the serving dish or on individual plates. You can cover the dish with plastic wrap pierced here and there with a sharp knife to further speed up cooking times and help the food cook evenly.

·CLASSIC SAUCES, DRESSINGS, AND STOCKS·

CITRONETTE

---

3 tbsp lemon juice

⅓ cup oil

salt and white pepper

Stir ½ tsp salt into the lemon juice in a small bowl until it has dissolved. Then gradually beat in the oil, adding a little at a time so that it blends in evenly. Season with freshly ground white pepper.

———•———

VINAIGRETTE

2½ tbsp wine vinegar

½ tsp dried mustard

½ cup olive oil

salt and pepper

Optional ingredients:

¼–½ clove garlic

1 tbsp finely chopped fresh herbs or 1 generous pinch dried herbs

---

Mix the vinegar with the dried mustard and ½ tsp salt in a small bowl, blending thoroughly. Beat continuously as you add the oil a little at a time. Season to taste, adding the extra flavoring if desired.

———•———

MAYONNAISE

---

3 egg yolks

1 tsp Dijon mustard

few drops Worcestershire sauce (optional)

2 cups light olive oil

1 lemon

wine vinegar

salt and pepper

Place the egg yolks in the blender with ½ tsp salt, the mustard, and a few drops of Worcestershire sauce if desired. Blend at high speed, pouring the oil very slowly through the small hole in the lid. After just 1 minute, the mayonnaise should increase in volume and become very thick. Add the lemon juice and a few drops of wine vinegar and blend for a few seconds. Taste and add more salt and lemon juice if needed and freshly ground white pepper to taste.

———•———

VELOUTÉ SAUCE

---

2 cups chicken stock (see page 37)

2 tbsp butter

¼ cup all-purpose flour

salt and white pepper

The above quantities will make a fairly thick, coating sauce to cover cooked vegetables and can be used to brown them in a shallow dish in the oven or under the broiler. Bring the stock to a gentle boil. Melt the butter in a small, heavy-bottomed saucepan over low heat. Add the flour and stir with a wooden spoon over the heat for about 2 minutes; do not allow to brown at all. Remove from the heat and pour in all the hot stock at once, beating energetically with a whisk to ensure that the sauce is smooth and free of lumps. Return to a slightly higher heat and keep stirring with the whisk or with a wooden spoon until the sauce comes to a boil. Turn down the heat and simmer while stirring for a further minute. Take off the heat and stir in salt and freshly ground pepper to taste.

35

BÉCHAMEL SAUCE

2 cups milk

1 carrot

½ celery stalk

1 onion stuck with 2 cloves

1 bay leaf

few peppercorns

2 tbsp butter

¼ cup all-purpose flour

nutmeg

salt and white pepper

This is a useful sauce for coating vegetables, particularly prior to browning in the oven or under the broiler in gratin dishes. Bring the milk to a boil with the carrot, celery, onion, bay leaf, and peppercorns. Remove and discard the vegetables. Melt the butter in a small pan, add the flour, and stir for 1–2 minutes. Gradually add the flavored milk. Bring to a boil, stirring continuously, and cook for 1–2 minutes until thick. Season with nutmeg, salt, and white pepper.

———— • ————

CRÊPES OR PANCAKES

1¼ cups cold milk

1 cup sifted all-purpose flour

4 eggs

4 tbsp melted butter

2 tbsp oil

salt

The above quantities will yield enough batter for 18 crêpes or pancakes, each about 7in in diameter. Pour the milk in the blender; add the same volume of iced water and add the sifted flour with ½ tsp salt. Break in the eggs and add the butter. Cover and blend at high speed for 1 minute. Uncover and scrape off the flour that has stuck to the sides, replace the lid, and blend for another ½ minute. Chill the batter in the refrigerator for at least 6 hours before using. Lightly oil the inside of a nonstick skillet before cooking the first pancake (there is no need to oil again). Heat the skillet over moderately high heat; when very hot pour 3 tbsp of the batter into the center of the skillet and immediately tip this way and that so the batter spreads out to coat the entire bottom of the skillet thinly. Use a nonstick spatula to loosen the edges of the pancake. Turn or toss the pancake when the underside is pale golden brown and cook the other side, then slide out onto a heated plate. The first pancake is usually the least successful. Repeat this process with the remaining batter, stacking the cooked pancakes one on top of the other. Keep to moderate heat and make sure the skillet has time to reheat sufficiently before you pour in the next quantity of batter. The pancakes can be made in advance and kept in the refrigerator, covered with plastic wrap; they also freeze well.

———— • ————

GHEE OR CLARIFIED BUTTER

Melt butter over low heat in a small, heavy-bottomed saucepan and leave over this gentle heat for 10 minutes to allow the white whey or casein to coagulate at the bottom of the saucepan. Remove from the heat. Let it stand for a few minutes and then pour through a sieve lined with a piece of cheesecloth into a storage jar. Seal very tightly and store in the refrigerator or, for shorter periods, at room temperature. Indian recipes frequently call for ghee and it is also used in other cuisines for frying at temperatures at which unclarified butter would burn, since it is the casein content that burns. Ghee is also very useful for greasing oven-proof dishes before cooking foods at high temperatures.

COURT-BOUILLON

3 shallots, peeled and sliced

20 parsley stalks, cut in sections

2 sprigs thyme

1 bay leaf

2¼ cups white wine vinegar

½ cup best virgin olive oil

coarse sea salt and 5 each black and white peppercorns

This *court-bouillon* is particularly suitable for pickling vegetables or for cooking ready for canning in oil. To make a straightforward aromatic cooking stock, however, use just ¼ cup fresh, strained lemon juice or 2 tbsp white wine vinegar for every 1½ cups cold water for the liquid content, otherwise the flavors of the *court-bouillon* will be too sharp.

Place all the listed ingredients in a large saucepan with 1¾ cups cold water (tie the thyme and peppercorns in a piece of cheesecloth). Add 1 tbsp coarse sea salt. Bring slowly to a boil over very gentle heat, add the vegetables and simmer until they are tender but still have a little bite left in them. If they are to be canned, drain and arrange in layers in suitable jars and completely cover with the best cold-pressed virgin olive oil. Seal tightly and keep in a cupboard at room temperature or in the larder.

—— • ——

BEEF STOCK

1½ lb lean stewing or braising beef

1 medium-sized leek

1 small onion

1 medium-sized carrot

1 baby turnip

1 green celery stalk with its leaves

few parsley stalks

1 bay leaf

2 cloves

5 black peppercorns

1 tbsp coarse sea salt

Peel the vegetables and rinse them well. Make a deep cross cut in both ends of the onion and stud it with the cloves. Then place all the ingredients in a very large pot with 2 quarts cold water. Cover and bring to a boil. Skim the surface and simmer for 3 hours.

Remove the beef; strain the stock through a piece of cheesecloth placed in a sieve or through a very fine sieve held over a large bowl. Add a little more salt if necessary and when cold, place in the refrigerator to chill overnight. Take the stock out of the refrigerator and immediately remove the layer of solidified fat from the surface. The beef will not be wasted: eat it hot with a sharp, vinegary sauce or serve it cold the next day.

—— • ——

CHICKEN STOCK

1 chicken carcass

1 leek

2 carrots

1 onion

1 celery stalk

few parsley stalks

1 sprig thyme

few black peppercorns

1 tsp salt

Break up the chicken carcass and place with any leftover trimmings in a large pot with the vegetables and herbs. Add 2 quarts cold water, cover, and bring to a boil. Skim the surface then simmer for about 2 hours. Strain through a fine sieve and cool. Chill, then remove the solidified layer of fat before using.

VEGETABLE STOCK

2 green celery stalks
2 carrots
1 large onion
1 leek
1 potato
few parsley stalks
1 bay leaf
few peppercorns
1 tsp salt

Slice all the vegetables and place with the herbs in a large pot. Pour in 1½ quarts cold water, cover, and bring to a boil. Simmer for 15 minutes, then strain.

FISH STOCK (FUMET)

2–2½ lb fresh fish trimmings, heads, bones, etc.
3 large shallots or 1 medium-sized mild onion
1 small carrot
2 celery stalks
2 sprigs thyme
1 bay leaf
generous 2 cups dry white wine
5 each white and black peppercorns
salt

Rinse the fish trimmings well under running cold water, break or cut up any large pieces, and place in a very large pot with 1½ quarts cold water and ½ tsp salt. Bring slowly to a boil over gentle heat, skimming off any scum as it rises to the surface. Peel the shallots, and cut into thin rings; slice the peeled carrot and the celery finely. Add these and the remaining ingredients to the water and fish and simmer uncovered for 30 minutes.

Rinse a clean cheesecloth in cold water and wring out; use it to line a sieve or colander and slowly pour the fish stock through it into a large bowl to strain. Add more salt if necessary; leave to cool at room temperature and refrigerate in an airtight container.

CRUSTACEAN STOCK

1 large scallion
1 medium-sized carrot
1 small celery stalk
20 parsley stalks
1 small green chili pepper (optional)
2–2¼ lb shrimp heads and shells
2 sprigs thyme
1 bay leaf
small pinch fennel seeds (optional)
¼ tsp mild curry powder (optional)
scant 1 cup dry vermouth or dry white wine
10 white peppercorns
coarse sea salt

Peel and trim the scallion, carrot, and celery where necessary and slice finely; cut the parsley stalks into small pieces; remove the seeds and stalk from the green chili pepper and slice into thin rings. Place all the ingredients in a very large pot together with 1 quart cold water and 1 tsp coarse sea salt. Bring to a boil very slowly over gentle heat and simmer gently, uncovered, for 20 minutes. Strain through a large, fine-mesh sieve.

·STORING AND PRESERVING·

*T*here is nothing to equal freshly picked vegetables, but you can ensure plenty of taste and vitamin content if you buy the vegetables you require as frequently as possible and store for the minimum length of time in the vegetable compartment of your refrigerator. Supermarkets have found the best way of keeping vegetables in moderately good condition for several days by arranging them in single layers in polystyrene trays, covered with plastic wrap. Even when cooked, vegetables should be refrigerated in airtight containers or they will deteriorate, dry out, and become discolored. Deep-frozen vegetables will, of course, keep for months and many will retain most of their taste and texture.

Those of you who have vegetable gardens will find you occasionally have a glut of certain vegetables. These are best deep-frozen to avoid waste or the prospect of serving them with monotonous regularity. You can freeze some vegetables raw (peeled, chopped onion, celery, mushrooms, tomatoes, and peppers) but most must be pre-cooked or blanched for about 5 minutes. You should then transfer them to a sieve or colander and plunge them into iced water to cool quickly. After thoroughly draining and drying the vegetables (a salad spinner is a useful piece of equipment for this purpose), pack them in special foil containers or plastic freezer bags and suck out most of the air with a straw inserted into the bag, the plastic gathered tightly around it with your hand. Label the bag with the contents and freezing date, then freeze. You should eat these vegetables within 8 months, preferably sooner.

Deep-freezing raw vegetables

Mushrooms: for cultivated mushrooms, simply trim off the ends of the stalks and wipe with a damp cloth; for wild mushrooms, brush to get rid of grit, scrape off any discolored sections with a peeling knife, and trim off the root end. Enclose in plastic freezer bags, seal tightly after extracting as much air as possible, label, and date. Use within 3 months.

Tomatoes: choose firm, ripe tomatoes. Wash and dry them thoroughly. Place in a single, well spaced layer on a tray and quick-freeze; transfer them to a plastic freezer bag, seal, label, and date. Use within 8 months of freezing date.

Peppers: wash, dry, cut in half, and remove the stem, seeds, and white membrane inside. Quick freeze and store as you would tomatoes.

Celery: cut off the base and run a swivel-blade potato peeler along the outer surfaces to get rid of any strings; brush or wipe off any traces of soil. Freeze the stalks whole and use only for cooked dishes (casseroles, flavorings for sauces, etc.).

It is better not to thaw vegetables before cooking them but to cook them frozen; if you must thaw them, allow them to thaw inside their unopened packaging until just before using them as they deteriorate rapidly once exposed to the air.

You can place deep-frozen cooked vegetable dishes directly in preheated conventional, convection and microwave ovens and reheat them provided you have used containers that are safe in the freezer and in the oven or microwave.

- Wash vegetables thoroughly but quickly; do not leave them soaking in water for a long time. Cook them in as little water as possible to prevent excessive loss of mineral salts and vitamins.

- When you have used water for boiling vegetables, keep it for making delicious soups or sauces, or add to casseroles, risottos, and similar dishes; vegetable stock is full of precious nutrition and is too good to waste.

- Once trimmed and sliced, the cut surfaces of globe artichokes discolor and darken as oxidization takes place; to prevent this from happening, drop them into a bowl of cold water acidulated with lemon juice immediately after slicing or cutting them.

- The smell of onions and garlic is penetrating and persistent; keep a separate, small chopping board exclusively for these vegetables. The only totally effective method to prevent your eyes from watering while slicing onions or grating horseradish is to wear goggles!

- Never slice or chop your onions until just before you intend to cook them; if chopped and left for a long time, they acquire an unpleasant smell. You can, however, chop them (on their own or mixed with chopped carrot and celery as a ready-to-use mirepoix mixture) and then immediately store them in the freezer, either in an ice container designed for the purpose or in the ice cube tray. Remove the cubes when frozen solid and store in plastic bags. Take out the number of cubes you require when you need them.

- It is always a good idea to keep plenty of chopped parsley in the freezer: store it in a hermetically sealed container and take out as much as you need at a time, breaking and crumbling it off the block and putting the rest back in the freezer.

- To get rid of the bitterness sometimes present in eggplants and cucumbers (although most of today's varieties have been cultivated to minimize this), slice and spread out on a clean cloth or paper towels in a single layer, sprinkle with salt, and leave for 30 minutes to 1 hour. Then cover with another cloth and press down firmly to absorb the liquid. Alternatively, place the slices in a colander, sprinkle each layer with a little salt, and place a plate and a weight on top; dry the slices with paper towels.

- When removing seeds from tomato slices or when hollowing out whole tomatoes for stuffing, keep all the seeds, caps, and pieces of flesh; strain them and use the liquid to flavor sauces and soups.

- Cauliflower and cabbage produce an unpleasant, pervasive smell when cooking due to their sulphur content; to lessen this, add a few drops of vinegar or a cube of stale bread to the cooking water.

- To preserve the bright green color of certain vegetables, such as string beans, and to prevent them from turning a darker, dull color when done, cook them until just tender but still crisp in an uncovered pot containing plenty of lightly salted water. Many chefs freshen crisp vegetables by running under cold water and then steam to reheat them. This is effective, and also practical when serving large numbers but it depletes many of the mineral salts and vitamins.

·LEAF AND STALK VEGETABLES·

CELERY AND GREEN APPLE JUICE

p. 49

Preparation: 15 minutes
Difficulty: very easy
Appetizer

HAM AND CRESS SANDWICHES

p. 49

Preparation: 15 minutes
Difficulty: very easy
Snack

INDIAN BREAD WITH CABBAGE STUFFING

Paratha

p. 49

Preparation: 25 minutes
Cooking time: 3–4 minutes
Difficulty: easy
Snack

SALAD APPETIZER WITH POMEGRANATE SEEDS AND FOIE GRAS

p. 50

Preparation: 25 minutes
Difficulty: very easy
Appetizer

ITALIAN EASTER PIE

p. 50

Preparation: 1½ hours
Cooking time: 30 minutes
Difficulty: fairly easy
Snack or appetizer

STUFFED PE T'SAI ROLLS

p. 51

Preparation: 30 minutes
Cooking time: 20 minutes
Difficulty: fairly easy
Snack or appetizer

CELERY CANAPÉS WITH ROQUEFORT STUFFING

p. 52

Preparation: 20 minutes
Difficulty: very easy
Appetizer

CAESAR SALAD WITH QUAIL'S EGGS

p. 52

Preparation: 20 minutes
Difficulty: very easy
Appetizer

SALADE GOURMANDE WITH SHERRY VINAIGRETTE

p. 52

Preparation: 20 minutes
Difficulty: very easy
Appetizer

AIDA SALAD

p. 53

Preparation: 40 minutes
Difficulty: easy
Appetizer

ENDIVE AND SCAMPI

p. 53

Preparation: 45 minutes
Difficulty: easy
Appetizer

TURKISH STUFFED VINE LEAVES

p. 53

Preparation: 1 hour
Cooking time: 1 hour
Difficulty: fairly easy
Appetizer

ENDIVE, FENNEL, AND PEAR SALAD

p. 54

Preparation: 20 minutes
Difficulty: very easy
Appetizer

GREEN SALAD WITH ROQUEFORT DRESSING

p. 54

Preparation: 10 minutes
Difficulty: very easy
Appetizer

FENNEL, ORANGE, AND WALNUT SALAD

p. 55

Preparation: 20 minutes
Difficulty: easy
Appetizer

Pasta with turnip tops

Red Chioggia endive
(radicchio)
Cichorium intybus

Red Treviso endive (radicchio)
Cichorium intybus

Cress
Lepidium sativum

Friar's beard chicory
Salsola soda

Dandelion
Taraxacum vulgare

Arugula
Eruca sativa

Fennel
Foeniculum vulgare

Cardoon
Cynara cardunculus

Borage
Borago officinalis

Watercress
Nasturtium officinale

Vine leaves
Vitis vinifera

Crisp head lettuce
Lactuca sativa

Escarole
Cichorium endivia

Belgian endive
Cichorium intybus var. foliosum

Wild chicory
Cichorium intybus

Chicory
Cichorium endivia var. crispa

Romaine lettuce
Lactuca sativa

Wild lamb's lettuce or corn salad
Valerianella locusta

Boston lettuce
Lactuca sativa

*Cabbage and
ham rolls*

HERB AND RICE SOUP WITH PESTO

p. 61

*Preparation: 35 minutes
Difficulty: very easy
First course*

CREAM OF CAULIFLOWER SOUP

p. 61

*Preparation: 15 minutes
Cooking time: 30 minutes
Difficulty: very easy
First course*

CURRIED CREAM OF CAULIFLOWER SOUP

p. 62

*Preparation: 15 minutes
Cooking time: 30 minutes
Difficulty: very easy
First course*

PICKLED CABBAGE SOUP

Sauerkrautsuppe

p. 62

*Preparation: 5 minutes
Cooking time: 30 minutes
Difficulty: very easy
First course*

TUSCAN BEAN SOUP

Ribollita

p. 62

*Preparation: 30 minutes
Cooking time: 2 hours 45 minutes
Difficulty: very easy
First course*

CREAM OF CABBAGE SOUP

p. 63

*Preparation: 25 minutes
Cooking time: 50 minutes
Difficulty: very easy
First course*

RISOTTO WITH ENDIVE

p. 64

*Preparation: 15 minutes
Cooking time: 20 minutes
Difficulty: easy
First course*

NETTLE RISOTTO

p. 64

*Preparation: 15 minutes
Cooking time: 20 minutes
Difficulty: very easy
First course*

ENDIVE SALAD WITH PORT AND RAISINS

p. 55

*Preparation: 15 minutes
Difficulty: very easy
Appetizer*

SPINACH SALAD WITH RAISINS AND PINE NUTS

p. 55

*Preparation: 15 minutes
Difficulty: very easy
Appetizer*

PERSIAN SPINACH SALAD WITH YOGHURT DRESSING

Borani

p. 56

*Preparation: 25 minutes
Difficulty: very easy
Appetizer*

SPINACH JAPANESE STYLE

Horenso no Ohitashi

p. 56

*Preparation: 15 minutes
Cooking time: 2 minutes
Difficulty: very easy
Appetizer*

SPINACH MOLDS

p. 56

*Preparation: 15 minutes
Cooking time: 30 minutes
Difficulty: very easy
Appetizer*

BRUSSELS SPROUTS WITH FRESH CORIANDER AND LIME JUICE

p. 57

*Preparation: 12 minutes
Cooking time: 20 minutes
Difficulty: very easy
Appetizer*

ARUGULA, ARTICHOKE, AND EGG SALAD

p. 57

*Preparation: 15 minutes
Difficulty: very easy
Appetizer*

DANDELION AND SCAMPI SOUP

p. 58

*Preparation: 30 minutes
Cooking time: 5 minutes
Difficulty: easy
First course*

LETTUCE AND PEA SOUP

Endiviensuppe

p. 58

*Preparation: 10 minutes
Cooking time: 30 minutes
Difficulty: very easy
First course*

CHINESE WATERCRESS SOUP

p. 58

*Preparation: 15 minutes
Cooking time: 2 minutes
Difficulty: very easy
First course*

CREAM OF WATERCRESS SOUP

p. 59

*Preparation: 15 minutes
Cooking time: 15 minutes
Difficulty: easy
First course*

RUSSIAN SPINACH AND SORREL SOUP WITH SMETANA

Šči

p. 59

*Preparation: 40 minutes
Cooking time: 40 minutes
Difficulty: easy
First course*

JAPANESE SPINACH AND EGG SOUP

p. 60

*Preparation: 15 minutes
Cooking time: 15 minutes
Difficulty: easy
First course*

SWISS CHARD AND CABBAGE SOUP

p. 60

*Preparation: 15 minutes
Cooking time: 20 minutes
Difficulty: very easy
First course*

NETTLE SOUP

p. 61

*Preparation: 20 minutes
Cooking time: 30 minutes
Difficulty: very easy
First course*

CAULIFLOWER AND RICE AU GRATIN

p. 64

Preparation: 20 minutes
Cooking time: 40 minutes
Difficulty: easy
First course

CAULIFLOWER AND MACARONI PIE

p. 65

Preparation: 15 minutes
Cooking time: 15 minutes
Difficulty: easy
First course

PASTA WITH TURNIP TOPS

p. 65

Preparation: 15 minutes
Cooking time: 30 minutes
Difficulty: very easy
First course

SPINACH NOODLES WITH WALNUT SAUCE

p. 66

Preparation: 1 hour + 1 hour for drying the pasta
Cooking time: 2–3 minutes
Difficulty: fairly easy
First course

PASTA WITH SPINACH AND ANCHOVIES

p. 66

Preparation: 20 minutes
Cooking time: 12 minutes
Difficulty: very easy
First course

SPINACH AND CHEESE DUMPLINGS WITH SAGE BUTTER

p. 67

Preparation: 45 minutes
Cooking time: 15 minutes
Difficulty: easy
First course

PASTA WITH BROCCOLI IN HOT SAUCE

p. 67

Preparation: 25 minutes
Cooking time: 20 minutes
Difficulty: very easy
First course

RIBBON NOODLES WITH ARUGULA, TOMATO, AND CHEESE

p. 68

Preparation: 10 minutes
Cooking time: 3 minutes
Difficulty: very easy
First course

STUFFED LETTUCE ROLLS

p. 68

Preparation: 1 hour
Cooking time: 5 minutes
Difficulty: easy
First course

GREEN GNOOCHI

p. 69

Preparation: 1 hour 15 minutes
Cooking time: 20 minutes
Difficulty: fairly easy
First course

GREEN RAVIOLI WITH WALNUT SAUCE

p. 70

Preparation: 1 hour
Cooking time: 10 minutes
Difficulty: fairly easy
First course

LETTUCE AND BLACK OLIVE PIZZA

p. 70

Preparation: 35 minutes
Cooking time: 15 minutes
Difficulty: easy
First course

APULIAN PIZZA PIE

p. 71

Preparation: 35 minutes
Cooking time: 30 minutes
Difficulty: easy
First course

CHICKEN AND ENDIVE GALETTE

p. 72

Preparation: 1 hour + overnight chilling time
Cooking time: 15 minutes
Difficulty: fairly easy
First course

STUFFED LETTUCE

p. 73

Preparation: 40 minutes
Cooking time: 6 minutes
Difficulty: easy
Main course

BELGIAN ENDIVE AND CHEESE MOLD WITH TOMATO SAUCE

p. 74

Preparation: 30 minutes
Cooking time: 40 minutes
Difficulty: easy
Main course

ESCAROLE WITH BEANS AND OLIVE OIL

p. 75

Preparation: 10 minutes
Cooking time: 2 hours 30 minutes
Difficulty: very easy
Main course

BROILED ENDIVE WITH CHEESE AND CHIVE DUMPLINGS

p. 75

Preparation: 20 minutes
Cooking time: 15 minutes
Difficulty: very easy
Main course

SWISS CHARD AU GRATIN

p. 76

Preparation: 35 minutes
Cooking time: 30 minutes
Difficulty: easy
Main course

FENNEL AND LETTUCE AU GRATIN

p. 76

Preparation: 25 minutes
Cooking time: 20 minutes
Difficulty: easy
Main course

SPINACH, ARTICHOKE, AND FENNEL PIE

p. 77

Preparation: 35 minutes
Cooking time: 25 minutes
Difficulty: easy
Main course

FENNEL, SPINACH, AND BEEF SALAD

p. 78

Preparation: 25 minutes
Difficulty: very easy
Main course

SPINACH SOUFFLÉ

p. 78

Preparation: 45 minutes
Cooking time: 35 minutes
Difficulty: fairly easy
Main course

SPINACH STIR-FRY WITH PORK, EGGS, AND GINGER

p. 79

Preparation: 25 minutes
Cooking time: 5 minutes
Difficulty: easy
Main course

GREEK SPINACH AND FETA CHEESE PIE

Spanakópita

p. 80

Preparation: 20 minutes
Cooking time: 20 minutes
Difficulty: easy
Main course

INDIAN SPICED CABBAGE AND POTATOES

Masala Gobhi

p. 81

Preparation: 20 minutes
Cooking time: 20 minutes
Difficulty: easy
Main course

CABBAGE AND HAM ROLLS

p. 82

Preparation: 50 minutes
Cooking time: 40 minutes
Difficulty: fairly easy
Main course

CAULIFLOWER FU-YUNG

p. 82

Preparation: 25 minutes
Cooking time: 5 minutes
Difficulty: easy
Main course

CAULIFLOWER POLONAIS

p. 83

Preparation: 15 minutes
Cooking time: 20 minutes
Difficulty: very easy
Main course

CAULIFLOWER TIMBALE

p. 83

Preparation: 40 minutes
Cooking time: 35 minutes
Difficulty: easy
Main course

BROCCOLI AND SHRIMP TERRINE

p. 84

Preparation: 1 hour 15 minutes
Cooking time: 40 minutes
Difficulty: fairly easy
Main course

STEAMED BROCCOLI WITH SAUCE MALTAISE

p. 84

Preparation: 30 minutes
Cooking time: 15 minutes
Difficulty: easy
Main course

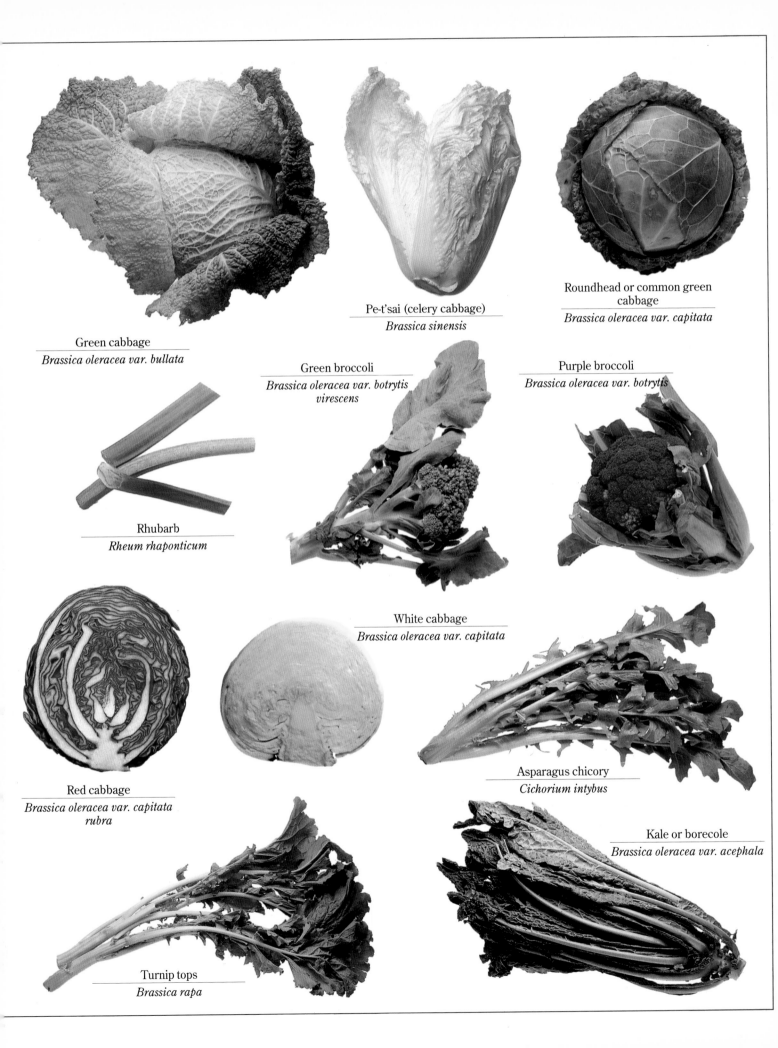

Green cabbage
Brassica oleracea var. bullata

Pe-t'sai (celery cabbage)
Brassica sinensis

Roundhead or common green cabbage
Brassica oleracea var. capitata

Rhubarb
Rheum rhaponticum

Green broccoli
Brassica oleracea var. botrytis virescens

Purple broccoli
Brassica oleracea var. botrytis

White cabbage
Brassica oleracea var. capitata

Red cabbage
Brassica oleracea var. capitata rubra

Asparagus chicory
Cichorium intybus

Kale or borecole
Brassica oleracea var. acephala

Turnip tops
Brassica rapa

BRUSSELS SPROUTS WITH HOLLANDAISE SAUCE

p. 85

Preparation: 30 minutes
Cooking time: 10–12 minutes
Difficulty: fairly easy
Main course

BRAISED LETTUCE

p. 85

Preparation: 15 minutes
Cooking time: 20 minutes
Difficulty: very easy
Accompaniment

STIR-FRIED LETTUCE WITH OYSTER SAUCE

p. 86

Preparation: 10 minutes
Cooking time: 5 minutes
Difficulty: very easy
Accompaniment

FRIED CHICORY ROMAN STYLE

p. 86

Preparation: 10 minutes
Cooking time: 10 minutes
Difficulty: very easy
Accompaniment

ENDIVE WITH LEMON FLEMISH STYLE

p. 86

Preparation: 5 minutes
Cooking time: 20 minutes
Difficulty: very easy
Accompaniment

ESCAROLE WITH OLIVES

p. 87

Preparation: 15 minutes
Cooking time: 1 hour
Difficulty: very easy
Accompaniment

BRAISED ENDIVE

p. 87

Preparation: 10 minutes
Cooking time: 15–20 minutes
Difficulty: very easy
Accompaniment

ASPARAGUS CHICORY AND BEANS APULIAN STYLE

p. 87

Preparation: 15 minutes
Cooking time: 30 minutes
Difficulty: very easy
Accompaniment

BUTTERED SPINACH INDIAN STYLE

p. 88

Preparation: 25 minutes
Cooking time: 15 minutes
Difficulty: very easy
Accompaniment

SPINACH WITH SHALLOTS

p. 88

Preparation: 25 minutes
Cooking time: 5 minutes
Difficulty: very easy
Accompaniment

SPINACH AND RICE INDIAN STYLE

p. 88

Preparation: 25 minutes
Cooking time: 30 minutes
Difficulty: easy
Accompaniment

FRIAR'S BEARD CHICORY WITH LEMON DRESSING

p. 89

Preparation: 10 minutes
Cooking time: 8–10 minutes
Difficulty: very easy
Accompaniment

FRIAR'S BEARD CHICORY WITH GARLIC AND CHILI

p. 89

Preparation: 10 minutes
Cooking time: 5 minutes
Difficulty: very easy
Accompaniment

POACHED TURNIP TOPS

p. 89

Preparation: 15 minutes
Cooking time: 20 minutes
Difficulty: very easy
Accompaniment

CELERY AU GRATIN

p. 90

Preparation: 20 minutes
Cooking time: 20 minutes
Difficulty: very easy
Accompaniment

CELERY JAPANESE STYLE

Selori no Kimpira

p. 90

Preparation: 15 minutes
Cooking time: 3 minutes
Difficulty: very easy
Accompaniment

FENNEL AU GRATIN

p. 90

Preparation: 20 minutes
Cooking time: 25 minutes
Difficulty: very easy
Accompaniment

CURRIED FENNEL

p. 91

Preparation: 15 minutes
Cooking time: 25 minutes
Difficulty: very easy
Accompaniment

CAULIFLOWER AND POTATOES INDIAN STYLE

p. 91

Preparation: 20 minutes
Cooking time: 10 minutes
Difficulty: easy
Accompaniment

CAULIFLOWER SALAD

p. 92

Preparation: 15 minutes
Cooking time: 20 minutes
Difficulty: very easy
Accompaniment

SWEET-SOUR CELERY CABBAGE PEKING STYLE

p. 92

Preparation: 10 minutes
Cooking time: 8 minutes
Difficulty: very easy
Accompaniment

BRAISED RED CABBAGE WITH APPLES

Rotkraut

p. 93

Preparation: 15 minutes
Cooking time: 40 minutes
Difficulty: very easy
Accompaniment

BRUSSELS SPROUTS WITH ROASTED SESAME SEEDS

p. 93

Preparation: 5 minutes
Cooking time: 20 minutes
Difficulty: very easy
Accompaniment

BROCCOLI PURÉE

p. 93

Preparation: 25 minutes
Cooking time: 15 minutes
Difficulty: easy
Accompaniment

RICE WITH GREEN CABBAGE

p. 94

Preparation: 25 minutes
Cooking time: 2 hours
Difficulty: very easy
Accompaniment

CARDOONS AU GRATIN

p. 94

Preparation: 25 minutes
Cooking time: 2 hours
Difficulty: very easy
Accompaniment

SWEET SPINACH PIE

p. 95

Preparation: 35 minutes
Cooking time: 20–25 minutes
Difficulty: easy
Dessert

PERSIAN RHUBARD FRAPPÉ

Sharbate Rivas

p. 95

Preparation: 20 minutes
Cooking time: 30 minutes
Difficulty: very easy
Dessert

RHUBARB CAKE

p. 96

Preparation: 30 minutes + overnight
standing time
Cooking time: 45 minutes
Difficulty: easy
Dessert

Fennel and lettuce au gratin

Spinach
Spinacia oleracea

Beet greens
Beta vulgaris

Green celery
Apium graveolens

Swiss chard
Beta vulgaris var. cycla

Nettle
Urtica urens

White celery
Apium graveolens

Cauliflower
Brassica oleracea var. botrytis

White hearting broccoli
Brassica oleracea var. italica

Brussels sprouts
Brassica oleracea var. gemmifera

CELERY AND GREEN APPLE JUICE

4 crisp green celery stalks

4 green apples

4 cups unsweetened apple juice

Chop the celery. Wash, quarter, and core the apples but do not peel them; cut the quarters into fairly small pieces. Pour the apple juice into the blender, add the celery and apples, and process at high speed until smooth. Unless you have a large-capacity blender or a juicer, you will need to process the ingredients in 2 batches. If the juice is too sharp for your taste, add a generous pinch of sugar. Alternatively, add a small pinch of celery salt and a dash of Tabasco.

————•————

HAM AND WATERCRESS SANDWICHES

1–2 bunches watercress

8 thin slices white or whole wheat bread

butter, softened at room temperature

Dijon mustard

½ lb very thinly sliced ham or cold roast pork

salt and pepper

Rinse the watercress thoroughly, dry, and take all the leaves off the stalks.

Trim off the crusts from the bread. Spread 4 slices generously with butter, the other 4 with a thin spread of mustard. Place sliced ham or pork on the 4 slices spread with mustard and season lightly with salt and freshly ground pepper. Place plenty of watercress on top of the meat, heaping it up in the center. Sprinkle with a pinch of salt and cover with the buttered slices, pressing down firmly.

INDIAN BREAD WITH CABBAGE STUFFING
Paratha

2¼ cups whole wheat flour

½ cup melted ghee (clarified butter, see page 36)

1 cup shredded green cabbage

¾ cup cream cheese

1–2 small green chili peppers, finely chopped

4 tbsp chopped coriander leaves

salt

Sift the flour into a large bowl with 1 tsp salt, stir in scant 2½ tbsp of the ghee (clarified butter), and gradually add ¾–1 cup cold water; the dough should be soft but not too moist to knead easily. Knead until smooth and fairly elastic, shape into a ball, wrap in foil or plastic wrap, and chill for 20 minutes.

Cut out all the large ribs from the cabbage; wash the leaves, dry thoroughly, and shred. Place in a large bowl and sprinkle with 3 tbsp salt. Stir, then leave to stand for 15 minutes. Transfer to a sieve or colander and rinse thoroughly under running cold water, drain, and squeeze out excess moisture. Place in a bowl and mix with the cheese, the chili peppers, coriander, and a little salt. Divide the dough into 10 evenly sized pieces; roll each one into a ball between your palms, keeping the ones you are not working on covered with foil or plastic wrap as they dry out quickly. On a lightly floured board, roll these balls out one at a time into thin disks about 4½–5 in in diameter. Place equal amounts of the cabbage and cheese stuffing on 5 disks, cover with the remaining 5, and seal the edges by pinching together (moisten the edges if necessary). These *parathas* must be fried at once: lightly coat the inside of a large cast-iron skillet with ghee and then heat until very hot; cook the *parathas* for 2 minutes on each side. Keep the fried *parathas* hot in the oven, uncovered, while you finish cooking the rest. Coat the skillet with more ghee whenever necessary. Serve at once.

————•————

SALAD APPETIZER WITH POMEGRANATE SEEDS AND FOIE GRAS

1 ripe pomegranate

selection of small salad leaves to cover 4 plates

4 slices bread

8 butter curls

light olive oil

1 lemon

12 thin slices rare beef tenderloin

4 medium-sized Boston lettuce leaves

4 slices canned truffle-flavored foie gras

salt

Peel the pomegranate and take out the seeds. Cut the bread slices diagonally in half, into two triangles, and toast them lightly. Place the butter curls in a small bowl of iced water

Mix ¼ cup olive oil, 1½ tsp lemon juice, and a pinch of salt and dress the salad with this. Heap an equal quantity of salad in the center of 4 plates, place 3 slices of beef on top, and sprinkle the pomegranate seeds on top.

Place a Boston lettuce leaf beside each of these mounds, and put a slice of *foie gras* and 2 butter curls on each leaf. Lastly, arrange 2 slices of toast to one side of the *foie gras*. Drizzle 1½ tsp olive oil over each portion of pomegranate seeds and beef and serve.

———— • ————

ITALIAN EASTER PIE
Torta Pasqualina

For the pie dough:

generous 2½ cups unbleached white flour

scant ½ cup olive oil

For the filling:

1 lb spinach

1 lb Swiss chard or silver beet

1 medium-sized onion

1½ tbsp butter + 1½ tbsp oil

9 very fresh small eggs

½ cup grated Parmesan cheese

1 small clove garlic, finely chopped

1½ tbsp finely chopped fresh marjoram

generous 1 cup ricotta cheese

extra oil for greasing

salt and pepper

Begin by making the olive oil pie dough: sift the flour with a pinch of salt into a large mixing bowl. Make a well in the center and pour 1 cup boiling water and the oil into it. Gradually stir all this liquid into the flour then knead for about 10 minutes, until the dough is smooth. Divide into 6 pieces of equal size and shape each into a ball; place these dough balls in a single layer in the mixing bowl and cover with a clean, damp cloth. Leave to stand for 1 hour.

While the pie dough is resting, make the filling: remove the stalks, wash the spinach and Swiss chard or beet leaves well, and blanch both types of leaf vegetable in a large pot of boiling, salted water for 5 minutes. Drain, leave to cool, then squeeze out all the moisture and chop finely. Peel the onion and chop it very finely; sweat in the butter and 1½ tbsp olive oil in a covered saucepan over low heat until wilted and tender. Add the chopped spinach and chard leaves and cook, uncovered, for 5 minutes, stirring continuously with a wooden spoon. Remove from the heat and leave to cool.

Break 5 of the eggs into a large mixing bowl and beat lightly, seasoning with a little salt and freshly ground pepper. Stir in the Parmesan cheese, the garlic, marjoram, ricotta, and the spinach mixture. Preheat the oven to 475°F.

Grease a 10-in fairly shallow cake pan with a little olive oil. Have a couple of drinking straws handy to "inflate" the pie shell. Take a dough ball out of the bowl and keep the rest covered. Flatten it with the palms of your hands, carefully pushing and pulling it into a very large, thin circle and use this to line the base of the pan, allowing for an extra ¾ in up the sides. Press the dough against the lightly oiled surface and brush lightly with olive oil.

Repeat this operation with a second dough ball, placing the circle on top of the first one, pressing the spare pie dough against the sides of the cake pan, and brushing its surface with oil. Do the same with a third ball, but do not brush the surface with oil. Spread all the filling in a very thick, even layer on top of the third layer of pie dough. Press the back of a large wooden spoon gently down into the filling to make 4 evenly spaced shallow hollows and break an egg into each one. Season each egg with a little salt and freshly ground pepper. Make another circle out of the fourth dough ball, place it carefully on top of the eggs and filling, and press the margin against the sides of the cake pan just as you did before but this time insert a drinking straw at one point as you do so. Brush the surface lightly with oil. Repeat this process with the last two portions of pie dough, but do not oil the surface of the last, topmost layer. Blow gently down the straw to introduce as much air as possible between the filling and the layered pie lid. Seal the little aperture by pressing it against the inside of the cake pan as you remove the straw. Run a knife between the inside of the pan and the upper edges of the pastry layers, then roll these over, away from the edge of the tin; pinch them or press with the tines of a fork to seal.

Place in the oven and bake for 30 minutes.

———— • ————

STUFFED PE-T'SAI ROLLS

¼ lb skinned, boned chicken breast

¼ lb cooked ham

pinch monosodium glutamate (optional)

1½ tbsp Chinese rice wine or dry sherry

For the sauce:

2 tbsp cornstarch or potato flour

8 large leaves pe-t'sai (celery cabbage)

2 large leaves nori *(black Japanese seaweed sheets)*

salt and white pepper

1 tbsp potato flour or cornstarch

1 cup chicken stock (see page 37)

pinch ground ginger

1½ tbsp Chinese rice wine or dry sherry

salt and white pepper

Chop the chicken and ham finely and transfer to a bowl. Season with a pinch of salt, a small pinch of monosodium glutamate if used, a pinch of pepper, and 1½ tbsp rice wine or sherry; stir well.

Blanch the pe-t'sai (celery cabbage) leaves one at a time for 30 seconds in a large pot of boiling salted water, spreading out each one flat to dry on a cloth or paper towels as you take it out of the water. Cut a V-shaped section out of the bottom of the stalk of each leaf with a small, sharp knife, removing enough to make rolling the leaves easy when it comes to wrapping the stuffing. Pile 4 leaves one on top of the other "head to tail"; do the same with the remaining 4 leaves in a separate pile.

Sprinkle a thin layer of cornstarch or potato flour over the exposed surface of the 2 topmost leaves, spread out a piece of *nori* seaweed on top of both, and then cover each with half the filling spread out in a thick layer. Carefully roll the leaves and filling into 2 long sausage shapes (they will not be very neat, but this does not matter). Transfer these to a lightly oiled heatproof plate and steam over fast boiling water for 20 minutes.

Make the sauce. Mix the cornstarch or potato flour with 2 tbsp cold water and stir into the saucepan containing the chicken stock. Season with a pinch each of salt and pepper, the ginger, rice wine or dry sherry. Bring to a boil, stirring continuously. Simmer for 1–2 minutes.

Cut the pe-t'sai rolls into slices and hand round the sauce separately.

———— • ————

CELERY CANAPÉS WITH ROQUEFORT STUFFING

1 head celery

½ lb Roquefort cheese

½ cup soft cream cheese

salt and pepper

Separate the celery stalks and cut off the parts too narrow to stuff at the ends. Rinse thoroughly, dry, and run a potato peeler over their outer sides to remove any strings. Cut the stalks into sections about 1½ in long. Make the filling: use a fork to break up the Roquefort and work it into a paste; gradually work in the cream cheese until smoothly blended. Season to taste with salt and freshly ground white pepper.

Fill the celery canapés with this mixture. Cover with plastic wrap and refrigerate until required.

———•———

CAESAR SALAD WITH QUAIL'S EGGS

1 Romaine or crisphead lettuce

5 canned anchovy fillets

6 small slices whole wheat bread

best-quality olive oil

24 cooked quail's eggs, shelled

For the anchovy-flavored citronette dressing:

½ clove garlic

3 tbsp lemon juice

½ canned anchovy fillet, very finely chopped

⅓ cup freshly grated Parmesan cheese

½ cup extra-virgin olive oil

freshly ground pepper

Wash the lettuce, drain and dry, then shred or simply tear the leaves into small pieces. Cut the anchovy fillets into very small sections. Make croutons by cutting the bread into squares and frying them in a little olive oil in a nonstick skillet; sprinkle them with a little salt and set aside. Place all the dressing ingredients in a blender and process until smooth. Just before serving, spread out the shredded lettuce on 4 individual plates, space out an equal number of anchovy pieces, quail's eggs, and croutons on top of each bed of lettuce, and sprinkle with the citronette dressing. It is unlikely that you will want to add salt as both the anchovies and the Parmesan add a salty flavor to the salad.

———•———

SALADE GOURMANDE WITH SHERRY VINAIGRETTE

2 small Belgian endives

1 small head radicchio

2 Boston lettuces (heart only)

4 slices canned truffled foie gras, chilled

½ lb cold, cooked French beans

For the sherry vinaigrette:

3 tbsp sherry vinegar

1 tsp Dijon mustard

½ cup light olive oil

salt and pepper

Trim, rinse, and drain all the salad leaves. Dry with paper towels or in a salad spinner. Slice the Belgian endive lengthwise into thin strips; cut or tear the larger leaves of the radicchio and the lettuce hearts into small pieces. Run a butter curler over the chilled *foie gras* very lightly to take off thin, curling strips, rather like butter curls. Mix the salad leaves

with the French beans in a wide salad bowl.
Make the sherry vinaigrette: blend the mustard and
salt with the vinegar in a small bowl using a wooden
spoon; gradually add the olive oil, beating continuously. Stir in freshly ground white pepper. Dress the
salad with this and serve with *foie gras* curls on top.

———— • ————

AIDA SALAD

| 2 heads chicory |
| 2 hard-boiled eggs |
| 2 red peppers |
| 1 melon |
| 2 canned artichoke hearts |
| citronette dressing (see page 35) |

Shred the chicory. Push the hard-boiled eggs
through a sieve into a bowl. Skin the red peppers
(see page 16) and shred. Cut the melon in half,
remove the seeds, and use a melon scoop to scoop
out small balls of flesh. Cut the artichoke hearts into
strips.
Arrange a bed of chicory on each plate, sprinkle
with the artichoke pieces, place the melon balls in
the center, surrounded by the red pepper strips, and
sprinkle with the sieved hard-boiled egg. Hand
round the citronette dressing separately.

———— • ————

ENDIVE AND SCAMPI SALAD

| 14 oz scampi or shrimp |
| 3–4 heads Belgian endive |

| For the walnut citronette dressing: |
| ⅓ cup shelled walnuts |
| ¼ cup lemon juice |
| ½ cup olive oil |
| 1 tsp walnut oil (optional) |
| salt and white pepper |

If you have bought raw scampi or large shrimp,
steam them for 3–4 minutes, then peel them. Prepare the endive, rinse briefly, and dry. Carefully
remove 12 undamaged outer leaves and set aside.
Make the dressing. Add 1 tbsp of the walnuts to boiling water and simmer for 2 minutes, drain, and peel
off the thin inner skin if present. Chop them coarsely and place in the food processor with the lemon
juice, freshly ground pepper, and 1 tsp salt. Process
at high speed for a few seconds, then continue to
process while trickling in the oil. Pour into a small
jug or sauceboat. Chop the remaining walnuts very
coarsely and set aside.
Fan out 3 of the reserved outer leaves on each of the
4 plates, to one side. Shred the remaining endive
leaves and place in the center of each plate; arrange
an equal number of scampi on each bed and sprinkle with some of the dressing. Decorate with the
chopped walnuts and serve at once.

———— • ————

TURKISH STUFFED VINE LEAVES

| 1 small onion |
| 2 tbsp pine nuts |
| ½ cup rice |
| generous pinch ground cinnamon |
| generous pinch ground allspice |
| 2 tbsp seedless white raisins, soaked in luke-warm water |
| 40 small vine leaves, fresh or vacuum-packed |
| olive oil |

2 lemons, cut into wedges

salt

These quantities should yield about 30 stuffed rolls, allowing for a few spoiled ones. Preheat the oven to 300°F. Chop the onion finely. Heat 6 tbsp olive oil in a medium-sized saucepan and fry the onion gently, stirring frequently, until lightly browned, then add the pine nuts and cook over low heat for 2 minutes, stirring continuously.

Add the rice, ½ tsp salt, the cinnamon, and a generous pinch of ground allspice. Drain the seedless white raisins and squeeze out excess moisture, add to the saucepan, and continue cooking and stirring for a further 1 minute.

Add 1 cup boiling water to the rice and other ingredients. Cover the saucepan tightly, turn down the heat to very low, and cook for 15 minutes or until all the water has been absorbed and the rice is done. Stir in a little more salt if needed and leave to cool.

If using fresh vine leaves, wash them and blanch for 30 seconds in a large pot of boiling water. Drain and spread them out carefully to lie flat, veined side downward on clean cloths or paper towels. If you are using vacuum-packed or canned vine leaves, just rinse, drain well and spread out.

Grease a wide, fairly shallow ovenproof dish with oil and spread out 10 of the vine leaves to cover the bottom. Place 1½ tsp of the rice mixture toward the stem end of each leaf and roll up in neat parcels, tucking in the spare uncovered leaf space on either side as you do so to prevent the rice filling from oozing out. Pack snugly into the dish, preferably in a single layer. Sprinkle with 3 tbsp oil and 6 tbsp cold water. Cover the dish with a sheet of foil and bake in the oven for 50 minutes. Serve warm or cold as an appetizer, with lemon wedges for each person to squeeze over them.

---·---

ENDIVE, FENNEL, AND PEAR SALAD

¼ lb arugula

¼ lb chicory

2 heads Belgian endive

2 small bulbs fennel

2 tomatoes

16 shelled walnuts

2 pears

1 lemon

mustard-flavored citronette dressing (see page 78, half quantities)

Trim, wash, and dry all the vegetables; mix the salad greens together in a large bowl. Shred the Belgian endives; repeat this operation with the fennel bulbs, having first removed their outermost layer, and mix them together in another bowl. Slice each of the tomatoes into quarters or 8 wedges. Chop the walnuts very coarsely. Peel the pears, cut them lengthwise into quarters, remove the cores, and slice; place in a bowl and sprinkle with lemon juice to prevent them discoloring.

Arrange all these prepared ingredients attractively in a bowl or in a large fan shape on a serving plate. Sprinkle with the walnuts. Make sure the citronette dressing is well blended and sprinkle all over the salad just before serving.

---·---

GREEN SALAD WITH ROQUEFORT DRESSING

2 Boston lettuces

For the Roquefort dressing:

2½ tbsp wine vinegar

2 oz Roquefort cheese, crumbled into small dice

½ cup light olive oil

3 tbsp light cream

salt and freshly ground pepper

Separate the lettuce leaves. Wash and spin then shake or pat dry. Tear the larger leaves across in half and place on 4 individual plates or in a large bowl. Place the smaller leaves on top. Make the dressing: mix 1 tsp salt with the vinegar in a bowl. Add the crumbled Roquefort and work into the vinegar with a fork until fairly smooth. Mix in 1½ tbsp cold water and some freshly ground pepper. Add the oil in a thin trickle while beating with a balloon whisk. Beat in the cream. Sprinkle the Roquefort dressing over the lettuce and serve at once.

———— • ————

FENNEL, ORANGE, AND WALNUT SALAD

4 bulbs fennel

extra-virgin olive oil

2 large oranges

8 shelled walnuts

salt and white pepper

Remove and discard the outermost layer from the fennel bulbs. Trim, wash, and dry the bulbs and cut them lengthwise into quarters, then cut these pieces into matchstick strips.

Arrange the fennel strips on individual plates in a fan shape; sprinkle with a little salt, freshly ground white pepper, and a little olive oil.

Peel the oranges with a knife, cutting away the inner skin as well as the peel and pith; take the segments out of their thin membranous pockets, remove any seeds, and fan out the segments on top of the fennel. Reserve 4 walnuts for decoration; chop the others

coarsely and sprinkle around the orange segments. Place the reserved walnuts at the pointed, lower end of each fan and serve.

———— • ————

ENDIVE SALAD WITH PORT AND RAISINS

¼ cup seedless white raisins

½ cup port

4 medium-sized heads Belgian endive

6 walnuts

citronette dressing (see page 35)

Soak the seedless white raisins in the port. Rinse the Belgian endive briefly and dry thoroughly. Cut off the tough end then slice lengthwise in quarters; cut these pieces into 1½-in lengths. Chop the walnuts coarsely.

Arrange the salad in individual bowls and sprinkle with the drained port-flavored seedless white raisins. Make the citronette dressing and sprinkle all over the salad. Serve at once.

———— • ————

SPINACH SALAD WITH RAISINS AND PINE NUTS

1–1¼ lb spinach

3 tbsp seedless white raisins

3 tbsp pine nuts

For the dressing:

3 tbsp lemon juice

½ cup olive or sunflower oil

salt and white pepper

Pick out all the smallest, crispest spinach leaves, wash, and dry in a salad spinner. Soak the seedless white raisins briefly in warm water to soften; drain. Arrange the spinach leaves on 4 individual plates and sprinkle seedless white raisins and pine nuts over each portion.

Mix 1 tsp salt with the lemon juice until it has dissolved. Beat in the oil with a small balloon whisk, adding a little at a time. Season with freshly ground pepper, sprinkle over the salad, and serve at once.

———— • ————

PERSIAN SPINACH SALAD WITH YOGHURT DRESSING
Borani

1¼ lb very fresh young spinach
1 medium-sized shallot or ½ small Bermuda onion
1 cup natural yoghurt
3 tbsp fresh lime juice
3 tsp chopped, fresh mint or 1 tsp dried mint
salt and black pepper

Bring a large saucepan of salted water to a boil. Trim and wash the spinach, add the leaves to the saucepan of boiling water, and blanch for 2 minutes. Drain well, leave to cool, and then squeeze out as much moisture as possible from the spinach a handful at a time. Chop very finely and place in a mixing bowl. Slice the peeled shallot or Bermuda onion very finely and mix with the yoghurt in a small bowl, adding 1 tsp salt, the lime juice, and a little freshly ground pepper. Stir this yoghurt dressing into the spinach, cover, and chill for 2 hours in the refrigerator. Turn into a chilled salad bowl just before serving and sprinkle with the mint.

———— • ————

SPINACH JAPANESE STYLE
Horenso no Ohitashi

1¼ lb very fresh spinach
½ cup Japanese soy sauce
1 tsp sugar
pinch monosodium glutamate (optional)
6 tbsp katzuobushi *(Japanese fish scales) or 6 tbsp white sesame seeds*

The Japanese ingredients in this recipe are available from Oriental grocery shops, gourmet markets, and health food shops. Trim the spinach leaves, picking out the small leaves with thin, tender stalks, and wash them thoroughly. You should allow about 1¼ lb for this dish. Blanch in a large saucepan of boiling salted water for a maximum of 2 minutes. Drain very thoroughly and leave to cool. Do not squeeze out the moisture.

Make the sauce: mix the soy sauce in a small saucepan with 3 tbsp water, the sugar, and the monosodium glutamate if used. Heat gently, stirring, until the sugar has completely dissolved. Leave to cool. Arrange the spinach on 4 individual plates, moisten with the soy dressing, and sprinkle with the fish scales or the sesame seeds.

———— • ————

SPINACH MOLDS

1¼ lb blanched spinach leaves
1 shallot or ½ small Bermuda onion
1 tbsp butter
4 large eggs
½ cup heavy cream
½ cup milk
¼ cup grated Parmesan cheese
salt and pepper

These quantities are sufficient to fill four 1-cup capacity or six 1/2-cup capacity ovenproof timbale molds or ramekins. Preheat the oven to 350°F. Drain the spinach and squeeze out all excess moisture. Peel the shallot or onion and chop very finely; sweat gently in the butter until soft, add the spinach and cook, stirring, for 2 minutes. Process to a fine purée in the food processor.

Grease the inside of the molds all over with butter. Heat plenty of water in the kettle. Beat 2 whole eggs and 2 yolks and beat lightly with a pinch of salt and a generous pinch of white pepper. Beat in the cream, milk, freshly grated cheese, and spinach. Pour into the molds.

Place the molds in a roasting pan and pour enough boiling water from the kettle into it to come about two-thirds of the way up their sides. Cook in the oven for 30 minutes. Take out of the oven, take the molds out of the water and allow to stand for a minute. Turn out onto individual heated plates. Serve hot or cold.

———— • ————

BRUSSELS SPROUTS WITH FRESH CORIANDER AND LIME JUICE

1½ lb Brussels sprouts
2 shallots or 1 small onion
1 tbsp butter
½ cup heavy cream
¼ beef or chicken bouillon cube
1 small red chili pepper
1 juicy lime
3 tbsp chopped fresh coriander leaves
salt and black pepper

Heat plenty of salted water in a large saucepan. Trim and wash the Brussels sprouts but do not make the usual cross-cut in their solid, stalk ends. Add to the boiling water and cook until tender but still a little crisp (8–15 minutes, depending on their size). Drain and cut lengthwise in half. Chop the peeled shallots or onion and sweat in the butter over very low heat until soft, using the pot in which the sprouts were cooked. Add the cream, the crumbled ¼ bouillon cube and stir. Leave to simmer gently. Meanwhile, wash the red chili pepper, remove the stalk and seeds, and chop fairly coarsely or cut into very thin rings. Stir into the cream and then add the Brussels sprouts. Sprinkle with the lime juice, stir once more, and cover the pan. Remove from the heat and leave to stand for 5 minutes, then transfer to a heated serving dish and garnish with the chopped coriander leaves.

———— • ————

ARUGULA, ARTICHOKE, AND EGG SALAD

4 very fresh eggs
½ lb arugula or any small-leaved salad greens
½ cup extra-virgin olive oil
juice of 1 lemon
8 cooked or canned artichoke hearts
1 tsp finely chopped scallion or chives + 1½ tbsp olive oil
salt and pepper

Have the eggs at room temperature, place them in a saucepan, and add sufficient cold water to cover them. Heat very slowly to boiling point and as soon as the water reaches a full boil remove from the heat. Take the eggs out and leave to cool; they will be deliciously soft-boiled. If you prefer your eggs almost hard-boiled, boil for 5 minutes after the water has reached boiling point and cool quickly in cold water. Wash the arugula.

Make a citronette dressing (see page 35), gradually beating the olive oil into the mixture of lemon juice,

½ tsp salt, and freshly ground pepper. Drain the artichoke hearts if using canned ones, cut into strips, and mix gently with the citronette dressing. Lift the artichoke strips out of the dressing with a slotted spoon, place in the center of a shallow bowl, and surround with the arugula.

Shell the eggs carefully, cut in half, and place on top of the artichoke strips. Stir the chopped scallion or chives into the extra 1½ tbsp of olive oil with a pinch of salt and drizzle over the eggs.

Sprinkle the remaining citronette dressing over the arugula and serve at once.

———— • ————

DANDELION AND SCAMPI SOUP

½ lb very fresh young dandelion leaves

¾ lb scampi or jumbo shrimp

3 tbsp olive oil

2 cloves garlic

1 sprig fresh thyme

3¼–3½ cups crustacean stock (see page 38)

salt and pepper

Serve with:

garlic croutons (see method)

Heat the stock. Peel the scampi or shrimp if you have not already done so (to provide some of the heads and shells for making the stock) and remove the black vein or intestinal tract that runs down their backs.

Wash the dandelion leaves well and cut into thin strips (chicory may be used instead of dandelion leaves, if preferred). Fry the peeled, minced garlic gently in the oil with the thyme, then add the scampi or shrimp and cook for 1 minute. Add the dandelion leaves and cook, stirring, for a further 2 minutes. Pour in the hot stock and simmer for 1 minute.

Add a little more salt if necessary and white pepper to taste. Serve at once with bread croutons fried in butter with or without a little minced garlic.

LETTUCE AND PEA SOUP
Endiviensuppe (Germany)

1 quart chicken stock (see page 37)

½ lb fresh or frozen peas

2 medium-sized potatoes

1 small onion

1 head escarole

1½ tbsp butter

1½ tbsp oil

pinch pure saffron threads or powder

salt and pepper

Serve with:

toasted slices of French bread

You can use chicory or Belgian endive for this soup if escarole is unavailable. Bring the chicken stock to a boil, add the peas and boil gently until very tender; allow the liquid and peas to cool a little and then process in the blender in several batches or in the food processor. Peel and dice the potatoes; peel the onion and chop very finely. Shred the well washed and dried escarole or chicory leaves. Sweat the onion and lettuce in the butter and oil in a large saucepan until tender; add the potatoes and fry briefly, stirring. Pour in the puréed peas and stock, add salt to taste and the saffron. Simmer for 10 minutes. Season with a little pepper and serve.

———— • ————

CHINESE WATERCRESS SOUP

3½ cups beef stock (see page 37)

2 bunches watercress

2 small scallions

8 oz very thinly sliced beef tenderloin

1½ tbsp all-purpose flour

1½ tbsp light soy sauce

salt and pepper

Heat the stock. Wash and drain the watercress. Chop the scallions.

Place the raw tenderloin in the freezer for 10 minutes to make it easier to slice wafer-thin. Cut into fine strips; place in a bowl and sprinkle with 1 tsp salt and the flour; stir well.

When the stock has come to a boil, stir in the soy sauce, the beef strips, and the watercress. Simmer gently for 1 minute only. Add the scallion pieces and simmer for a further minute, then remove from the heat. Season with a pinch of freshly ground black pepper and serve.

———— • ————

CREAM OF WATERCRESS SOUP

¾ lb watercress
1 quart vegetable stock (see page 38) or chicken stock (see page 37)
3 shallots
2 tbsp butter
6 tbsp all-purpose flour
2 egg yolks
½ cup heavy cream
salt
Serve with:
slices of French bread fried in butter

Wash the watercress. Drain. Take off all the leaves and their little stalks and discard all the thicker stalks. Heat the stock.

Peel the shallots and chop very finely indeed. Fry in the butter for 3 minutes without browning at all. Add the watercress and a pinch of salt. Stir well, cover with a tight-fitting lid, and leave to sweat over gentle heat for 5 minutes, or until the watercress has turned dark and wilted. Sift the flour into the saucepan and stir well over low heat for 1 minute. Take off the heat and use a balloon whisk or wooden spoon to blend in the boiling hot stock, adding this gradually

to avoid any lumps forming.

Return to the heat, cover, and simmer for a further 5 minutes. Turn off the heat. Use a hand-held electric beater to beat the soup until it is very smooth. If you are not going to serve it immediately, stop the preparation at this point and leave the saucepan uncovered. Just before serving the soup, reheat it to boiling point and draw aside from the heat. Beat the egg yolks with the cream in a small bowl and whisk into the soup. Place over very low heat and continue beating for 2 minutes as the soup thickens a little. Do not allow to boil or else the eggs will curdle. Turn off the heat, add a little more salt if wished, stir, and ladle the soup into heated soup bowls.

———— • ————

RUSSIAN SPINACH AND SORREL SOUP WITH SMETANA
Šči

2½ lb spinach
½ lb small sorrel leaves
1 medium-sized carrot, peeled and very finely chopped
1 large Spanish onion, peeled and very finely chopped
3 tbsp finely chopped parsley
2 tbsp butter
¼ cup all-purpose flour
2 cups beef stock (see page 37)
2 bay leaves
4 hard-boiled eggs
salt and black peppercorns
For the smetana:
½ cup whipping cream, stiffly beaten
¼ cup sour cream
1 lemon

Wash the spinach and sorrel. If sorrel is unavailable, you can simply make up the weight with more spinach. Sweat the spinach in a covered saucepan for 2 minutes with a pinch of salt and no added water. Place the spinach in a sieve and refresh by holding the sieve under running cold water; squeeze out all the moisture by hand. Process to a smooth purée in the food processor.

Sweat the finely chopped onion, carrot, and parsley in the butter for 10 minutes over low heat in a large covered saucepan, stirring at frequent intervals. Mix in the flour and cook, stirring continuously, for about 30 seconds. Remove from the heat and pour in all the boiling stock at once; beat vigorously to prevent lumps forming. Stir in the puréed spinach, the bay leaf, and a few whole black peppercorns (tied in a small piece of cheesecloth). Simmer gently for 10 minutes; shred the sorrel leaves, add to the soup, and simmer for a further 10 minutes, stirring occasionally. Draw aside from the heat, remove the peppercorns, and add salt to taste.

Slice the hard-boiled eggs thinly; make the smetana by mixing the stiffly beaten cream, the sour cream, and a little lemon juice to taste. Ladle the soup into individual heated soup bowls; place 1-1/2 tbsp of the smetana in each and float the egg slices on top (one egg per serving). Serve at once.

———— • ————

JAPANESE SPINACH AND EGG SOUP

1 dashinomoto *(cube instant fish bouillon)*
3½ cups boiling water
¼ cup white wine vinegar
1 lb fresh young spinach
peel of ½ lemon
4 very fresh eggs
1 tsp Japanese light soy sauce
salt

Make the fish stock by pouring 3½ cups boiling water from the kettle into a saucepan containing the *dashinomoto* cube. Set 1 quart cold water mixed with

the vinegar to heat in another saucepan. Wash the spinach and add to a large saucepan of boiling salted water to blanch for only 30 seconds; drain in a sieve and immediately refresh by rinsing under running cold water. Shred the lemon rind. When the vinegar water comes to a very gentle boil, break the eggs on to a saucer one at a time and slide them into the water at the point where the water is actually bubbling; once the whites have set, remove them carefully with a slotted spoon or ladle and place each in a heated soup bowl. Divide the blanched spinach into 4 portions and place beside the egg in each bowl. Stir the soy sauce into the boiling hot stock and ladle into the bowls. Sprinkle the strips of lemon rind onto the surface and serve.

———— • ————

SWISS CHARD AND CABBAGE SOUP

5 cups chicken stock (see page 37)
5 oz Swiss chard
¼ lb green cabbage leaves
1 small onion
1 small bunch parsley
1 small bunch fresh basil or small pinch dried basil
3 tbsp extra-virgin olive oil
½ cup rice
1½ tbsp unsalted butter
½ cup finely grated Parmesan cheese
salt and pepper

Heat the chicken stock slowly in a large saucepan to just below boiling point. Wash the Swiss chard well once you have removed the stalks. Wash the cabbage leaves, removing the large, hard ribs, and check their weight after you have prepared them. Shred the leaves. Peel the onion and chop very finely and do likewise with the washed and dried parsley leaves and basil.

Fry the onion gently in the olive oil in a large sauce-pan until tender and pale golden brown. Add the shredded leaves, turn up the heat to fairly high, and cook, stirring, for 2 minutes. Add 1/2 cup of the hot stock, cover, and simmer gently for 5 minutes.

Add the rest of the hot stock, bring to a boil, and sprinkle in the rice. Stir stock, then leave to boil gently for about 10–15 minutes. When the rice is just tender, remove from the heat, stir in the chopped parsley and basil, the butter, and cheese. Season with salt and pepper to taste and serve.

———•———

NETTLE SOUP

¾ lb nettles or spinach

1 medium-sized leek

2 tbsp butter

6 tbsp all-purpose flour

4½ cups chicken stock (see page 37)

½ cup light cream

salt and pepper

Use only the tender tips of the nettles, gathered when they are young shoots. When you have ¾ lb, wash them well, add to a pot of boiling salted water, and blanch for 2 minutes; then drain and refresh under running cold water. Squeeze out as much moisture as possible and chop finely. Trim and wash the leek, and chop finely.

Melt the butter in a large saucepan over low heat and fry the leek gently for 5 minutes. Add the nettles and cook for 2 minutes while stirring. Sift in the flour, cook for 1 minute, stirring, then pour in the hot stock beating with a balloon whisk or stirring vigorously with the spoon to prevent any lumps forming. Turn down the heat as low as possible and simmer gently for 30 minutes.

Allow the soup to cool a little then process to a smooth, creamy texture in the blender before returning to the saucepan. Stir in the cream, add salt and pepper to taste, and serve.

HERB AND RICE SOUP WITH PESTO

approx. ½ lb mixed herbs and wild or cultivated salad leaves (e.g. fresh borage, basil, marjoram, a little fresh sage, Swiss chard, chicory, escarole)

⅓ cup pesto sauce (see page 177)

1 cup rice

3 tbsp extra-virgin olive oil

grated Parmesan cheese

salt

Rinse the borage and the other herbs and the leaf vegetables well, take the leaves off their stalks, and shred all but the smallest. Put them in a saucepan with 5 cups water and 1 tsp salt; cover and barely simmer over low heat for 30 minutes. While they are cooking, make the *pesto* sauce (if fresh basil is un-available, buy the *pesto* ready-made in a jar) and mix it with ¼ cup of the herb stock before adding the rice. Turn up the heat and allow the herb soup to come to a boil; sprinkle in the rice, stir, and boil over moderate heat, uncovered, for 10 minutes. When the rice is just tender, stir in half the *pesto* mixture and 1½ tbsp of the oil and boil for a further 2–3 min-utes. Draw aside from the heat, stir in the rest of the *pesto* and oil, add salt to taste, and serve sprinkled with grated Parmesan cheese.

———•———

CREAM OF CAULIFLOWER SOUP

1 large cauliflower

1 celery stalk

1 medium-sized onion

4 parsley stalks

3½ cups chicken stock (see page 37)

1 cup light cream

1 tsp potato flour or cornstarch

2 egg yolks

⅓ cup grated Swiss cheese (Gruyère)

salt and white pepper

Serve with:

croutons

Trim and wash the cauliflower and cut it into small pieces; you will need about 1¼ lb net weight. Trim and prepare the celery and peel the onion; slice them both thinly.

Rinse the parsley stalks and chop coarsely. Place all these vegetables in the stock in a large saucepan. Cover, bring to a boil, and simmer for 25 minutes. Draw aside from the heat and beat with a hand-held electric beater until smooth, or cool a little, then process briefly in the blender.

Mix the potato flour or cornstarch with ½ cup of the cream and stir into the soup; place over low heat and continue stirring until it boils and thickens slightly. You can prepare the soup up to this point in advance and set aside or refrigerate until shortly before serving.

Just before serving, reheat to just below boiling point. Beat the egg yolks lightly with the remaining cream and the Swiss cheese and then beat a little at a time into the soup with a balloon whisk. Do not allow to boil. Keep beating over gentle heat until the soup thickens, take off the heat, and continue beating for about 30 seconds. Add salt and pepper to taste and serve with tiny bread croutons fried in butter.

———•———

CURRIED CREAM OF CAULIFLOWER SOUP

Use the same ingredients and quantities as for the previous recipe, Cream of Cauliflower Soup, and add:

3 tsp mild curry powder

1 dried chili pepper, seeded (optional)

3 tbsp chopped coriander

garlic

Follow the Cream of Cauliflower Soup method, adding the curry powder and the crumbled chili pepper to the chicken stock. Sprinkle the finished soup with the coriander just before serving. Rub the slices of bread with the cut surfaces of a garlic clove, cut into small squares, and fry in butter in a nonstick skillet. Curry is the perfect flavoring for cauliflower.

———•———

PICKLED CABBAGE SOUP
Sauerkrautsuppe

1 quart chicken stock (see page 37)

1½ tbsp butter

¼ cup all-purpose flour

½ lb canned or bottled Sauerkraut

1 tsp cumin seeds

salt and pepper

Heat the stock. Melt the butter in a large saucepan, stir in the flour, and cook over low heat for 2 minutes. Remove from the heat and pour in the stock all at once, beating energetically with a balloon whisk to prevent lumps forming. Return to the heat, add the Sauerkraut and the cumin, cover, and leave to simmer gently for about 30 minutes. Add salt and pepper to taste and serve very hot.

———•———

TUSCAN BEAN SOUP
Ribollita

3 cloves garlic

1 medium-sized onion

2 leeks, trimmed

1 carrot

1 green celery stalk

2 large canned tomatoes, drained

1¼ lb dried white haricot beans, pre-soaked

1 ham bone or knuckle

1 sprig rosemary

1 sprig thyme

extra-virgin olive oil

¾ lb kohlrabi or cabbage

4 slices 2-day old whole wheat bread

salt and pepper

Peel 2 garlic cloves and the onion and chop them finely; fry gently in ⅛ cup olive oil in a large fireproof casserole dish, preferably earthenware or enameled cast iron. Prepare and clean all the other vegetables; chop the heart of the leek having discarded the tough outer layers, the carrot, the celery, and the tomatoes and add to the casserole dish. Stir, cover with a tight-fitting lid, and sweat over low heat for 15 minutes, stirring at intervals. Add the beans, the ham bone, and enough cold water to completely cover the contents of the casserole. Cover and simmer for 1½ hours. Add salt to taste after 1¼ hours.

Gently fry a peeled, minced garlic clove with the rosemary and thyme in 3 tbsp oil for a few minutes, then pour the oil through a small sieve and set aside. Prepare and wash the cabbage then shred. Remove the ham bone from the casserole dish; transfer about 1½ large ladlefuls of beans to a bowl and reserve. Purée the remaining soup in a food processor, or push through a sieve, and transfer this purée to the casserole dish; stir in the flavored oil and the shredded cabbage. Bring to a boil and then simmer for 1 hour. Add salt to taste, add the reserved whole beans, stir and simmer for a further 5 minutes.

Remove from the heat and add plenty of freshly ground pepper. Place slices of 2-day old or lightly toasted whole wheat bread into heated individual bowls and ladle the soup onto these.

CREAM OF CABBAGE SOUP

1½ lb green cabbage

3½ cups beef, or chicken stock (see page 37)

1 clove garlic

2 sprigs parsley

2 sprigs fresh basil

1 leek

1 medium-sized potato

¼ lb spicy sliceable sausage

¼ cup light cream

2–3 tbsp grated Swiss cheese (Gruyère or Emmenthal)

3 tbsp finely chopped parsley

salt and black peppercorns

Serve with:

croutons

Trim, prepare, and wash the cabbage, cut it lengthwise into quarters, take out the hard core and largest ribs, rinse, and drain well. Shred the leaves and place in a large cooking pot with the stock, the herbs, garlic, and 5 black peppercorns. Trim and wash the leek; slice it into rings; peel the potato and cut into small pieces. Add both to the soup together with ½ tsp salt, cover, bring to a boil, and simmer for 50 minutes.

Remove the skin of the sausage (a highly flavored and rather fatty sausage made with plenty of paprika is best) and dice. Heat a nonstick skillet and when it is very hot, fry the sausage for 2–3 minutes. The sausage should be crisp; drain off and discard all the fat it has released.

When the cabbage is very tender, use a hand-held electric beater to reduce the soup to a creamy consistency (alternatively, allow to cool a little and process in batches in the blender, then return to the saucepan and reheat to just below boiling point). Draw aside from the heat; add salt to taste, stir in the Swiss cheese, cream, parsley, and sausage. Serve sprinkled with croutons.

RISOTTO WITH ENDIVE

*1½ quarts chicken stock (see page 37) or vegetable stock
(see page 38)*

½ lb red Treviso endive or Belgian endive

1 small onion

¼ cup butter

1½ cups risotto rice (e.g. arborio)

½ cup dry white wine

⅓ cup grated Parmesan cheese

salt and pepper

Bring the stock slowly to a boil. Shred the endive. If
you cannot buy the red Treviso endive, use Belgian
endive instead. Peel the onion and chop very finely;
sweat it in 2 tbsp of the butter in a wide, fairly shallow saucepan. Add the endive and cook gently, stirring until tender.

Add the rice, stir for a minute or two, then add the
wine. Pour in about 1 cup of the boiling hot stock
and leave to cook, stirring occasionally; once the
rice has absorbed most of the liquid, add more. Test
the rice; when it is just tender (after about 14 minutes), draw aside from the heat, add the remaining
butter and freshly grated Parmesan cheese. Stir
gently, season, and serve.

———— • ————

NETTLE RISOTTO

1½ quarts chicken stock (see page 37)

¼ lb tender young nettle tops or spinach leaves

3 large shallots or 1 small onion

¼ cup butter

1½ cups risotto rice (e.g. arborio)

½ cup dry white wine

2–3 tbsp grated Parmesan cheese

salt and pepper

Heat the stock in a large saucepan. Wash the nettle
leaves thoroughly; these are best gathered in the
spring or very early summer when they are young
shoots. Blanch for 2 minutes in boiling salted water,
drain and chop finely. Peel the shallots and chop
very finely and cook gently in half the butter in a
large saucepan for a few minutes until tender. Add
the chopped nettles and cook for 2–3 minutes.
Sprinkle in the rice and cook for 2 minutes, stirring.
Pour in the wine.

Cook, uncovered, until all the wine has evaporated,
then add 1 cup boiling hot stock; leave the risotto to
cook, stirring occasionally and adding about ½ cup
boiling stock at a time as the rice absorbs the liquid.
The risotto should be very moist when cooked and
not at all stiff or sticky. After about 14 minutes' cooking time the rice will be tender but still have a little
"bite" left in when tested; take off the heat and stir in
the remaining butter; sprinkle with the Parmesan
cheese. Season and serve.

———— • ————

CAULIFLOWER AND RICE AU GRATIN

1 small cauliflower

1 tsp baking soda

⅔ cup rice

½ cup butter

¾ cup grated Parmesan cheese

3 cups milk

¼ cup all-purpose flour

nutmeg

salt and pepper

Clean the cauliflower, separate all the florets, and
cut the larger pieces of stalk into small pieces. Cook
in a large saucepan of boiling salted water for 10
minutes with the baking soda; drain well.

Place the rice in a fairly small saucepan with 1½ cups

water and 1/2 tsp salt. Once the water has come to a boil, cover the pan and cook over low heat for 15 minutes by which time the rice should be tender and should have absorbed all the water. Stir in 1-1/2 tbsp of the butter and 1/4 cup of the grated Parmesan cheese. Add a little more salt if wished. Preheat the oven to 400°F.

Make a fairly thick béchamel sauce: heat the milk slowly to boiling point; melt 2 tbsp of the butter in a small saucepan, add the flour, and cook, stirring, for 2 minutes over gentle heat. Draw aside from the heat and add all the hot milk at once, beating vigorously with a whisk to prevent lumps forming. Replace the saucepan on the heat and keep beating as it reheats and boils very gently for 2 minutes. Season with a little salt, freshly ground white pepper, and a pinch of nutmeg. Stir in 1/4 cup of the Parmesan cheese.

Grease a shallow, ovenproof dish with 1 tbsp of the remaining butter and spread the rice out in it evenly. Cover with a layer of the cauliflower; combine the two without breaking up the cauliflower too much and completely cover with the béchamel sauce. Sprinkle the remaining Parmesan cheese over the surface and dot with the remaining butter in small flakes. Place in the oven for 15 minutes, after which the surface should be crisp and golden brown. Turn off the oven and leave to stand with the door ajar for 10 minutes before serving.

———— • ————

CAULIFLOWER AND MACARONI PIE

1 lb cauliflower florets

½ cup extra-virgin olive oil

3 large cloves garlic

½ dried chili pepper or pinch cayenne pepper

¾ lb macaroni

freshly grated Parmesan cheese

salt and pepper

Bring a large saucepan three-quarters full of salted water to a boil. Clean the cauliflower and divide into small florets. Rinse these and drain well.

Heat ½ cup oil in a very wide skillet or wok and stir-fry the minced cloves of garlic with the cauliflower florets over high heat for 2 minutes. Reduce the heat to low, add ¼ cup of the boiling salted water heating for the pasta, the seeded and crumbled chili pepper, cover, and simmer for about 10 minutes.

Cook the pasta in the fast boiling salted water until barely tender and still with a good deal of bite to it; drain but reserve 1 cup of the water in a measuring jug. Add half this water and all the pasta to the cauliflower, stir gently, and leave to simmer uncovered until the pasta has finished cooking, adding more of the reserved water when necessary. The finished dish should be very moist. Turn off the heat. Season to taste and sprinkle the grated cheese on top. Serve immediately.

———— • ————

PASTA WITH TURNIP TOPS

1 lb turnip tops or broccoli

1 lb pasta shells

4 cloves garlic

1 small red chili pepper

extra-virgin olive oil

salt and freshly ground pepper

Bring a large saucepan of salted water to a boil ready for the pasta. Trim the larger stems and ribs from the turnip tops and discard, together with any wilted leaves. (Substitute broccoli, if wished.) Wash well, drain, and cut into small pieces. Sprinkle the pasta into the fast boiling water to prevent it coming off the boil and cook for 8 minutes before adding your chosen green vegetable, the peeled, whole garlic cloves, and the whole chili pepper. Cook until the green vegetable is only just tender and still a little crisp (about 8 minutes). Reserve a little of the cooking liquid when draining the contents of the saucepan. Remove and discard the garlic and chili pepper. Transfer to a large, heated serving dish and sprinkle with a little of the reserved liquid, the olive oil, and freshly ground pepper. Serve at once.

———— • ————

SPINACH NOODLES WITH WALNUT SAUCE

¼ lb spinach leaves
4½ cups unbleached white flour
2 eggs
⅔ cup freshly grated Parmesan cheese
salt
For the walnut sauce:
1 large slice 2-day old white bread
½ cup milk
¾ cup shelled walnuts
1 small clove garlic
1½ tbsp fresh marjoram leaves
2 tbsp pine nuts, toasted
¼ cup extra-virgin olive oil
¾ cup heavy cream
salt

Wash the spinach leaves, and cook for a few minutes with no added water in a saucepan until wilted. Cool, then squeeze out excess moisture; chop finely or process to a purée.

Make the dough: make a well in the flour in a large mixing bowl or on a pastry board, break the eggs into it, and add the spinach. Gradually mix these into the flour; knead well; the dough should be smooth and soft. Wrap in plastic wrap and chill in the refrigerator for at least 30 minutes, then roll out into a thin square sheet. Fold the top and bottom edges over to make 3-in "hems" or turnings. Keep turning both sides over until the 2 flattish rolls meet

in the center. Take one folded side and place it on top of the other by making a final fold, placing the two flattish rolls on top of one another. Cut firmly down across this roll with a very sharp, large knife making ⅛-in wide slices, squashing the layers together as little as possible and unraveling each slice once cut. Spread out these very thin ribbon noodles (taglierini) to dry for at least 1 hour. If you prefer, buy ready-made green tagliatelle.

Make the walnut sauce. Soak the bread in the milk, squeeze out excess moisture, and place in the food processor with the walnuts, peeled garlic clove, marjoram, a pinch of salt, and the pine nuts. Process briefly. Turn into a bowl and gradually mix in the oil and cream.

Bring a very large saucepan three-quarters full of salted water to a boil, add the noodles, and keep at a fast boil for 2–3 minutes (longer if using commercially prepared pasta), or until tender with just a little bite to it. Drain and stir in the walnut sauce. Serve at once with the Parmesan cheese.

———— • ————

PASTA WITH SPINACH AND ANCHOVIES

½ lb spinach leaves
4 slices Mortadella
1 cup fairly thick béchamel sauce (see page 36)
1 clove garlic, finely chopped
1 piece fresh ginger, peeled and finely chopped
3 canned tomatoes
¾ lb pasta shapes (e.g. quills or shells)
2 tbsp butter
¼ cup oil
3 small canned anchovy fillets, finely chopped
¼ cup freshly grated Parmesan cheese
½ cup heavy cream
salt and pepper

Bring a saucepan of salted water to a boil and blanch the spinach leaves in it; drain and leave to cool, squeeze out the moisture, and chop finely.

Bring plenty of salted water to a boil in a large saucepan. Meanwhile, chop the Mortadella very finely by hand or in the food processor; stir it into the béchamel sauce which should be of pouring consistency. Chop the garlic and ginger finely together. Dice the tomatoes. Add the pasta to the fast boiling salted water.

While the pasta is cooking, make the sauce: cook the garlic and ginger gently in the butter and oil in a very wide nonstick skillet until they start to release a good aroma; draw aside from the heat and stir in the chopped anchovy fillets, crushing these and working them into the oil with a wooden spoon. Return to a slightly higher heat, add the spinach, and cook for 1 minute while stirring. Stir in the béchamel sauce, the Parmesan, and the cream. Drain the pasta when just tender and add to the sauce; stir over low heat for 1 minute. Add salt and pepper to taste, then stir in the diced cold tomato gently, and serve at once.

SPINACH AND CHEESE DUMPLINGS WITH SAGE BUTTER

2¾ lb spinach
2 eggs
nutmeg
1¼ cups ricotta cheese
⅓ cup grated Pecorino Classico or any semi-hard cheese
1½ tbsp all-purpose flour
¼ cup butter
1 tsp finely chopped sage leaves
salt and freshly ground black pepper

Bring a very large, wide saucepan of salted water to a boil. Trim and wash the spinach; blanch for 1 minute in the boiling water and drain. Reserve the water. Squeeze out all the moisture and chop. Place in a large skillet and cook over low heat for about 2 minutes, stirring to eliminate any remaining moisture. Transfer to a bowl and leave to cool.

Beat 1 whole egg and 1 extra yolk in a mixing bowl with 1 tsp salt, a little pepper, and a pinch of nutmeg. (Beat the spare egg white lightly in a small bowl and set aside). Blend in both types of cheese, followed by the spinach. Sprinkle in the flour and stir very thoroughly, adding salt to taste. If the mixture is very soft, add an extra 1 tbsp flour. Cover the bowl with plastic wrap and chill in the refrigerator for at least 2 hours, preferably overnight.

Bring the reserved blanching water back to a boil. Take the mixture out of the refrigerator and use a tablespoon dipped in cold water repeatedly to make slightly flattened, egg-shaped dumplings. Dip your fingers in the spare egg white to prevent the mixture sticking to your hands. Space out the shaped dumplings on a large plate or board as you prepare them. Melt the butter gently in a small saucepan and add the sage. With the water at a very gentle boil, barely more than a simmer, carefully add the dumplings one at a time. They must have plenty of room; cook in batches if necessary. At first they will sink to the bottom, and after about 3 minutes they will bob up to the surface; as they do so, remove them with a slotted spoon and drain in a colander. Handle them with care as they tend to break up easily. Place an even number of these dumplings in very hot individual plates and sprinkle with the sage butter.

PASTA WITH BROCCOLI IN HOT SAUCE

1 lb broccoli
1–2 chili peppers, dried or fresh
½ cup extra-virgin olive oil

4 garlic cloves, minced
¾ lb pasta shapes (e.g. shells or quills)
salt

Set a large saucepan with plenty of salted water in it to come to a boil. Prepare the broccoli, discarding the thicker part of the stalks and dividing into small florets. Rinse well and drain.

Remove the seeds from the chili pepper(s) if using fresh, and cut into thin rings; if using dried chilis, crumble them. Heat the oil in a large, nonstick skillet and cook the garlic very briefly (do not allow to color). Add the chilis and broccoli and sprinkle with a pinch of salt. Stir-fry over high heat for 1–2 minutes, then turn down the heat to very low, cover, and cook for 10–15 minutes, adding a little water at intervals to prevent the contents catching and burning. The vegetables should be tender but still crisp.

Add the pasta to the saucepan of fast boiling water and cook until just tender. Reserve ½ cup of the cooking water, drain off the rest, and add the pasta to the broccoli, together with some or all of the reserved water. Cook, uncovered, for 1 minute, gently mixing the broccoli into the pasta. Serve immediately on heated plates.

———— • ————

RIBBON NOODLES WITH ARUGULA, TOMATO, AND CHEESE

½ lb fresh taglierini (thin ribbon noodles, see recipe for Spinach Noodles with Walnut Sauce on page 66)
¼ lb arugula
2 medium-sized firm, ripe tomatoes
1 small clove garlic
¼ cup grated Pecorino Romano cheese
½ cup extra-virgin olive oil
salt and black pepper

Buy commercially-prepared thin ribbon noodles if you prefer. Heat plenty of salted water in a large saucepan in which to cook the pasta. Make the sauce: weigh the ¼ lb arugula after removing the stalks, wash well, and dry in a salad spinner. Pour boiling water onto the tomatoes in a bowl, wait 10 seconds, then drain and fill the bowl with cold water; drain again and peel, remove the seeds and any hard stalk sections, and chop the flesh coarsely. Place the arugula leaves in the food processor with the garlic (use half a small skinned clove) and grated cheese; process just enough to chop finely, not to turn into a paste. Turn into a mixing bowl, stir in the oil and the drained, chopped tomatoes. Add a little salt and plenty of freshly ground pepper. Add the pasta to the boiling salted water, stir then leave to cook for 3 minutes if fresh noodles are used; packaged pasta will take longer. Drain when just tender, reserving ½ cup of the liquid. Return the pasta to the saucepan, sprinkle with some of the reserved hot water so that it is very moist, and then stir in the sauce; serve at once.

———— • ————

STUFFED LETTUCE ROLLS

1 quart beef stock (see page 37)
2 large Boston lettuces
1½ cups closed cap mushrooms
3 very thin veal escalopes
butter
1 clove garlic
3 slices lean cooked ham
3 slices prosciutto (Italian cured ham)
few fresh marjoram leaves
⅓ cup freshly grated hard cheese
1 egg yolk
salt and pepper

Serve with:

croutons

Heat the beef stock slowly. Bring a large saucepan of salted water to a boil. Prepare the lettuce, selecting the middle, fairly large leaves. Reserve the outermost, old leaves and the very small ones for use in other dishes.

Spread out 3 clean dry cloths side by side on a work surface. Blanch the lettuce leaves one by one for 30 seconds in the saucepan of boiling salted water, remove each one with a slotted ladle, and spread out flat on the cloths, with the greener, inner side uppermost, taking care not to tear them. Fry the veal escalopes lightly in a little butter with a peeled whole clove of garlic; sprinkle with a little salt and pepper when cooked, chop into fairly small pieces, and set aside. Cut both types of ham into strips, then chop finely. Process the veal, ham, and marjoram leaves briefly in the food processor, just enough to chop further and mix well without turning the mixture into a paste.

Transfer this stuffing to a large mixing bowl, stir in the egg yolk, the grated cheese, a little salt, and plenty of pepper. Stir until the mixture is evenly blended and homogenous.

Place 1½–2½ tsp of the stuffing in the center of each lettuce leaf, fold the rib end of the leaf over the filling, fold over the two sides, and then roll up, forming a neat parcel; press the edge of the leaf gently to seal.

When all the leaves and stuffing have been used up, arrange the parcels in a single layer, with the joins downward, in a shallow fireproof casserole dish; sprinkle with 1 cup of the boiling hot stock, cover, and simmer gently for 5 minutes.

Fry the bread croutons in butter until golden brown and very crisp.

Carefully arrange an equal number of the lettuce parcels side by side in 4 wide, heated, soup dishes; sprinkle some croutons on top of each serving, ladle in the remaining hot stock, and serve at once.

———— • ————

GREEN GNOCCHI

1 lb potatoes

1 lb arugula

½ lb watercress leaves

1 scallion

1 clove garlic

1 large bunch basil

olive oil

2¼ cups all-purpose flour

3 eggs

nutmeg

salt and white pepper

For the vegetable and herb sauce:

1 small clove garlic

1 small carrot

1 small green celery stalk

1 scallion

1 sprig parsley

1 sprig basil

1 bay leaf

1 sprig sage

1 3-in white section of leek

1 small bunch arugula

1 small bunch corn salad or mâche

2 medium-sized ripe tomatoes

extra-virgin olive oil

6 tbsp freshly grated Parmesan cheese

salt and pepper

Steam and mash the potatoes. Blanch the arugula and watercress in boiling salted water for 1 minute; drain, squeeze out all moisture, and chop finely.

Trim and peel the scallion and garlic and chop finely together with 1 tbsp of the fresh basil leaves; fry gently in 3 tbsp olive oil for a few minutes. Add the chopped arugula and watercress, a little salt, pepper, and a pinch of nutmeg. Cook over low heat for a few minutes while stirring, then turn up the heat a little to dry out the mixture. Transfer to the food processor and process to a paste.

Stir the vegetable purée into the potatoes, sift in the flour, and work this in, gradually blending in 1 whole egg and 2 extra yolks, a generous pinch of salt, freshly ground white pepper, and nutmeg. Knead well until very smooth and homogenous; divide the dough up into several even portions and roll each out into a long sausage shape of even thickness, cut into 3/4-in sections and press each section against the tines of a fork. Space out the gnocchi on a clean floured cloth or board.

Make the sauce: Chop the garlic, carrot, celery, scallion, parsley, and basil together finely. Cook this mixture gently in 1/4 cup olive oil in a very large skillet for 10 minutes, stirring frequently. Tie up the bay leaf, sage, and sliced leek in a cheesecloth bag. Wash all the leaves of the green salad vegetables, dry well, and chop finely. Peel the tomato and remove the seeds; chop the flesh.

Add the chopped salad greens and tomatoes to the skillet together with the cheesecloth bag, salt, and pepper; cover and cook for 15 minutes.

Add the gnocchi in several batches to a very large saucepan of boiling salted water; when each batch bobs up to the surface, the gnocchi are cooked. Remove with a slotted ladle and place in a colander then cook the next batch. Place the gnocchi in the hot sauce and stir. Turn off the heat, sprinkle with Parmesan, dot with pieces of butter, and serve. Serves 6.

———— • ————

GREEN RAVIOLI WITH WALNUT SAUCE

1 lb escarole
1½ lb mixture Swiss chard or silver beet or other mild beet leaves and fresh borage
2 eggs

1 small clove garlic
½ cup grated Parmesan cheese
¼ lb ricotta cheese
1¾ cups unbleached white flour
¼ cup dry white wine
walnut sauce (see page 66)
salt

Trim the escarole, Swiss chard, and borage. Trim and wash all the vegetables very well, cook in a covered pot with no added liquid until wilted and tender. Drain and leave to cool then squeeze out all the moisture. Chop finely in the food processor, transfer to a mixing bowl, and combine very thoroughly with the lightly beaten eggs, the garlic, ⅛ cup Parmesan cheese, the ricotta, and salt. Cover with plastic wrap and refrigerate.

Make the pasta dough: make a well in the center of the flour and pour the wine and approximately ¾ cup water into it, adding a generous pinch of salt. Mix, adding a little water if necessary, just enough to make the dough hold together. Knead well until smooth and elastic. Roll out into a very thin sheet on a lightly floured board and cut into rectangles measuring approximately 2¾ × 2 in. Place 1 tbsp of the chilled stuffing in the center of each piece and fold over, pinching the edges to seal well (moisten if necessary).

Cook these large ravioli in the usual way, in a very large pot of fast boiling salted water and drain after about 5 minutes. Cook in 2 batches if necessary. Serve with walnut sauce and a sprinkling of the remaining Parmesan.

———— • ————

LETTUCE AND BLACK OLIVE PIZZA

1 oz compressed yeast
pinch sugar
4½ cups unbleached white flour

light olive oil

2 large heads escarole

2 cloves garlic

1–2 dried chili peppers

15 large black olives, pitted

salt and pepper

Crumble the yeast into ½ cup lukewarm water with a pinch of sugar. Stir gently. Leave to stand for 10 minutes at warm room temperature or until the surface is covered with foam. Sift the flour into a large mixing bowl with 1 tsp salt, make a well in the center, and pour in the yeast mixture and 3 tbsp olive oil.

Use your hand or a wooden spoon to stir the yeast and oil, gradually incorporating the flour. When it is all combined, transfer to a lightly floured pastry board and knead energetically for about 5 minutes or until the dough is very light, smooth, and elastic. Shape into a ball and return to the mixing bowl; place a damp cloth over the bowl and leave to rise in a warm place for 30 minutes or longer, until doubled in bulk.

Prepare the escarole. Heat 2 tbsp oil in a very large saucepan and gently fry the lightly minced garlic cloves for a minute to flavor the oil, then remove them. Crumble the chili pepper into the oil, add the escarole, a pinch of salt, and some freshly ground pepper, and cook over fairly high heat for 2 minutes, stirring continuously. Cover tightly, reduce the heat to very low, and leave to cook without any added liquid for 10 minutes. Take the lid off and add the olives, and cook over a higher heat for a few minutes to allow all the liquid produced by the escarole to evaporate. Season to taste and remove from the heat.

Check the dough after 1 hour and when it has doubled in bulk, preheat the oven to 475°F. Flatten the dough with a lightly floured rolling pin on a floured board and roll out into a 12-in disk.

Press into a well oiled tart pan. Spread the escarole and olive mixture evenly over the pizza dough, sprinkle with a very little olive oil and bake in the oven for 15 minutes or until the exposed dough round the edge is crisp and golden.

APULIAN PIZZA PIE

4½ cups unbleached white flour

1 oz compressed yeast

pinch sugar

2 large heads escarole

¾ cup chicken stock (see page 37)

olive oil

2 cloves garlic

15 firm black olives, pitted

2 tbsp capers

3 tbsp seedless white raisins

butter for greasing

¼ cup pine nuts

salt and pepper

Make the pizza dough as described in the previous recipe and leave it to rise. Make the filling: trim and wash the escarole, discarding the outermost leaves and separating the rest. Cook these in the stock in a large, covered saucepan over low heat for 20 minutes by which time they will be wilted and tender. Heat 3 tbsp olive oil in a large skillet, add the whole, peeled cloves of garlic and cook gently until pale golden brown, then discard the garlic. Turn up the heat a little and add the contents of the saucepan without draining off any liquid. Stir well and cook for 5 minutes. Sprinkle with a pinch of salt, add some freshly ground pepper and the olives, capers, and seedless white raisins. Stir, cover, and simmer over low heat for a further 5 minutes.

Grease a nonstick skillet with a little butter and cook the pine nuts gently, stirring continuously until they are lightly colored. Stir into the escarole mixture and take off the heat; taste for seasoning. Preheat the oven to 360°F when the dough has doubled in bulk. Divide the dough in half and roll out these portions into thin disks on a lightly floured board; roll out one to a diameter of 12 in and use to line a cake pan or pie plate approximately 10 in in diameter and ideally 1½ in deep. Dip a pastry brush in oil and brush a thin film over the surface of the dough lining. Fill with the escarole mixture, leveling out the surface, and drizzle over a few drops of oil. Roll out the second disk to a diameter of about 10-1/2 in, and place over the pie. Press the edges together, folding them over, away from the rim in a narrow fold all round the edge and press with the tines of a fork to make a decorative border. Brush the surface lightly with a little oil and bake for 30 minutes, until golden brown. Serve hot or warm.

— • —

CHICKEN AND ENDIVE GALETTE

For the aromatic chicken stock:

3–3¼-lb free-range chicken

1 leek

2 cloves

1 celery stalk

1 baby turnip

1 carrot

few parsley stalks

1 freshly peeled tomato skin

1 sprig fresh thyme

1 bay leaf

salt and peppercorns

For the pancakes:

see recipe on page 36

For the filling:

meat from the chicken, taken off the bone

1 head radicchio

3 heads red Treviso endive or Belgian endive

1 medium-sized onion

2 tbsp butter

1 cup creamy velouté sauce (see method)

½ cup reduced chicken stock

salt and pepper

For the creamy velouté sauce:

¼ cup butter

½ cup all-purpose flour

4 cups (1 quart) aromatic chicken stock (see above)

½ cup heavy cream or crème fraîche

nutmeg

salt and white pepper

For the topping:

5–6 thin slices Swiss cheese (Gruyère or Emmenthal)

1 cup coarsely grated Parmesan cheese

3 tbsp butter

Use 2 soufflé dishes or similar deep, ovenproof receptacles about 10 in in diameter for this recipe. You will need to cook the chicken, take the flesh off the bone, and make the pancakes the day before you plan to serve this dish.

Place the chicken in a very large saucepan with 5 cups water, a pinch of salt, the well washed and trimmed white part of the leek stuck with the 2 cloves, and all the other ingredients listed above under stock. Cover and bring slowly to a boil; remove the lid, turn up the heat a little, and remove any scum as it rises to the surface. Simmer for about 30 minutes or until the chicken is cooked (test by inserting a pointed knife deep into the thigh; if the juices run clear it is done). Take the flesh off the bone when cool enough to handle. Strain the stock and refrigerate overnight. Refrigerate the chicken meat.

Make the pancake batter following the recipe on page 36, cover, and refrigerate for at least 6 hours before making the pancakes; use more batter for each one than usual as you will need fairly thick pancakes; make them in a skillet of a slightly smaller diameter than the soufflé dishes in which you will be layering them with the filling. This batter should yield ten, eight of which should be good enough to use. Pile them up on top of one another, cover with plastic wrap, and refrigerate overnight.

The next day remove all the fat that will have solidified in a layer on top of the chicken stock. Strain the stock again, this time through a piece of cheesecloth placed in a sieve. Measure out 1 cup, add salt to taste, and heat, reducing to 3/4 cup. Preheat the oven to 400°F. Cut the chicken flesh, without the skin, into thin strips. Trim the radicchio and Treviso endive or Belgian endive, rinse, drain, and shred. Peel the onion and chop very finely; cook gently in 2 tbsp of the butter in a large saucepan until tender but not browned. Add the radicchio and endive leaves, sprinkle with a pinch of salt, and cook, stirring, for 3 or 4 minutes. Pour in 1/2 cup of the reduced stock, cover tightly, and cook over moderate heat for 5 minutes, until the vegetables are wilted and tender; continue cooking briefly, uncovered, to let the remaining liquid evaporate. Draw aside from the heat and stir in the chicken strips.

Make the sauce: melt the butter, stir in the flour, and cook over low heat for 2 minutes; add all the 4 cups of hot stock at once off the heat, beating quickly. Return to the heat and simmer for a few minutes, stirring, until the sauce is very smooth and glossy. Stir in the cream or *crème fraîche* and add a little salt, freshly ground white pepper, and nutmeg to taste; simmer for a further 2 minutes then take off the heat. Add 1 cup of this sauce to the chicken mixture and fold in gently.

Grease 2 soufflé dishes about 10 in in diameter with butter and trim the pancakes down to size if necessary. Cover the bottom of the ovenproof dishes with a thin layer of the remaining sauce; cover this with a pancake, laid flat. Spread a thin layer of sauce on top, sprinkle with a little reduced chicken stock; divide one-third of the chicken and vegetable mixture between the 2 dishes, spreading it out evenly and stopping about 1 in short of the edges. For the next layer, divide a quarter of each cheese between the 2 dishes. Place the second pancake on top; spread another thin layer of sauce, sprinkle with chicken stock, cover with the same quantity as before of the chicken and vegetable filling, a cheese layer, followed by the third pancake. Repeat this sequence one more time; when the last 2 pancakes are in place, spread a generous layer of the remaining sauce on top; use up the remaining quarter of the cheeses for a last, even layer. Place a very small piece of butter in the center of each of the lattice spaces. Place in the oven for 15–20 minutes or until the sauce bubbles and the surface has browned. Turn off the oven, leave to stand for a few minutes with the door ajar, and then serve, cutting down through the layers as if slicing a cake.

———•———

STUFFED LETTUCE

2 or 3 large Boston lettuces
1 large boneless chicken breast, skin removed
2 slices white bread, crusts removed
1 cup milk
¼ cup dried mushrooms, pre-soaked
2 slices cooked lean ham
2 slices Mortadella
1 clove garlic
1½ tsp fresh marjoram leaves
½ cup ricotta cheese
⅓ cup grated Parmesan cheese
3 eggs
1¾ cups crushed tomatoes
olive oil
salt and pepper

Bring plenty of salted water to a boil in a large saucepan. Use the outermost lettuce leaves provided they are undamaged as well as the next couple of layers of large leaves: you will need at least 20 leaves. Save the rest of the lettuce for salads. Rinse the leaves well. Spread 3 or 4 clean cloths flat on the work surface. Blanch the leaves one at a time for 30 seconds, take out with a spatula and spread flat on the cloths with their inner, greener sides facing uppermost, handling them with great care.

Fry the chicken lightly in a little oil; sprinkle with a

73

little salt and pepper when done, slice into thin strips and set aside. Soak the bread in the milk, then squeeze out all excess moisture. Have the dried mushrooms ready soaked in a small bowl of warm water (use dried ceps if possible as these have plenty of flavor); squeeze out all excess moisture by hand when they have reconstituted. Shred the ham and Mortadella and process briefly with the chicken, mushrooms, peeled garlic, and marjoram to chop finely; do not over process into a paste. Transfer to a large mixing bowl and work in the ricotta, Parmesan, the eggs (one at a time) and the moistened bread. Season with salt and pepper.

Place about 2 tsp of this mixture in the middle of each lettuce leaf, adjusting the quantity to the size of the leaf. Fold the rib or stalk end of the leaf over the filling, fold the two sides over each end, and roll up, enclosing the filling neatly and pressing down gently on the joins. Pour two-thirds of the crushed tomatoes into a very wide skillet and season lightly if wished. Place the lettuce rolls in it, in a single layer. Cover with the remaining tomato. Heat gently to reach boiling point and then simmer gently for exactly 6 minutes with the lid on. Take the rolls out of the skillet and place on individual plates with some of the tomato liquid and serve warm rather than very hot.

— • —

BELGIAN ENDIVE AND CHEESE MOLD WITH TOMATO SAUCE

2½ lb Belgian endive
4 slices prosciutto (Italian cured ham) or cooked ham
2 tbsp butter
2 large shallots, finely chopped
2¼ cups milk
1 cup velouté sauce (see page 35)
2 eggs
¼ cup grated Parmesan cheese
2 tbsp grated Swiss cheese (Emmenthal)
1½ tbsp olive oil
fine dry breadcrumbs
salt and white pepper
For the tomato sauce:
2 cloves garlic
1 sprig thyme
2¼ cups crushed tomatoes (passato)
¼ cup dry white wine
pinch sugar
pinch oregano
1½ tbsp olive oil
salt and pepper

Preheat the oven to 350°F; grease a 5-cup soufflé dish lightly with oil and sprinkle in ⅓ cup breadcrumbs, to coat the oiled surfaces evenly; tip out excess.

Bring a large saucepan of salted water to a boil; prepare the Belgian endive (see page 21) and blanch the leaves in the boiling water for 5 minutes. Drain them well. Shred the *prosciutto*. Heat the butter in a very large nonstick skillet and fry the chopped shallots gently for 5 minutes, stirring. Add the *prosciutto* or ham and cook for a few seconds then add the Belgian endive. Continue stirring for 1 minute, then increase the heat a little and gradually stir in the milk. Simmer, uncovered, for 20 minutes, stirring at frequent intervals; the mixture should turn into a thick, creamy mixture. Season with a very little salt and some freshly ground pepper.

Make a thick velouté sauce; allow to cool for a few seconds before beating in the egg yolks one at a time with a wooden spoon, followed by both cheeses. Turn into a large mixing bowl and combine with the Belgian endive mixture. Season.

Pour boiling water from the kettle into a roasting pan and place in the oven; the water should be about 1½ in deep. Beat the egg whites with a small pinch of salt until very stiff but not dry or grainy; fold gently into the mixture in the mixing bowl, using a spatula.

Turn into the soufflé dish, tap the bottom of the dish on the work surface to make the mixture settle and level, sprinkle the surface with a fine layer of bread-crumbs, and place the dish in simmering water in the roasting pan. This *bain-marie* water should come just over halfway up the side of the dish; top up with more boiling water if necessary. Bake for about 40 minutes or until golden brown.

While it is cooking, make a mildly flavored, light tomato sauce: peel the garlic cloves, leave them whole but mince them slightly if you want a good garlic flavor, and fry gently over low heat with the thyme in 1-1/2 tbsp olive oil. Add the *passato*, sugar, 1/2 tsp salt, and 1/4 cup dry white wine. Cover and simmer for 20 minutes. Remove the thyme stalk and the garlic cloves, stir in 1/4 cup of water, add salt and pepper to taste and a pinch of oregano, and cook uncovered for 1 minute, stirring. Keep this pouring sauce hot.

Take the cooked mold out of the oven and leave to stand and settle for 5 minutes before running the sharp point of a knife round the inside edge of the dish. Turn a heated serving plate upside down on top of the soufflé dish and hold the two dishes together as you turn the serving dish the right way up again and release the mold onto it. Pour just a little sauce over the top of the mold, to trickle down the sides, top with a sprig of basil if wished, and hand round the remaining sauce separately.

———•———

ESCAROLE WITH BEANS AND OLIVE OIL

1 lb dried cannellini beans (see method)
1 tsp baking soda
2 cloves garlic
2 sage leaves
1¾ lb escarole
extra-virgin olive oil
salt and pepper

Soak the beans overnight in enough cold water to amply cover them mixed with the baking soda. The next day, drain and place the beans in a measuring jug to get an estimate of their volume and transfer to a fireproof earthenware or enameled cast-iron cas-serole dish with twice their volume of water. Add the unpeeled garlic cloves, sage leaves but no salt, and bring slowly to a gentle boil.

Simmer the beans over very low heat for 2½ hours or until they are tender, removing any scum from the surface at intervals. After just over 1 hour's cooking time, stir in 2½ tsp salt and 1½ tbsp olive oil. When they are cooked, remove and discard the sage leaves; squeeze the soft purée from the garlic into the beans and stir. Add a little salt if necessary.

Trim and thoroughly wash the escarole; blanch for 2 minutes in a large saucepan of boiling salted water. Drain well and place an equal amount in deep, heated soup bowls; ladle plenty of beans on top and sprinkle a little olive oil and freshly ground white or black pepper over them. Serves 6.

———•———

BROILED ENDIVE WITH CHEESE AND CHIVE DUMPLINGS

4 heads Belgian endive
1 cup cream cheese
1 egg yolk
1 small bunch chives
4 tbsp grated cheese (e.g. Emmenthal, Edam, Gouda, etc.)
16 slices bacon
¼ cup butter
¼ cup extra-virgin olive oil
salt and pepper

Preheat the oven to 400°F and have the broiler ready for use. Trim and rinse the endive then cut each head lengthwise in half and place cut side uppermost in a single layer in a lightly oiled gratin dish. Season and sprinkle with the olive oil.

Push the cream cheese through a sieve into a large mixing bowl and blend in the egg yolk with a wooden spoon; add about 1 tbsp chives snipped with the kitchen scissors, the grated cheese, salt, and pepper. Stir well then mold into flattened egg shapes by packing the mixture into a tablespoon and rounding the top off smoothly. Set aside.

Broil the bacon until crisp; drain off all the fat. Cook the endive by placing it under the very hot broiler for 8–10 minutes or remove from the gratin dish and dry-fry on the griddle; turn the endive halfway through the cooking time. Place in the preheated oven for 5 minutes in the gratin dish.

By now the endive will have wilted and softened; place a piece of crispy bacon on top of each endive half, then fill the spaces with the ricotta dumplings. Return the gratin dish to the oven for 4–5 minutes while you melt the butter. Pour the melted butter over the dumplings and serve.

——— • ———

SWISS CHARD AU GRATIN

1¼ lb Swiss chard

½ cup small button mushrooms

¼ cup butter

1½ tbsp oil

1 large shallot, finely chopped

1 tbsp finely chopped parsley

1¾ cups béchamel sauce (see page 36) or velouté sauce (see page 35)

2 egg yolks

¼ cup grated Swiss cheese (Gruyère or Emmenthal)

6 tbsp fine white breadcrumbs

½ cup freshly grated Parmesan cheese

salt and pepper

Preheat the oven to 400°F. Bring a large saucepan of salted water to a boil. Prepare and wash the Swiss chard, cutting off the green leaves and setting them aside for use in other dishes. Cut the wide, white stalks into 2½-in sections. Boil in the salted water for

15–20 minutes or until tender but still with a little bite left in them, then drain well. Trim and wipe the mushrooms; chop them very finely, then wrap in a clean cloth and twist the free ends hard to squeeze out all their moisture. Heat 1 tbsp of the butter in a nonstick skillet with 1½ tbsp oil and fry the shallot very gently. Add the mushrooms and cook, stirring, for 1–2 minutes. Add a pinch of salt and pepper and sprinkle in the parsley.

Make the béchamel or velouté sauce, ensuring that it is not very thick; once it has acquired a creamy consistency, remove from the heat and beat in the 2 egg yolks one at a time with a balloon whisk or wooden spoon, followed by the Swiss cheese and the shallot and mushroom mixture. Cover the bottom of a shallow ovenproof or gratin dish with a thin layer of this sauce, place the stalks evenly on top in one layer, overlapping one another and cover with the remaining sauce. Fry the fine breadcrumbs in 1½ tbsp butter; allow to cool before stirring in the Parmesan cheese and then sprinkle evenly over the surface. Melt the remaining 1½ tbsp of butter and drizzle over the topping. Place in the oven and bake for about 10 minutes or until the surface is crisp and golden brown.

——— • ———

FENNEL AND LETTUCE AU GRATIN

2 firm crisp head lettuces

1¾ lb fennel

3 tbsp all-purpose flour

¼ cup butter

1¾ cups milk

½ cup grated Parmesan cheese

1 egg yolk

breadcrumbs

nutmeg

salt and pepper

Preheat the oven to 400°F. Trim and wash the lettuce, cut in quarters lengthwise from the stalk and add these to a saucepan full of boiling salted water to blanch for 5 minutes. Drain well and set aside. Prepare the fennel (see page 19), cut lengthwise into thin slices, and boil in salted water for 15 minutes. Drain and set aside. Make a béchamel sauce with 3 tbsp all-purpose flour, 2 tbsp butter, 1 3/4 cups boiling milk (see method on page 36); season with a little salt, pepper, and nutmeg. Remove from the heat and stir in half the grated cheese and the egg yolk, beating with a whisk.

Grease a fairly deep, oval ovenproof dish and fill the dish with alternating deep sections of the vegetables, alternating paler broad bands of fennel with the brighter green of the lettuce. Cover evenly with the sauce and sprinkle the surface with the rest of the grated cheese and with a fairly thin layer of dry breadcrumbs. Dot flakes of the remaining butter over the surface.

Bake in the oven for 20 minutes, then finish with 3 minutes' browning under a very hot preheated broiler.

———•———

SPINACH, ARTICHOKE, AND FENNEL PIE

1¼ lb spinach
10 oz artichoke hearts
1 lb fennel
1 cup beef, chicken or vegetable stock (see pages 37–38)
1¾ cups ricotta cheese
½ cup fromage frais
½ cup grated hard cheese
2 eggs
12 oz frozen puff pastry, thawed
5 slices medium-hard cheese
1½ tbsp chopped chives
2 cloves garlic
¼ cup olive oil
nutmeg
salt and pepper

For the artichoke sauce:
2 large shallots or 1 small onion, finely chopped
4 thinly sliced artichoke hearts (see page 18)
1 sprig thyme
2 tbsp butter
olive oil
½ cup chicken or vegetable stock (see pages 37–38)
½ cup heavy cream
lemon juice
salt and pepper

Preheat the oven to 450°F. Heat a large saucepan two-thirds full of salted water. Trim and prepare the spinach, using only the leaves and more tender stalks; you should have about 12 oz net weight. Wash very thoroughly in cold water, drain, and blanch in the saucepan of boiling salted water for 2 minutes. Drain; squeeze out all the moisture when cool enough to handle. Heat 3 tbsp oil in a skillet with a peeled, whole clove of garlic, add the spinach, and fry very gently for 3–4 minutes while stirring. Remove the garlic clove and set the spinach aside to cool. Prepare the artichoke hearts, dropping each one into a bowl of acidulated water to prevent discoloration. Cut lengthwise into quarters, then halve each quarter lengthwise. If the artichokes are not very young and tender, use the disk-shaped bottoms and cook whole. Fry very gently in 1 tbsp olive oil with a peeled whole but lightly minced clove of garlic and a small pinch of salt for 2–3 minutes with the lid on. Drain off any liquid and remove the garlic clove. Set aside to cool. If only the bottoms are used, slice each one horizontally in half when cool enough to handle.

Trim the fennel, wash, dry, and shred. Simmer in the stock in a covered saucepan for 10 minutes until tender. Place the ricotta and the *fromage frais* in a mixing bowl with the grated cheese and work in 1 egg yolk, a pinch of salt, pepper, nutmeg, and the chopped chives.

Cut off two-thirds of the pastry (about 8 oz) and roll out on a lightly floured board into a circle 13–14 in in diameter and just over ⅛ in thick; use this to line a

77

greased 10-in cake pan, allowing the extra pastry to overlap the edges evenly all the way round the rim. Spread a layer of half the soft cheese mixture over the bottom and cover with all the spinach. For the next layer use the sliced cheese and cover this layer with the artichoke hearts; cover these with the remaining soft cheese mixture.

Roll out the remaining pastry to make a pie lid 10 in in diameter and cover the pie; bring the overlap over this lid and press the edges together against the inside rim of the pan with the tines of a fork. Glaze with beaten egg. Bake for 20–25 minutes or until golden brown.

Make the sauce: sweat the chopped shallots or small onion in the butter with 1 tbsp oil for 5 minutes, add the artichokes and the thyme, and cook over fairly low heat for 5 minutes while stirring. Add the stock, cover, and simmer gently for 15 minutes or until tender. Remove the thyme stalk and beat with a hand-held electric beater until quite smooth. Stir in the cream, season with salt and pepper to taste, and add a little lemon juice. Serves 10–12.

———•———

FENNEL, SPINACH, AND BEEF SALAD

1 lb trimmed beef tenderloin

½ lb spinach leaves

2 bulbs fennel

1½ cups fresh soft cheese (e.g. cream cheese or ricotta)

2½ tbsp chopped chives

salt and pepper

For the mustard-flavored citronette dressing:

1 cup extra-virgin olive oil

juice of 1 large, juicy lemon

2 tsp Dijon mustard

3 tbsp finely chopped parsley

salt and white pepper

Place the beef in the freezer for 10–15 minutes to make it firm enough to slice wafer-thin, or ask your butcher to do this for you. Wash the spinach leaves and dry in a salad spinner. Prepare the fennel as described on page 19.

Sieve the soft fresh cheese unless it is already very smooth and easy to spread; beat with a pinch of salt, plenty of freshly ground pepper, and the chives. Spread a beef slice flat on the chopping board and spread 1 tbsp evenly over it, taking care not to tear the very thin meat; place another slice on top and roll up. Repeat this until you have used up all the filling and slices, placing them on a freezer tray as you prepare them. Put the beef rolls in the freezer for 10 minutes, to prepare for the next slicing. Make the mustard dressing: mix the mustard, salt, and freshly ground pepper with the strained lemon juice and then beat in the olive oil a little at a time with a whisk (or use a blender). Stir in the parsley.

Slice the fennel lengthwise into very thin slices; mix these with about half the dressing in a bowl. Cut the beef rolls into slices about ½ in thick. Toss the spinach leaves in a little of the dressing in a separate bowl to coat all the leaves. Place the fennel slices in the center of a serving platter, surround them with the spinach leaves, and place the beef roll slices all around the spinach. Sprinkle the beef roll slices with a little olive oil and freshly ground white pepper. Serve at once.

———•———

SPINACH SOUFFLÉ

½ lb spinach leaves

1 large shallot

¼ cup butter

3 slices cooked ham, finely chopped

1 cup milk

½ cup all-purpose flour

5 eggs

1 tsp cream of tartar
¼ cup grated Swiss cheese (Emmenthal or Gruyère)
3 tbsp grated Parmesan cheese
nutmeg
salt and pepper
Serve with:
mousseline sauce (see page 107)

Have the mousseline sauce ready prepared. Preheat the oven to 400°F. Wash the spinach thoroughly, drain briefly, and cook in a covered saucepan with no added water for a few minutes until wilted and tender. Drain off any liquid and when cool enough to handle squeeze to expel all excess moisture. Chop finely.

Fry the finely chopped shallot in 1 tbsp butter, stirring, for 2–3 minutes, then add the ham and fry for a further 2–3 minutes. Add the chopped spinach, sprinkle with a small pinch of salt, and stir until all the excess moisture has evaporated. Remove from the heat.

Make the white sauce basis for the soufflé: heat the milk slowly; melt 2 tbsp butter in a fairly small saucepan, stir in the flour, and cook while continuing to stir for 2–3 minutes over low heat. Draw aside from the heat, add all the boiling milk at once, beating vigorously with a whisk to prevent lumps forming. Return to a slightly higher heat and keep stirring as the mixture comes to a gentle boil; simmer and stir until very thick.

Take off the heat, stir in a little salt, freshly ground white pepper, and a pinch of nutmeg to taste. Separate the egg whites carefully from the yolks, lightly beat the latter in a small bowl, and beat a little at a time into the hot sauce, then stir in the spinach mixture. You can prepare the soufflé up to this point several hours in advance but it is slightly easier to fold into the egg whites when freshly made and still warm. Cover the sauce to prevent a hard skin forming.

Before you are ready to proceed to the next stage of the soufflé, have the mousseline sauce ready and keep hot over simmering water; preheat the oven.

Beat the 5 egg whites until just frothy; add the cream of tartar; this will stabilize the protein and keep the egg whites stiff for longer. Beat until very stiff but do not overbeat (they would look dry or "grainy"). Fold about a quarter of the egg whites into the thick sauce, together with all but 1 tbsp of the finely grated Swiss cheese. Fold this mixture gently but thoroughly into the remaining egg whites.

Grease a 5-cup capacity soufflé dish with butter, sprinkle with the grated Parmesan cheese or very fine dry breadcrumbs, and pour the mixture into it. Tap the bottom on the work surface to settle and level the surface. Sprinkle with the remaining cheese, place in the oven, close the oven door and immediately turn down to 375°F. Bake for 25–30 minutes or until the soufflé has risen up well in the dish and is a good deep golden brown on top. Serve at once with the mousseline sauce.

——— • ———

SPINACH STIR-FRY WITH PORK, EGGS, AND GINGER

4 eggs
pinch baking powder
monosodium glutamate (optional)
2 tbsp cornstarch
2 tbsp Chinese rice wine or dry sherry
¼ lb thinly sliced pork tenderloin
½ lb prepared spinach
¼ cup dried Chinese tree ear mushrooms, pre-soaked
¼ cup canned bamboo shoots, drained
1 ¼–½-in piece fresh ginger
1 small trimmed leek, white part only
2 tbsp soy sauce
wine vinegar or cider vinegar
1½ tbsp sugar
1½ tbsp chicken stock (see page 37)

½ cup sunflower oil
salt and pepper
Serve with:
steamed rice

Make the marinade: beat 1 egg lightly and pour half of it into a bowl. Add a pinch each of salt, pepper, baking powder, monosodium glutamate (if used), 1½ tbsp cornstarch, and 1½ tsp rice wine or dry sherry. Shred the pork and mix well with the marinade. Leave to stand for 10 minutes.

Meanwhile, wash the spinach leaves and slice them lengthwise in half, along the central rib. Drain and rinse the mushrooms then trim off any rough parts. Cut the mushrooms and bamboo shoots into thin strips. Peel the ginger and chop finely together with the leek. Beat the 3 remaining eggs lightly with the reserved half egg, seasoning with a little salt and monosodium glutamate if wished.

Make the sauce: mix the remaining cornstarch with 1½ tbsp cold water in a bowl and add 2 tbsp soy sauce, 1½ tbsp rice wine or dry sherry, 1½ tbsp vinegar, 1½ tbsp sugar, 3 tbsp stock, and a pinch of monosodium glutamate (optional).

Twenty minutes before you intend serving the dish, steam the rice and keep hot while you cook the meat.

Heat ⅓ cup sunflower oil in a wok or large skillet. Heat a large serving plate in the oven. When the oil is very hot, add the pork and bamboo shoots and stir-fry over high heat for 1½ minutes, keeping the meat strips as separate as possible. Transfer to a heated serving plate. Add 1½ tbsp fresh oil to the oil remaining in the wok and when hot, but not smoking hot, pour in the beaten eggs and cook briefly until they have set. Remove with a ladle, draining well, and keep hot on the serving plate. Fry the leek and ginger for 30 seconds in the remaining oil. Add the spinach and stir-fry for a further 30 seconds; add the well-mixed sauce, stir to combine with spinach for 30 seconds as it thickens and comes to a boil. Return all the contents of the serving plate to the wok, stir briefly, then return to the serving plate and serve with the rice.

GREEK SPINACH AND FETA CHEESE PIE
Spanakópita

1¼ lb spinach leaves
2 medium-sized onions
½ lb feta cheese
5 eggs
6 tbsp olive oil
¼ cup chopped fresh dill or 1½ tsp dried dill
¼ cup finely chopped fresh parsley
⅓ cup heavy cream
2 packages puff pastry, each weighing 9 oz, thawed, or 1 package filo dough
salt and pepper

The above quantities will serve eight. Preheat the oven to 400°F. Wash the spinach thoroughly, drain briefly, and cook in a covered pan with no extra water for a few minutes, until wilted and tender. As soon as the spinach is cool enough to handle, squeeze out all excess moisture and chop very finely. Peel and finely chop the onions; finely chop or crumble the feta cheese. Beat 4 eggs lightly in a large mixing bowl with a pinch of salt and some freshly ground pepper.

Heat 6 tbsp oil in a saucepan and fry the onion gently until soft. Add the chopped spinach and a pinch of salt and stir well; cover and cook gently for 5 minutes. Stir in the dill and the parsley and cook for 2–3 minutes, stirring to allow any moisture to evaporate. Remove from the heat, stir in the cream, and leave to cool a little before stirring in the eggs and cheese, adding more seasoning.

Roll out each pastry piece on a lightly floured board large enough to line a large, shallow greased baking tray. If you prefer to use filo pastry, place half the sheets from the package one on top of the other to line the baking tray and brush each one with melted butter or oil; do likewise with the remaining sheets for the pie lid. Spread the spinach and cheese mixture over the pie shell and place the other half of the pastry on top, pinching the edges firmly together all the way round, or pressing with the tines of a fork to seal tightly. Beat the remaining egg lightly and brush all over the surface of the pastry to glaze.

Bake for about 20 minutes or until the pastry has puffed up and is crisp and golden brown.

———•———

INDIAN SPICED CABBAGE AND POTATOES
Aloo Gobhi

2¼–2½ lb green cabbage
4 potatoes
3 onions
3 small green chili peppers
1 1-in piece fresh ginger
2 tbsp coriander seeds
3 tbsp cumin seeds
2 dried chili peppers
1½ tsp ground dried green mango (amchur) (optional)
3 tbsp ghee (clarified butter, see page 36)
¼ cup chopped coriander leaves
salt
For the chapatis or Indian soft flat bread:
2¼ cups whole wheat flour or chapati flour
1 cup all-purpose flour
1½ tbsp ghee (clarified butter, see page 36)

Many shops sell ready-made fresh chapatis and nan bread, both of which are suitable for this dish. Heat the former in a hot oven (450°F) wrapped in foil for 10 minutes; heat the nan for no more than 5 minutes and brush it with the ghee as soon as you take it out of the oven. Packages of chapati mix are also available; follow the manufacturer's instructions.

Prepare the cabbage as described on page 20, wash, and shred. Peel the potatoes and onions and chop both coarsely. Trim the green chili peppers, remov-ing the stalk and seeds, and chop finely. Peel the ginger and chop very finely.

Grind 1 tbsp of the coriander seeds, 1 tbsp of the cumin seeds and the dried chili peppers (the latter can be seeded, as they are very hot) very finely in an electric grinder. Put the freshly ground spices through a sieve into a bowl and mix with the mango powder if used.

Heat 3 tbsp ghee in a large saucepan and fry the remaining 1½ tbsp whole coriander seeds and 2½ tsp cumin seeds gently for a few seconds, stirring with a wooden spoon.

Add the chopped green chili peppers and ginger and fry for 30 seconds while stirring. Add the ground spices, cook for a few seconds, then add the vegetables.

Stir well, turn down the heat to very low, cover, and sweat for about 25 minutes or until the vegetables are tender but keep their shape and are not at all mushy. Moisten with 1–2 tbsp of water or stock at intervals. When cooked, add a little salt, stir, and sprinkle with the chopped coriander leaves.

Make the chapatis: sift both types of flour into a large mixing bowl; make a well in the center and pour the ghee and ¾ cup lukewarm water into it. With your hand cupped and fingers held together, stir the butter and water, gradually incorporating all the flour. Knead the resulting soft dough for 6 minutes. Cover with a very damp cloth and leave to stand for 30 minutes.

Knead again and divide the dough into 15 balls of equal size. Return to the bowl and cover again with the dampened cloth. Grease a griddle or a large cast-iron skillet with a little ghee and heat until very hot. Take a ball of dough out of the bowl, flatten with your palms, and roll out on a lightly floured board to a disk 5½ in in diameter. Shake off excess flour and dry-fry for 1 minute, then turn and fry the other side for ½ minute. Provided the skillet is hot enough, the chapati will puff up; do not worry if this does not happen. Pile up the chapatis as you cook them, keeping them wrapped in a cloth. Serve immediately with the spiced vegetable mixture. Serves 6.

———•———

CABBAGE AND HAM ROLLS

1 large green cabbage

½ lb finely chopped lean roast pork

4 slices cooked ham, chopped

4 slices Italian coppa *or Westphalian ham, finely chopped*

⅓ cup grated Parmesan cheese

1 egg

1 clove garlic, minced

3 tbsp parsley, finely chopped

3–4 tbsp béchamel sauce (see page 36)

1 small onion, finely chopped

4 slices bacon, finely chopped

3 sage leaves

½ cup dry white wine

butter

nutmeg

salt and pepper

Select the largest undamaged cabbage leaves, wash them, and blanch one at a time in a large saucepan of boiling salted water for 2 minutes; spread the leaves out with the greener, inner side facing downward on the chopping board and use a small, sharp pointed knife to cut out the center rib neatly. Prepare 30 leaves in this way.

Mix all the ingredients for the stuffing very thoroughly together in a large mixing bowl, adding them in the order listed above, finishing up with the béchamel sauce. The mixture should be moist but firm enough to hold its shape. Season with salt, pepper, and nutmeg to taste. Mold by hand into oblong shapes and wrap these in the cabbage leaves, placing the stuffing on the greener, smoother side of the leaves and rolling up into neat parcels, closed at both ends. You may need more than one leaf to completely enclose the stuffing; secure with wooden cocktail sticks.

Fry the onion, bacon, and sage leaves in butter in a wide nonstick skillet until the onion is soft and transparent; add the cabbage rolls in a single layer and fry over moderate heat, turning them carefully after a few minutes. They should be very lightly browned on both sides. Sprinkle the wine all over them, cover tightly, and turn down the heat to very low. Simmer for 25–30 minutes. Serve with creamed potatoes.

·

CAULIFLOWER FU-YUNG

1 lb cauliflower florets

½ cup ground turkey

1½ tbsp all-purpose flour

pinch monosodium glutamate (optional)

2 egg whites

½ cup chicken stock (see page 37)

sunflower oil

2 thin slices cooked ham, shredded

green leaves of 2 scallions, shredded

salt and pepper

Served with:

steamed rice

Bring a large saucepan of water to a boil, add a generous pinch of salt, and blanch the cauliflower florets for 5 minutes. Drain well and leave to cool. Mix the turkey in a bowl with 1 tsp salt, the flour, monosodium glutamate (if used), the egg whites, and the stock.

Heat ¼ cup oil in a wok or large nonstick skillet and stir-fry the cauliflower florets over high heat for 2 minutes; sprinkle with a pinch of salt. Take the florets out of the wok with a slotted or mesh ladle and keep hot.

Add the turkey mixture to the remaining hot oil and keep stirring over lower heat until it thickens and becomes smooth and creamy. Return the cauliflower to the wok, reduce the heat further, and stir for 1–2 minutes. Transfer to a heated serving dish and garnish with the shredded ham and scallions. Serve at once with steamed rice.

———— • ————

CAULIFLOWER POLONAIS

1 medium-sized cauliflower

baking soda or cube of white bread

2 tbsp finely chopped parsley

4 slices cooked ham

4 hard-boiled eggs

1 lemon

¼ cup butter

½ cup fine fresh breadcrumbs

½ cup heavy cream

salt and pepper

Fill a deep saucepan wide enough to take the whole cauliflower half full of water; add salt and bring to a boil. Prepare the cauliflower, trimming off the leaves and the hard remains of the stalk, and cut a deep cross in the base. Rinse the cauliflower and place in the boiling water which should completely cover it. (Add a pinch of baking soda or a cube of white bread to help prevent a strong smell as it cooks, if wished.) Cover and boil gently for 17–20 minutes or until tender but still firm.

Meanwhile, make the sauce: chop the parsley, the ham, and the hard-boiled eggs, keeping them separate. Squeeze the juice out of the lemon and strain. Melt the butter in a small saucepan, add the breadcrumbs and fry for 1 or 2 minutes until lightly browned. Stir in the lemon juice and the cream and remove from the heat.

Drain the whole cauliflower and place carefully in a heated serving dish. Make sure the sauce is hot but not boiling, stir in the ham, eggs, and parsley, a pinch of salt, and plenty of freshly ground pepper. Pour this all over the cauliflower and serve at once on heated plates.

CAULIFLOWER TIMBALE

1 small cauliflower

1 cup fine fresh white breadcrumbs

pinch baking soda or 1 cube white bread

1 medium-sized onion

5 eggs

¼ cup grated Swiss cheese (Emmenthal or Gruyère)

1 cup milk

generous ¼ cup butter

nutmeg

salt and pepper

Serve with:

Fresh Tomato Sauce (see page 258)

Preheat the oven to 325°F. Lightly oil the inside of a 1½-quart capacity soufflé dish and sprinkle all over with 2 tbsp breadcrumbs, tipping out any excess. Bring a large saucepan of salted water to a boil, adding a pinch of baking soda or a cube of white bread to prevent the unpleasant cooking smell, if wished.

Wash the cauliflower and cut it into florets and the lower stalks into fairly small pieces. Boil for about 13 minutes or until tender; drain well. You will need just over 1 lb cooked cauliflower for this dish; place it in a large mixing bowl and crush finely with a potato masher or fork.

Peel and chop the onion finely and cook for about 10 minutes or until soft but not browned in 1 tbsp butter, stirring frequently with a wooden spoon. Scrape this out of the saucepan into another mixing bowl, season with salt, freshly ground white pepper, and nutmeg. Break the eggs into the bowl and beat lightly. Stir in the cheese and ⅔ cup of the fine fresh breadcrumbs. Bring the milk to a boil with ¼ cup of the butter and then add in a thin stream to the egg mixture, beating continuously. Stir in the cauliflower and add a little more salt and pepper if needed. The timbale can be prepared 2–3 hours in advance up to this point and left at cool room

83

temperature, then placed in a preheated oven to cook. Fill the soufflé dish with the mixture, sprinkle the remaining breadcrumbs on the surface, and bake for 35–40 minutes or until a skewer pushed deep into the timbale comes out clean and dry. Leave to stand and settle for 10 minutes before unmolding on to a heated serving plate. Pour a little of the sauce over the timbale. This is delicious served with peas mixed with small pieces of cooked ham or bacon and basil, and Fresh Tomato Sauce. Serves 6.

———— • ————

BROCCOLI AND SHRIMP TERRINE

1½ lb fresh or thawed frozen jumbo shrimp

¾ cup crustacean stock (see page 38)

½ lb trimmed broccoli

1 ¾ lb fillets of Dover sole, lemon sole or flounder

3 eggs

2½ cups light cream or half-and-half

pinch cayenne pepper or chili powder

lemon juice

salt and pepper

Peel the shrimp, remove the black intestinal tract running down their backs, and simmer them for 2 minutes in a covered saucepan with just enough of the crustacean stock to cover them. Drain.

Trim the broccoli spears and cook in plenty of boiling, salted water until tender but still crisp; drain in a colander and immediately refresh under running cold water. Drain again and leave to cool. Preheat the oven to 300°F. Cut the sole fillets into fairly small pieces and purée in the food processor; remove and leave to one side. Place 2 yolks and 3 whites in the food processor and process on maximum speed for 1 or 2 minutes until they become very frothy; add the cream a little at a time, processing after each addition. Add salt, pepper, and the cayenne pepper or chili powder. Process again. Pour half the contents of the food processor into a bowl. Replace the rest, add half the puréed sole, and process briefly. Add salt, pepper, and a little lemon juice. Process once more. Empty the contents of the food processor into a bowl, and repeat the operation with the reserved egg mixture and the remaining fish purée.

With all the mixture in one large bowl, taste and add more seasoning if necessary; cover with plastic wrap or foil and chill for 15 minutes.

You will need a 2½-quart oval or rectangular terrine dish, (or 2 terrines of half this capacity). Spoon the chilled mixture into a pastry bag fitted with a large-gage tip. Pipe the mixture neatly so that it lines the terrine (base and sides). Press the shrimp into the lined sides, tails uppermost, in a well spaced line. Pipe in more fish and egg mixture, then place the broccoli spears upside down in the terrine; pipe the remaining mixture all around the broccoli and fill the terrine completely, tapping the base gently on the work surface as you do so, to prevent air pockets; smooth the surface with a spatula. Place the terrine in a roasting pan, pour sufficient boiling water into the pan to come about halfway up the sides of the terrine, place in the oven, and cook for 40 minutes.

Unmold and serve hot. A velouté sauce, made with the crustacean stock or a homemade tomato sauce will go well with this terrine.

———— • ————

STEAMED BROCCOLI WITH SAUCE MALTAISE

1¾ lb broccoli

For the sauce:

1 cup butter

3 egg yolks

⅛ cup orange juice

1½ tbsp lemon juice

finely grated rind of 1 orange

salt

Trim off the larger, tougher ends of the broccoli; leave the tender leaves on the stalks. Sprinkle with a pinch of salt and steam for 12–15 minutes. (Microwave them if preferred.)

The sauce should be made just before serving or else it may separate. Have 2 tbsp of the butter cut into small cubes in the ice box. Melt the rest of the butter gently in a small saucepan; beat the egg yolks with a whisk or hand-held electric beater in the top of a double boiler or in a heatproof bowl until light and creamy. Beat in 1-1/2 tbsp each of lemon and orange juice and a pinch of salt. Add half the chilled cubes of butter and place over gently simmering water; beat continuously as the butter and egg yolks gradually heat and combine. As soon as the mixture starts to thicken appreciably, remove from the heat and add the remaining chilled butter, still beating continuously; when completely blended beat in the melted butter. Start by adding just a few drops at a time and then trickle it into the sauce while beating in a very thin stream. The sauce should increase in volume and thicken. Gradually stir in the remaining orange juice and grated rind. Serve the broccoli at once on individual plates, lightly coated with the hot sauce.

———•———

Prepare the Brussels sprouts, sprinkle with a small pinch of salt, and steam or microwave until just tender.

Make the hollandaise sauce: place the vinegar in a small saucepan with the chopped shallot (or ¼ mild onion), 4 peppercorns, and the thyme or bay leaf. Simmer gently, uncovered, to reduce to about 3 tbsp. Strain and set aside.

Have ¼ stick of the butter cut into small cubes and chilled in the ice box of the refrigerator. Beat the egg yolks with a whisk or hand-held electric beater until creamy in the top of a double boiler or in a heatproof bowl that will fit snugly over a saucepan. Gradually beat in 1½ tbsp iced water and a pinch of salt. Add half the chilled butter pieces and set the double boiler top or bowl over gently simmering water; beat continuously as the mixture gradually heats and the butter melts and combines with the egg yolks, adding the reduced vinegar a little at a time as you beat. When this mixture starts to thicken noticeably, remove from the heat immediately, add the remaining chilled butter without delay, and continue beating. Once this has melted and combined with the egg mixture, beat in the melted butter, starting by adding a few drops and then gradually increasing this to a thin trickle. The sauce should become light and thick. Add a little more salt and freshly ground white pepper if wished and then stir in lemon juice to taste.

———•———

BRUSSELS SPROUTS WITH HOLLANDAISE SAUCE

1 lb Brussels sprouts
For the hollandaise sauce:
½ cup wine vinegar
1 shallot, peeled and finely chopped
1 sprig fresh thyme or ½ bay leaf
1 cup butter
3 egg yolks
lemon juice
salt and white peppercorns

BRAISED LETTUCE

2 large Boston lettuces
2 large scallions
1 heaping tsp potato flour or cornstarch
½ cup milk
1 vegetable or chicken bouillon cube
¼ cup heavy cream
black pepper

Remove the older or damaged outermost leaves of the lettuces, trim off the very end of the base, wash the hearts well, then drain and cut lengthwise in quarters. Remove the roots, leaves, and outer layers of the scallions and slice the bulbs into thin rings. Mix the potato flour or cornstarch with the cold milk.

Place the lettuce and milk in a shallow saucepan, crumble in the bouillon cube, cover, and simmer gently for about 20 minutes over low heat or until tender and wilted. Stir in the cream and remove from the heat. Sprinkle with a little freshly ground black pepper and serve at once.

— • —

STIR-FRIED LETTUCE WITH OYSTER SAUCE

2 very fresh Romaine lettuces

¼ cup sunflower oil

For the oyster sauce:

¼ cup commercially prepared Chinese oyster sauce

2 tbsp light soy sauce

3 tbsp Chinese rice wine or dry sherry

6 tbsp cold chicken stock (see page 37)

1 tsp cornstarch or potato flour

pinch monosodium glutamate (optional)

Bring a large saucepan two-thirds full of salted water to a boil. Remove any damaged outer leaves from the lettuces, cut each whole lettuce lengthwise into quarters, wash, and drain very thoroughly, then cut each quarter across, in half. Add to the boiling water and blanch for only 3–4 seconds, removing the lettuce pieces quickly with a slotted ladle (or use a frying basket). Drain well. Prepare the sauce mixture by stirring the ingredients together in a small bowl. Heat the oil in a wok or large nonstick skillet; when it is very hot, add all the lettuce at once and stir-fry over high heat for a maximum of 2 minutes; the lettuce should still be fairly crunchy. Transfer the let-

tuce to a serving dish, reduce the heat, stir the sauce ingredients once more, and pour into the wok, stirring and scraping to loosen any cooking deposits with a wooden spoon. When the sauce has boiled and thickened, sprinkle all over the lettuce and serve immediately.

— • —

FRIED CHICORY ROMAN STYLE

2 lb chicory

¼ cup extra-virgin olive oil

2 cloves garlic

1 small dried chili pepper, seeds removed

salt

Bring a large saucepan of lightly salted water to a boil. Prepare the chicory, taking all the leaves off the hard remains of the stem, discarding the old, outer ones, and wash very thoroughly. Blanch the leaves in the boiling water for 2 minutes, drain well, then squeeze out excess moisture in a clean cloth.

Heat the oil in a large, nonstick skillet, add the peeled, lightly minced garlic cloves and the crumbled chili pepper. Fry gently for 30 seconds, then add the chicory and cook over low heat for 5 minutes, stirring 2 or 3 times. Serve hot.

— • —

ENDIVE WITH LEMON FLEMISH STYLE

8 heads Belgian endive

¼ cup butter

1¼ cups chicken stock (see page 37)

2½ tsp light brown cane sugar

2 lemons

salt and white pepper

Prepare the Belgian endive (see page 21), cutting out part of the tough base. Rinse, dry thoroughly, and cut lengthwise in half (if the heads are very fat, cut lengthwise into quarters). Divide the butter in half between 2 skillets with tightly-fitting lids and heat until it foams. Add the endive pieces, spreading them out in a single layer in the hot butter and fry over moderately high heat until lightly browned. Turn carefully and brown the other side.

Divide the stock equally between the two skillets, cover tightly, and reduce the heat to low. Simmer for 5 minutes. Mix the sugar with the strained juice of both lemons, remove the lids, and sprinkle the sweetened juice all over the endive. Turn up the heat a little to reduce the lemon juice and liquid and caramelize slightly, loosening the endive gently if it shows signs of sticking. Season and serve.

——— • ———

ESCAROLE WITH OLIVES

4½ lb escarole

4 cloves garlic

1 chicken or vegetable bouillon cube

24 large black Greek olives, pitted

6 tbsp extra-virgin olive oil

salt and black peppercorns

Place the escarole in a large saucepan with 2 cups water, 8 black peppercorns, and the crumbled bouillon cube. If there are too many leaves to fit in, add more as soon as the first batch has wilted and made room for them. Cover tightly and simmer gently for 50 minutes once the water has come to a boil, adding any remaining leaves as soon as possible. Stir at intervals, turning the leaves.

When this cooking time is up, add the oil, the olives, and a little salt. Cover again and continue cooking for a further 10 minutes.

Add a little more salt, stir, and cook, uncovered, for a final 15 minutes.

BRAISED ENDIVE

6 heads red Treviso endive or Belgian endive

3 tbsp butter

salt and pepper

4–6 tbsp chopped parsley

Rinse and drain the endive and cut lengthwise in quarters. Melt the butter in a wide nonstick saucepan with a very tight-fitting lid. Arrange the endive in a single layer and sprinkle with a little salt and pepper. Cover tightly, turn the heat down very low, and leave to cook for 10 minutes. Take the saucepan off the heat, wait a minute or two, then remove the lid and turn the endive. Cover again and leave to cook for a further 10 minutes, by which time the endive should be tender. Serve hot, sprinkled with parsley.

——— • ———

ASPARAGUS CHICORY AND BEANS APULIAN STYLE

2–2¼ lb asparagus chicory

2 cups freshly hulled fava beans

extra-virgin olive oil

salt and pepper

If you cannot buy asparagus chicory, use Belgian endives. Bring a large pan two-thirds full of salted water to a boil. Trim the asparagus chicory, using only the leaves and the younger, tender stalks. Cut into 2-in sections, wash, and drain. Add to the boiling water and cook for 15 minutes.

Add the fava beans and cook for a further 15 minutes, or until tender. Drain well and place on an unheated serving plate. Sprinkle with olive oil and freshly ground pepper while still quite hot and serve warm.

BUTTERED SPINACH INDIAN STYLE

2¾ lb spinach
2 medium-sized onions
1 heaping tsp finely chopped fresh ginger
1–2 small green chili peppers
⅓ cup ghee (clarified butter, see page 36)
1 tsp cane sugar
½ cup chicken stock (see page 37)
½ tsp garam masala
salt
Serve with:
pilau rice

Trim the spinach, using only the leaves for this dish; wash thoroughly. Drain in a salad spinner and shred.

Peel the onions and ginger and chop very finely; trim the stalks from the chili peppers, remove the seeds and chop the flesh finely; fry all these chopped ingredients gently in the ghee for 5 minutes in a large saucepan, then add the spinach, a pinch of salt, and the sugar. Cook, uncovered, over moderate heat, stirring continuously for 2–3 minutes. Pour in the stock, stir briefly, cover and simmer for 5 minutes. Take off the lid and cook while stirring for a few seconds to reduce excess liquid. Draw aside from the heat, sprinkle with the garam masala, and serve with pilau rice.

— • —

SPINACH WITH SHALLOTS

2¾ lb spinach
4 shallots or 1 Bermuda onion
3 tbsp butter
salt and pepper

Use only the young, tender spinach leaves for this recipe; the larger, older leaves and stalks can be reserved for use in other dishes (e.g. soups, etc). Wash them very well and do not bother to drain them thoroughly.

Peel and chop the shallots or Bermuda onion finely and cook in the butter until tender but not at all browned. Add the spinach leaves, fry them for a few seconds stirring and to coat with butter, then cover and cook over low heat for 2–3 minutes. They should still be fairly crisp. Sprinkle with a little salt and freshly ground pepper and serve.

— • —

SPINACH AND RICE INDIAN STYLE

2¾ lb spinach
1½ cups Basmati rice
2 medium-sized onions
½ cup ghee (clarified butter, see page 36)
1 tsp garam masala
salt

Bring plenty of water to a boil in a large saucepan and add a generous pinch of salt. Use only the tenderest leaves and blanch for just under 1 minute then drain; you should aim to have 1–1¼ lb net weight of washed leaves. (The other leaves and stalks can be used for recipes calling for longer cooking and need not be wasted.)

While the spinach is cooling, put the rice in a sieve and hold it under running cold water to rinse until the water runs clear. Drain. Squeeze all excess moisture out of the spinach by hand and chop finely. Peel the onions and chop very finely. Fry the onions gently in the ghee for 5 minutes, stirring until they are soft and pale golden brown. Add the spinach and the garam masala and stir-fry over moderate heat for 2–3 minutes. Take off the heat, add the rice with 2¼ cups cold water and 1 tsp salt. Stir, return to a higher heat, and bring to a boil. Cover, turning the heat down extremely low, and leave to cook for 15 minutes or until the rice is tender and has absorbed all the liquid. Stir with a fork to help separate the grains. If serving with hot spiced dishes, you will not need to add more salt as this dish provides a refreshingly bland, cooling contrast. Serves 6.

———•———

FRIAR'S BEARD CHICORY WITH LEMON DRESSING

1½ lb friar's beard chicory (buck's horn plantain) or substitute (see method)

extra-virgin olive oil

1 lemon

salt

Almost any type of edible leaf vegetable can be cooked to advantage in this way.
Rinse your chosen vegetable well after trimming, drain, and add to a large pot of boiling salted water, stirring with a fork or a long-handled spoon. Friar's beard chicory should be boiled hard for 8–10 minutes until tender but still with a little bite to it. Drain and serve hot or warm, sprinkled with a little olive oil mixed with lemon juice.

———•———

FRIAR'S BEARD CHICORY WITH GARLIC AND CHILI

1¼ lb friar's beard (buck's horn plantain, see method)

6 tbsp extra-virgin olive oil

4 cloves garlic

1 small dried chili pepper, seeds removed

salt

An almost infinite variety of green and leaf vegetables can be cooked in this way. Trim, prepare, and rinse your chosen vegetable well; dry in a salad spinner or pat gently dry with paper towels. Heat the oil in a large saucepan, add the whole, lightly minced garlic cloves and fry gently until browned. Crumble in the chili pepper, stir, then add the friar's beard chicory or substitute. Stir-fry over high heat for about 5 minutes, until slightly wilted and tender but still with a bit of crispness left. Sprinkle with a pinch of salt, stir once more, and serve piping hot.

———•———

POACHED TURNIP TOPS

4½ lb turnip tops or purple sprouting broccoli

1 large Spanish onion

1½ cups chicken stock (see page 37)

½ cup extra-virgin olive oil

salt and peppercorns

Serve with:

slices of mild, semihard cheese

crusty whole wheat bread

Trim the turnip tops, removing the larger stalks and ribs; if using sprouting broccoli, cut the larger pieces in half. Wash well. Peel the onion, cut lengthwise in half and slice both halves as thinly as possible.
Place the turnip tops or broccoli in a large, heavy-bottomed fireproof casserole dish (preferably, earthenware or enameled cast iron), add the onion, stock, a pinch of salt, and 3–4 black or white peppercorns. Bring to a boil, cover, and simmer gently for 15 minutes (10 minutes for broccoli).
Sprinkle the olive oil all over the turnip tops, turn up the heat and cook uncovered for up to 10 minutes (broccoli will need only about 5 minutes), or until the turnip tops are done but still have a little bite left to them. Serve with the reduced cooking liquid as a

sauce, and eat with slices of semihard cheese and whole wheat bread.

———•———

CELERY AU GRATIN

2 heads celery weighing approx. 2 lb in total

1 cup chicken stock (see page 37)

3 tbsp dry vermouth or dry white wine

1 shallot, peeled and finely chopped

1 sprig fresh thyme

½ tsp cornstarch or potato flour

¼ cup butter

¼ cup grated Swiss cheese (Emmenthal)

¼ cup grated Parmesan cheese

salt and white pepper

Preheat the oven to 400°F. Cut off the hard, woody bases and the leaves of the celery bunches, separate the stalks, and run a potato peeler down the outside of each stalk to get rid of any strings. Cut the stalks into 1½-in lengths, place in a wide, fairly shallow saucepan or skillet, add the stock, cover tightly, and cook over moderate heat for about 7 minutes or until they are tender but still quite crisp.

Drain off the liquid into a small saucepan and add the vermouth or wine, the shallot, and thyme; boil until the liquid has reduced by half. Mix the cornstarch or potato flour with a little cold water, stir into the liquid, then simmer for 1 minute while stirring the slightly thickened liquid.

Grease a gratin dish or any shallow ovenproof dish with butter and spread the celery pieces out in a single layer, slightly overlapping one another.

Melt 3 tbsp butter in a small saucepan and sprinkle it all over the celery. Sprinkle with the grated cheeses and a little freshly ground pepper. Drizzle the slightly thickened stock all over and place in the oven for about 10 minutes to melt the cheeses and brown lightly on top. Serve very hot.

CELERY JAPANESE STYLE
Selori no Kimpira

2 large, wide celery stalks

2 tbsp sunflower oil

1½ tbsp Japanese soy sauce

3 tbsp saké or dry sherry

pinch sugar

pinch ajinimoto (Japanese) taste powder (optional)

Trim off the very ends of the celery stalks and their leaves. Scrub well under running cold water. Use a potato peeler to remove the outer layer, making sure there are no strings left. Cut lengthwise in 4 equal strips, then cut these into large matchstick pieces about 1¾ in long and just over ⅛ in wide.

Heat the oil in a nonstick skillet and stir-fry the celery for 1½ minutes. Sprinkle with the soy sauce and saké, turn up the heat a little more, and reduce what little liquid there is so that the celery is just moist. Add a pinch of sugar and a pinch of the taste powder, if wished. Stir once more and serve on 4 individual plates. This is very good cold or hot, served with plain boiled or steamed rice.

———•———

FENNEL AU GRATIN

6 bulbs fennel

1 clove garlic

1 sprig thyme

¼ cup grated Parmesan or other hard cheese

1 cup fine fresh breadcrumbs

¼ cup butter

salt and pepper

Serve with:

Rösti Potatoes with Shallots (see page 196)

Broccoli Purée (see page 93)

Preheat the oven to 400°F. Heat a little water in the lower compartment of a steamer. Wash the fennel and cut lengthwise into wide, thick slices. Steam for 10 minutes or until tender but firm.

Melt 3 tbsp butter in a wide, fairly shallow fireproof casserole dish and fry the fennel slices lightly with a clove of garlic and a sprig of thyme. Season with a little salt and freshly ground pepper and brown lightly on both sides. Sprinkle with the grated cheese and a thin covering of breadcrumbs, dot pieces of butter over the surface, and place in the oven for 5 minutes.

Serve at once as a main dish vegetable accompaniment to fish or poultry dishes.

———•———

CURRIED FENNEL

6 bulbs fennel

1 clove garlic

1 sprig fresh thyme

1 cup chicken stock (see page 37)

2 tbsp mild curry powder

¼ cup grated Parmesan or other hard cheese

1 cup fine breadcrumbs

¼ cup butter

salt and pepper

Set up a steamer ready for use, with the water heating in the lower compartment. Preheat the oven to 350°F. Rinse the fennel, dry, and slice lengthwise into wide, fairly thick slices. Steam for approximately 10 minutes or until just tender. Alternatively, boil until tender. Melt 2 tbsp butter in a skillet and fry the fennel gently on both sides with the peeled garlic clove and thyme sprig for about 5 minutes to brown lightly. Add a little salt and freshly ground pepper and the stock mixed with the curry powder. Cover and simmer for 5 minutes.

Transfer to a greased gratin dish, sprinkle with the cheese and breadcrumbs, dot a few flakes of butter here and there, and brown in the oven for 10 minutes.

This dish can be prepared in advance, sprinkled with the topping shortly before placing in the oven, and given a little longer to heat through and brown.

———•———

CAULIFLOWER AND POTATOES INDIAN STYLE

1 small cauliflower

2 cold boiled potatoes

2 tsp ground cumin

1 tsp cumin seeds

1 small fresh green chili pepper, seeds removed

1 tsp ground coriander

½ tsp turmeric powder

⅓ cup sunflower oil

salt and pepper

Serve with:

chapatis (see page 81) or nan (Indian flat bread)

Divide the cauliflower into florets, discarding the root. Cut the potatoes into small cubes. Roast the ground cumin in a nonstick saucepan with no added oil over low heat for 1 minute while stirring; set aside. Heat the oil in a large nonstick skillet and cook the cumin seeds in it for about 3 seconds before adding the cauliflower florets and stir-frying them over moderately high heat. Reduce the heat to very low, cover, and cook for 7 minutes by which time the florets should be just tender. Mix 1 tsp salt with the chopped chili pepper, the roasted ground cumin, ground coriander, turmeric, and some freshly ground black pepper in a small bowl.

Add the potatoes to the cauliflower, stir in the spice

mixture, and continue stirring carefully for 4 minutes over low heat. Serve at once, with chapatis, or heated nan bread brushed with melted butter.

———— • ————

CAULIFLOWER SALAD

1 cauliflower

6 gherkins pickled in mild vinegar

6 small anchovy fillets

1 pickled or canned sweet pepper

9 green olives, pitted

9 black olives, pitted

2 tbsp capers

extra-virgin olive oil

wine or cider vinegar

salt and pepper

Bring a large saucepan of salted water to a boil. Cut the cauliflower (see page 19) into quarters and add to the boiling water to cook for 15–20 minutes or until tender but still firm. Drain immediately and leave to cool.

Cut the gherkins into thick slices, the anchovies into small pieces, and the drained sweet pepper into short, thin strips. Cut off the larger stalks and "core" of the cauliflower and divide the head into individual florets. Place these in a large salad bowl with all the other ingredients and sprinkle with a dressing made with a pinch of salt and freshly ground pepper mixed with 2 tbsp good wine vinegar or cider vinegar and ¼ cup best-quality olive oil. Stir carefully to dress without breaking up the florets.

Do not use gherkins, sweet peppers or capers that have been pickled in acetic acid as this is so sharp and strong that it will effectively drown every other flavor.

———— • ————

SWEET-SOUR CELERY CABBAGE PEKING STYLE

1 head pe-t'sai (celery cabbage) weighing about 2 lb

1 small piece fresh ginger, peeled and finely chopped

1 dried red chili pepper

6 tbsp sunflower or peanut oil

½ cup vegetable stock (see page 38)

salt

For the sauce:

2 tbsp all-purpose flour

2 tbsp light soy sauce

¼ cup wine vinegar or cider vinegar

¼ cup Chinese rice wine, saké or dry sherry

3 tbsp sugar

6 tbsp unsweetened tangerine or orange juice

½ chicken or vegetable bouillon cube, crumbled

Serve with:

steamed rice (see page 126)

Cut off the hard base of the celery cabbage and peel off the outer, wilted leaves. You will need about 1¾ lb of crisp, tender leaves. Rinse the leaves well, drain, and cut across into pieces about 1½ wide.

Make the sauce: mix the flour with the soy sauce in a small bowl; stir in all the other sauce ingredients as listed above.

Heat the oil in a large wok or nonstick skillet and fry the chopped ginger for a few seconds before crumbling in the chili pepper and adding the celery cabbage. Stir-fry over high heat for 2–3 minutes.

Add the stock, stir, cover, and reduce the heat to very low. Simmer for 5 minutes, stir the sauce ingredients once more, and then mix into the contents of the wok. Turn up the heat a little and stir for 1 minute or until the sauce has thickened and coated all the leaves; add a very little salt if wished.

Serve at once. This is delicious with plain steamed rice.

BRAISED RED CABBAGE WITH APPLES
Rotkraut

1 red cabbage

4 firm cooking apples

¼ cup red wine vinegar

1 small onion

2 tbsp butter

1½ tbsp oil

1 cup red wine

salt and black peppercorns

Wash and trim the cabbage and cut into quarters, removing and discarding the tough remains of the stem and the very large ribs. Shred. Peel, quarter, core, and slice the apples and mix the cabbage and apples in a large mixing bowl, sprinkling them with the vinegar and a generous pinch of salt. Stir.

Peel the onion and chop finely; fry gently in the butter and oil in a large, heavy-bottomed saucepan or casserole dish (preferably stainless steel or enameled cast iron). When the onion is tender but not browned, add the shredded cabbage and the apple slices. Stir well and add the wine. Cover with a tight-fitting lid and simmer gently for 40 minutes, adding a little water if necessary. When the cabbage is tender, add salt if needed and plenty of freshly ground black pepper, stir once more, and serve very hot.

———— • ————

BRUSSELS SPROUTS WITH ROASTED SESAME SEEDS

1¼ lb Brussels sprouts

3 tbsp white sesame seeds

2 tbsp butter

2 tbsp olive oil

salt and white pepper

Bring a large saucepan of water to a boil. Add a generous pinch of salt. Clean the Brussels sprouts and add to the boiling water to boil fast, uncovered, for up to 15 minutes or until they are just tender. Drain in a colander.

Roast the sesame seeds without any oil or fat, spread out in a large nonstick skillet for 1–2 minutes over moderate heat, stirring continuously to prevent them overcooking or burning. Remove from the heat. Heat the olive oil in the emptied saucepan used to boil the Brussels sprouts, add the sprouts, and stir gently over low heat for 2–3 minutes. Transfer to a heated serving dish, sprinkle with the sesame seeds, and serve at once.

———— • ————

BROCCOLI PURÉE

2¾ lb broccoli

3 shallots or 1 medium-sized Bermuda onion

½ clove garlic

3 tbsp butter

½ cup vegetable stock (see page 38) or chicken stock (see page 37)

nutmeg

2 tbsp grated Swiss cheese (Emmenthal or Gruyère)

½ cup heavy cream or sour cream

salt and pepper

Trim and prepare the broccoli, discarding all the larger, older stalks. Cut off the budding flower heads or small florets; chop the stalks coarsely.

Peel the shallots or onion and chop finely together with the garlic; fry these gently in 1 tbsp of the butter until soft, add the chopped small stems, and cook while stirring for 2–3 minutes. Pour in the stock, cover, and simmer over low heat for 10 minutes or until the stalks are very tender and the liquid has completely evaporated. Reduce to a purée in the food processor, adding a pinch each of salt, pepper, and freshly grated nutmeg.

Boil the florets in salted water until just tender, turn into a sieve, and refresh under running cold water. Drain well before chopping finely.

Heat the purée just before you plan to serve it, stirring in the chopped florets. Draw aside from the heat when hot, stir in 2 tbsp butter, the grated Swiss cheese, a little more salt and pepper if needed, and the cream.

RICE WITH GREEN CABBAGE

8 large, crisp green cabbage leaves
3 medium-sized waxy potatoes
1 cup risotto rice (e.g. arborio)
4 fairly thick bacon slices
3–4 sage leaves
2 tbsp butter
1½ tbsp oil
⅓ cup grated Parmesan or other hard cheese
salt and pepper

Bring a large saucepan two-thirds full of water to a boil with a pinch of salt. Rinse the cabbage leaves and shred. Peel the potatoes and dice. Add the cabbage and potatoes to the boiling water and cook fast for 2 minutes. Sprinkle in the rice and cook for about 14 minutes or until it is tender but still has a little bite to it. Drain and leave to cool. This first stage of preparation is best completed several hours in advance, as the mixture is easier to handle if chilled in the refrigerator; take it out and leave at room temperature for about 20 minutes before completing the recipe.

Chop the bacon and fry in the hot butter and oil in a large saucepan with the sage leaves for 2–3 minutes while stirring, until the bacon is crisp and brown. Do not allow the butter to burn. Add the rice and vege-

table mixture, turn up the heat, and fry while stirring for 2–3 minutes to reheat and flavor. Remove from the heat, sprinkle with the grated cheese, and serve at once on hot plates. It is excellent with very fresh white (non-oily) fish that has been rolled in seasoned flour and shallow-fried in a little oil and butter.

CARDOONS AU GRATIN

approx. 2 lb cardoons or celery
1 lemon
1½ quarts vegetable stock (see page 38)
¼ cup small slivers of Parmesan cheese
¼ cup small slivers of Swiss cheese (Emmenthal)
¼ cup butter
white truffle, fresh or canned (optional)
salt and white pepper

Have the stock ready prepared in advance. Reheat it slowly over a low heat, stirring now and then. Trim and wash the cardoons; you will need to remove the outer, stringy layer of the stalks with a potato peeler. Cut into 2-in sections and drop them straight into a bowl of iced water mixed with the juice of 1 lemon as you prepare them or else they will discolor. When the cooking liquid has come to a boil, drain the cardoons and add to it. Regardless of what vegetable you choose, the cooking liquid should completely cover it. Simmer for anything from 1–2 hours for cardoons, a fraction of this for celery. Drain.

Preheat the oven to 400°F. Grease a wide, shallow ovenproof dish with butter and place a layer of the cooked vegetables in it. Cover with half the mixed cheeses. Dot half the remaining butter in flakes over the surface. Cover with a second layer of the vegetables, followed by the cheeses and remaining butter. If you are using a truffle or canned truffle peelings, sprinkle these on top and bake in the oven for 10 minutes.

SWEET SPINACH PIE

2–2½ lb spinach

1 piece fresh ginger, very finely chopped

3 tbsp butter

1½ tbsp seedless white raisins soaked in ½ cup dark rum

5 eggs

½ cup light brown cane sugar

4 allspice berries, ground

nutmeg

1 small lemon

2 tbsp pine nuts

1¼ cups heavy cream

2 8-oz packages frozen puff pastry

2 tbsp sesame seeds

Serve with:

¾ cup whipping cream, stiffly beaten

Trim off the spinach stalks and keep them for use in other dishes; wash the leaves, do not dry or shake them, but place directly in a nonstick saucepan with no additional water.

Cover and cook over moderate heat for 2 minutes, turning the spinach once. Transfer to a colander, refresh under running cold water; drain and squeeze very tightly to eliminate as much moisture as possible. Chop finely.

Peel and chop the ginger; heat 1½ tbsp of the butter in a nonstick skillet and cook the ginger for a few seconds, stirring; add the spinach and cook for 2 minutes, stirring continuously. Set aside to cool.

Have the seedless white raisins ready soaked in the rum for 15 minutes.

Beat 4 of the eggs in a very large bowl with the sugar; add the allspice, a pinch of grated nutmeg, and the finely grated rind of the lemon.

Cook the pine nuts in 1½ tbsp butter in a nonstick saucepan for a few seconds, shaking the pan; they should be a pale golden brown. Add the cream to the egg mixture in the bowl, followed by the spinach, the raisins and rum, and the pine nuts. Preheat the oven to 400°F. Roll out the two thawed pieces of pie dough into rectangles and use one to line a nonstick jelly roll pan or shallow baking tray. Spread all the spinach mixture evenly over the surface, leaving an uncovered border all the way round the edges. Cover with the second rectangle of pie dough and seal the edges tightly, pressing them down all the way round with the tines of a fork. Beat the remaining egg lightly and brush all over the surface of the pie. Sprinkle with the sesame seeds and bake in the oven for 20–25 minutes, until crisp and pale golden brown. Leave to stand for 10 minutes before serving with whipped cream.

———— • ————

PERSIAN RHUBARB FRAPPÉ
Sharbate Rivas

2–2½ lb tender, pink rhubarb stalks

2½ cups superfine sugar

finely chopped ice

Decorate with:

cocktail cherries

mint leaves

Wash the rhubarb stalks well and trim off both ends; if the stalks are not very thin and young, run the potato peeler down them to remove any tough fibers. Cut into small pieces. Place in a heavy-bottomed enameled fireproof casserole dish or stainless steel saucepan with 2¼ cups water. Bring to a boil, cover, and simmer gently for 25 minutes over low heat. Purée in a food processor, pour into a measuring jug, and add a little water if necessary to make the volume up to 1¾ cups. Pour back into an enameled or stainless steel casserole dish or saucepan, add the sugar, and stir over moderate heat until it has totally dissolved. Boil the rhubarb purée gently, uncovered, for about 6 minutes or until a candy thermometer registers 425°F (or until a drop added to a saucer of iced water forms a malleable ball). Draw aside from the heat and leave to cool, then chill.

Just before serving, pour 1/2 cup of the rhubarb into large, tall glasses and fill up with chopped ice. Decorate with the cocktail cherries and mint; place the glasses on small plates with long-handled ice cream soda spoons. Alternatively, freeze the cooled rhubarb mixture in an ice-cream maker or in a shallow bowl, stirring at intervals as it freezes and thickens.

—— • ——

RHUBARB CAKE

1¼ lb rhubarb stalks
1½ cups sugar
generous pinch cinnamon
generous ½ cup butter
3 eggs
2 cups all-purpose flour
1½ tsp baking powder
6 tbsp milk
1 lemon
salt

The day before you plan to make this cake, trim and rinse the rhubarb, running a potato peeler over the surface to take off the outermost layer unless it is very young and tender. Any thick stalks should be cut lengthwise in half, so that all the pieces are of an even thickness. Cut them into 1½-in sections and place in layers in a large nonmetallic bowl (plastic or glass are ideal). Mix 1 cup of the sugar with the cinnamon and sprinkle a little onto each layer, reserving most of it for sprinkling over the top layer. Leave to stand in a cool place overnight.

When it is time to make the cake, preheat the oven to 475°F. Work the butter, softened at room temperature, with ½ cup of the sugar; beat until pale and fluffy; beat in the eggs one at a time. Sift in the flour with the baking powder and a pinch of salt; stir in well, adding the milk and the juice and finely grated rind of the lemon.

Grease a 10-in pie dish or shallow springform cake pan with butter; drain off all the liquid from the rhubarb and chop 7 oz of the pieces finely; fold into the cake batter and turn into the pie dish or cake pan; smooth the surface level with a spatula. Cover with the remaining pieces of rhubarb, arranging these neatly side by side in concentric circles, working from the outer edge inward.

Bake in the oven for 45 minutes; leave to settle and cool slightly before taking it out of the cake pan (if baked in a ceramic pie dish you need not take it out). Serve warm or at room temperature.

·SHOOTS AND FRUITS·

GLOBE ARTICHOKES IN OIL

p. 105

Preparation: 1 hour + 1–2 weeks
maturing time
Cooking time: 12–15 minutes
Difficulty: easy
Snack

SUN-DRIED TOMATOES IN OIL WITH BREAD AND CHEESE

p. 105

Preparation: 30 minutes + 2 hours
soaking time + 2 weeks maturing time
Difficulty: very easy
Snack

BEAN SPROUT, CUCUMBER, AND CRAB SALAD

Kani to Mamiyashi

p. 105

Preparation: 10 minutes + 1 hour chilling
time
Cooking time: 2–3 minutes
Difficulty: easy
Appetizer

ASPARAGUS WITH BÉARNAISE SAUCE

p. 106

Preparation: 30 minutes
Cooking time: 10–12 minutes
Difficulty: fairly easy
Appetizer

STEAMED ASPARAGUS WITH MOUSSELINE SAUCE

p. 107

Preparation: 30 minutes
Cooking time: 10–12 minutes
Difficulty: fairly easy
Appetizer

JAPANESE ASPARAGUS SALAD

p. 107

Preparation: 15 minutes
Cooking time: 5 minutes
Difficulty: very easy
Appetizer

EGGPLANT TURKISH STYLE

Imam Bayildi

p. 107

Preparation: 15 minutes + 30 minutes
standing time
Cooking time: 1 hour 15 minutes
Difficulty: easy
Appetizer

EGGPLANT MOCK PIZZAS

p. 108

Preparation: 15 minutes
Cooking time: 20 minutes
Difficulty: very easy
Appetizer

TURKISH EGGPLANT

Patlican Ezmesi

p. 108

Preparation: 20 minutes
Cooking time: 30 minutes
Difficulty: very easy
Appetizer

EGGPLANT JAPANESE STYLE

Yaki Nasubi

p. 109

Preparation: 15 minutes
Cooking time: 30 minutes
Difficulty: very easy
Appetizer

DEEP-FRIED EGGPLANT TURNOVERS

Melitzano Burakakia

p. 109

Preparation: 30 minutes
Cooking time: 15 minutes
Difficulty: easy
Appetizer

GREEK EGGPLANT DIP

Melitzano Salata

p. 110

Preparation: 20 minutes + 30 minutes
chilling time
Cooking time: 30 minutes
Difficulty: very easy
Appetizer

DEEP-FRIED EGGPLANT WITH YOHURT

Patlican Kizartmasi

p. 110

Preparation: 10 minutes + 30 minutes
standing time
Cooking time: 20 minutes
Difficulty: easy
Appetizer

GREEK SALAD

p. 110

Preparation: 15 minutes
Difficulty: very easy
Appetizer

Provençal stuffed
vegetables

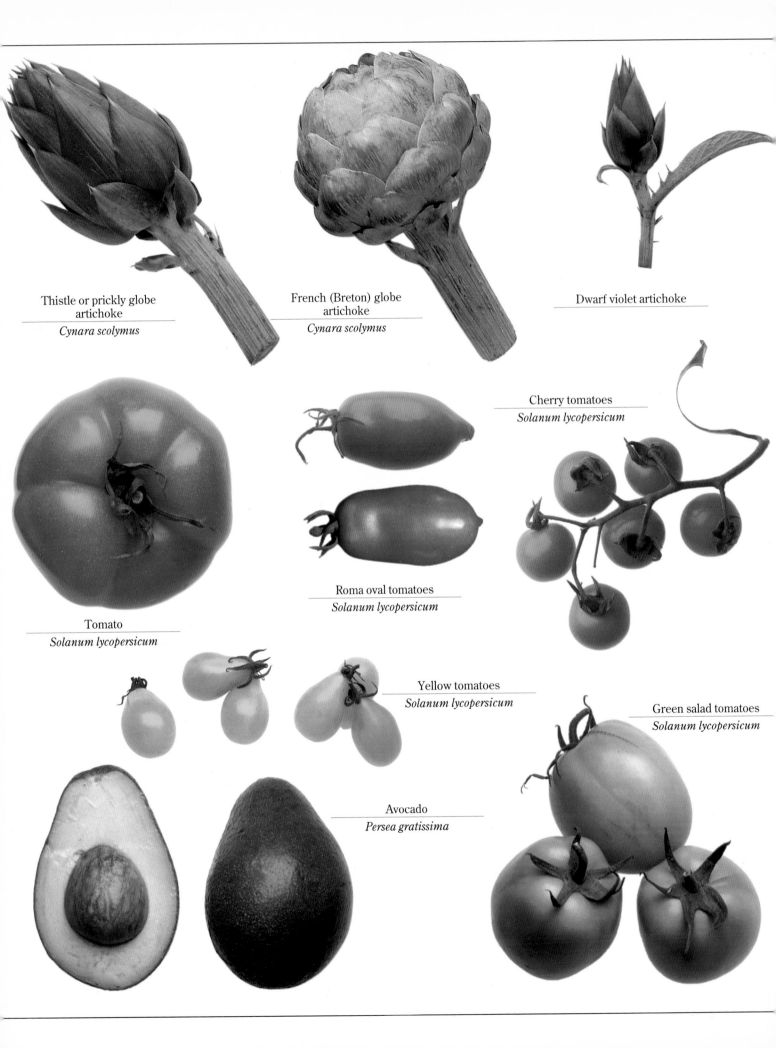

Thistle or prickly globe
artichoke
Cynara scolymus

French (Breton) globe
artichoke
Cynara scolymus

Dwarf violet artichoke

Cherry tomatoes
Solanum lycopersicum

Roma oval tomatoes
Solanum lycopersicum

Tomato
Solanum lycopersicum

Yellow tomatoes
Solanum lycopersicum

Green salad tomatoes
Solanum lycopersicum

Avocado
Persea gratissima

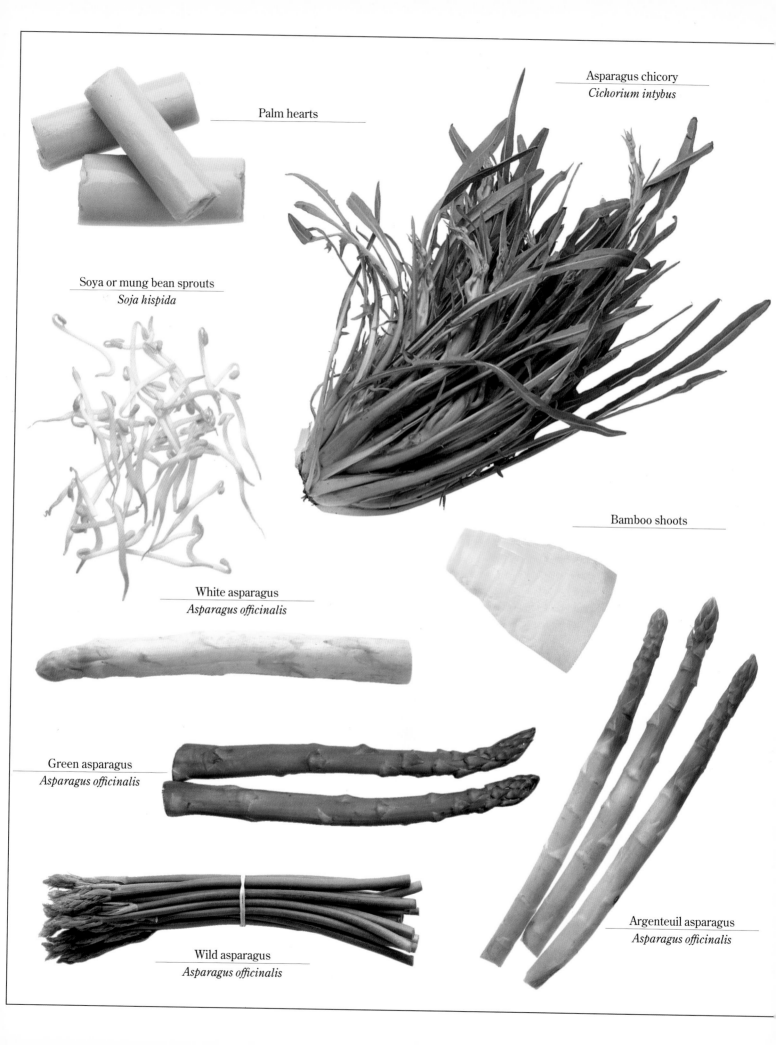

Palm hearts

Asparagus chicory
Cichorium intybus

Soya or mung bean sprouts
Soja hispida

Bamboo shoots

White asparagus
Asparagus officinalis

Green asparagus
Asparagus officinalis

Argenteuil asparagus
Asparagus officinalis

Wild asparagus
Asparagus officinalis

STUFFED BAKED PEPPER AND ANCHOVY ROLLS

p. 111

Preparation: 30 minutes
Cooking time: 35–40 minutes
Difficulty: very easy
Appetizer

AVOCADO, TOMATO, AND GRAPEFRUIT SALAD WITH CITRONETTE DRESSING

p. 111

Preparation: 20 minutes
Difficulty: easy
Appetizer

AVOCADO AND SHRIMP IN TUNA SAUCE

p. 111

Preparation: 15 minutes
Cooking time: 2 minutes
Difficulty: very easy
Appetizer

JAPANESE CUCUMBER AND HAM SALAD

Sunomono

p. 112

Preparation: 10 minutes + 1 hour chilling time
Difficulty: very easy
Appetizer

CUCUMBER WITH SALMON ROE AND HARD-BOILED EGGS

Ikura to Kyuri

p. 112

Preparation: 10 minutes + 20 minutes standing time
Cooking time: 10 minutes
Difficulty: easy
Appetizer

GUACAMOLE

p. 113

Preparation: 15 minutes
Difficulty: very easy
Appetizer

PILI-PILI TOMATO SALAD

p. 113

Preparation: 15 minutes + 3 hours soaking time
Difficulty: very easy
Appetizer or accompaniment

STUFFED TOMATOES AU GRATIN

p. 114

Preparation: 30 minutes
Cooking time: 15 minutes
Difficulty: very easy
Appetizer or accompaniment

ZUCCHINI ESCABECHE

p. 114

Preparation: 15 minutes + a few hours chilling time
Cooking time: 20 minutes
Difficulty: easy
Appetizer or accompaniment

DEEP-FRIED OKRA

p. 115

Preparation: 30 minutes
Cooking time: 20 minutes
Difficulty: easy
Appetizer or accompaniment

SPRING VEGETABLE SOUP

p. 115

Preparation: 20 minutes
Cooking time: 1 hour
Difficulty: very easy
First course

CREAM OF ASPARAGUS SOUP

p. 115

Preparation: 20 minutes
Cooking time: 40 minutes
Difficulty: easy
First course

CASTILIAN CREAM OF ARTICHOKE SOUP

p. 116

Preparation: 20 minutes
Cooking time: 40 minutes
Difficulty: very easy
First course

CHILLED TOMATO SOUP WITH SCAMPI

Crème de tomates Bec Fin

p. 117

Preparation: 30 minutes + 3 hours chilling time
Cooking time: 10 minutes
Difficulty: very easy
First course

CREAM OF TOMATO SOUP

p. 117

Preparation: 20 minutes
Cooking time: 30 minutes
Difficulty: very easy
First course

PUMPKIN SOUP WITH AMARETTI

p. 118

Preparation: 25 minutes
Cooking time: 30 minutes
Difficulty: very easy
First course

ARGENTINE ICED PUMPKIN SOUP

Fonda del Sol

p. 118

Preparation: 20 minutes + 2 hours chilling time
Cooking time: 1 hour
Difficulty: very easy
First course

BAMBOO SHOOT AND ASPARAGUS SOUP

p. 118

Preparation: 20 minutes + 30 minutes soaking time
Cooking time: 40 minutes
Difficulty: very easy
First course

AVOCADO SOUP

Sopa de Aguacate

p. 119

Preparation: 15 minutes
Cooking time: 5 minutes
Difficulty: very easy
First course

CHINESE CUCUMBER, PORK, AND CHICKEN SOUP

p. 119

Preparation: 15 minutes
Cooking time: 8 minutes
Difficulty: very easy
First course

ZUCCHINI SOUP WITH BASIL AND LEMON

p. 120

Preparation: 25 minutes
Cooking time: 40 minutes
Difficulty: very easy
First course

GAZPACHO

p. 120

Preparation: 30 minutes + 2–6 hours chilling time
Difficulty: very easy
First course

RISOTTO WITH ZUCCHINI FLOWERS

p. 121

Preparation: 15 minutes
Cooking time: 20 minutes
Difficulty: easy
First course

ARTICHOKE RISOTTO

p. 121

Preparation: 20 minutes
Cooking time: 20 minutes
Difficulty: easy
First course

SPAGHETTI WITH TOMATOES AND CHILI

p. 122

Preparation: 15 minutes
Cooking time: 10 minutes
Difficulty: very easy
First course

SPAGHETTI WITH MEDITERRANEAN SAUCE

p. 122

Preparation: 15 minutes
Cooking time:
Difficulty: very easy
First course

SPAGHETTI WITH EGGPLANT

p. 122

Preparation: 15 minutes
Cooking time: 20 minutes
Difficulty: easy
First course

SPAGHETTI WITH ZUCCHINI FLOWERS, CREAM, AND SAFFRON

p. 123

Preparation: 15 minutes
Cooking time: 10 minutes
Difficulty: very easy
First course

PASTA CAPRI

p. 123

Preparation: 30 minutes
Cooking time: 10 minutes
Difficulty: easy
First course

TURKISH STUFFED PEPPERS

Zeytinyagli Biber Dolmasi

p. 124

Preparation: 45 minutes
Cooking time: 45 minutes
Difficulty: easy
First course

PALM HEART, SHRIMP, AND AVOCADO SALAD

p. 125

Preparation: 15 minutes
Cooking time: 2 minutes
Difficulty: very easy
Lunch/Supper dish

CHICKEN, TOMATO, AND AVOCADO SALAD WITH TUNA SAUCE

p. 125

Preparation: 30 minutes
Cooking time: 6 minutes
Difficulty: easy
Lunch/Supper dish

BEAN SPROUTS WITH PORK AND STEAMED RICE

Momyashi to Gyuniko

p. 126

Preparation: 25 minutes + 20 minutes soaking time
Cooking time: 25 minutes
Difficulty: easy
Main course

WHITE ASPARAGUS WITH SPICY SAUCE

p. 127

Preparation: 20 minutes
Cooking time: 12–14 minutes
Difficulty: very easy
Main course

ASPARAGUS WITH EGGS AND PARMESAN

p. 127

Preparation: 20 minutes
Cooking time: 12 minutes
Difficulty: very easy
Main course

ASPARAGUS MOULDS

p. 128

Preparation: 15 minutes
Cooking time: 30 minutes
Difficulty: very easy
Main course

GLOBE ARTICHOKES ROMAN STYLE

p. 128

Preparation: 30 minutes
Cooking time: 20 minutes
Difficulty: easy
Main course

ARTICHOKE AND EGG PIE

p. 129

Preparation: 1 hour 30 minutes
Cooking time: 40 minutes
Difficulty: fairly easy
Main course

ARTICHOKE FRICASSÉE

p. 129

Preparation: 40 minutes
Cooking time: 20 minutes
Difficulty: easy
Main course

BAKED VEGETABLE OMELET

p. 130

Preparation: 45 minutes
Cooking time: 30 minutes
Difficulty: easy
Main course

DEEP-FRIED EGGPLANT AND CHEESE SANDWICHES

p. 130

Preparation: 10 minutes
Cooking time: 15 minutes
Difficulty: easy
Main course

STUFFED EGGPLANTS

p. 131

Preparation: 30 minutes
Cooking time: 15 minutes
Difficulty: easy
Main course

EGGPLANT, TOMATO, AND CHEESE BAKE

p. 131

Preparation: 25 minutes
Cooking time: 15 minutes
Difficulty: easy
Main course

NEAPOLITAN STUFFED EGGPLANTS

p. 132

Preparation: 25 minutes
Cooking time: 15 minutes
Difficulty: easy
Main course

STUFFED EGGPLANTS GREEK STYLE

Papuzakia

p. 132

Preparation: 20 minutes
Cooking time: 45 minutes
Difficulty: easy
Main course

EGGPLANT CUTLETS WITH TOMATO RELISH

p. 133

Preparation: 20 minutes
Cooking time: 5 minutes
Difficulty: easy
Main course

EGGPLANT TIMBALES WITH FRESH MINT SAUCE

p. 134

Preparation: 15 minutes
Cooking time: 40 minutes
Difficulty: easy
Main course

EGGPLANT PIE

p. 134

Preparation: 20 minutes
Cooking time: 1 hour 10 minutes
Difficulty: easy
Main course

CAPONATA

p. 135

Preparation: 25 minutes + 1 hour standing time
Cooking time: 40 minutes
Difficulty: easy
Main course

PROVENÇAL STUFFED VEGETABLES

Petits Farcis

p. 136

Preparation: 1 hour
Cooking time: 30 minutes
Difficulty: easy
Main course

RATATOUILLE

p. 137

Preparation: 30 minutes + 1 hour standing time
Cooking time: 40 minutes
Difficulty: easy
Main course

PEPERONATA

p. 137

Preparation: 25 minutes
Difficulty: easy
Main course

STUFFED PEPPERS

p. 138

Preparation: 40 minutes
Cooking time: 1 hour 10 minutes
Difficulty: easy
Main course

MEXICAN PEPPERS WITH POMEGRANATE SAUCE

Chiles en Nogada

p. 138

Preparation: 30 minutes
Cooking time: 30 minutes
Difficulty: easy
Main course

PEPPER, ZUCCHINI, AND TOMATO FRICASSÉE

Pisto Manchego

p. 139

Preparation: 30 minutes
Cooking time: 45 minutes
Difficulty: easy
Main course

Eggplant mock pizzas

Winter squash
Cucurbita maxima

Pumpkin or winter squash
Cucurbita moschata

Baby zucchini with flowers
Cucurbita pepo

Okra
Hibiscus esculentum

Zucchini
Cucurbita pepo

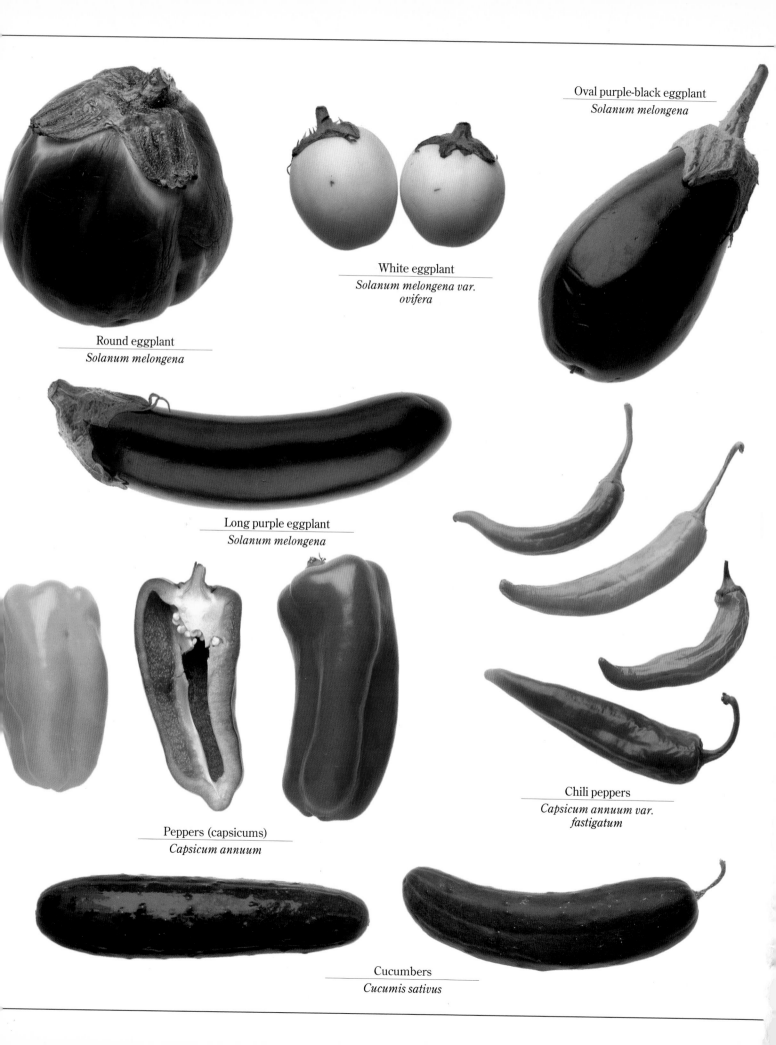

Oval purple-black eggplant
Solanum melongena

White eggplant
*Solanum melongena var.
ovifera*

Round eggplant
Solanum melongena

Long purple eggplant
Solanum melongena

Chili peppers
*Capsicum annuum var.
fastigatum*

Peppers (capsicums)
Capsicum annuum

Cucumbers
Cucumis sativus

Cream of tomato soup

LOBSTER CATALAN

p. 140

Preparation: 30 minutes + defrosting time
Cooking time: 10 minutes
Difficulty: easy
Main course

AVOCADO AND CHICKEN SALAD WITH BLUE CHEESE DRESSING

p. 141

Preparation: 1 hour
Cooking time: 1 hour
Difficulty: easy
Main course

TEMPURA WITH TENTSUYU SAUCE

p. 141

Preparation: 30 minutes + 1 hour chilling time
Cooking time: 30 minutes
Difficulty: fairly easy
Main course

ZUCCHINI WITH DILL SAUCE

p. 142

Preparation: 40 minutes
Cooking time: 50 minutes
Difficulty: easy
Main course

PERSIAN STUFFED ZUCCHINI

Dolmeh Kadu

p. 143

Preparation: 30 minutes
Cooking time: 20–25 minutes
Difficulty: easy
Main course

ZUCCHINI FRITTERS

Mücver

p. 143

Preparation: 25 minutes
Cooking time: few minutes
Difficulty: easy
Main course

BRAISED ARTICHOKES

p. 144

Preparation: 30 minutes
Cooking time: 15 minutes
Difficulty: very easy
Accompaniment

ARTICHOKE BOTTOMS IN CHEESE SAUCE

p. 144

Preparation: 30 minutes
Cooking time: 50 minutes
Difficulty: easy
Accompaniment

CRISPY GLOBE ARTICHOKES

p. 145

Preparation: 20 minutes
Cooking time: 15 minutes
Difficulty: easy
Accompaniment

ARTICHOKE BOTTOMS VENETIAN STYLE

p. 145

Preparation: 40 minutes
Cooking time: 20 minutes
Difficulty: easy
Accompaniment

BAKED TOMATOES WITH MINT

p. 145

Preparation: 10 minutes
Cooking time: 50 minutes
Difficulty: very easy
Accompaniment

TOMATOES AU GRATIN

p. 146

Preparation: 10 minutes
Cooking time: 30 minutes
Difficulty: very easy
Accompaniment

BAKED EGGPLANTS

p. 146

Preparation: 10 minutes
Cooking time: 35 minutes
Difficulty: very easy
Accompaniment

NEAPOLITAN EGGPLANT

p. 146

Preparation: 20 minutes + 30 minutes standing time
Cooking time: 2 minutes
Difficulty: very easy
Accompaniment

EGGPLANT AND MINT RAITA

p. 147

Preparation: 15 minutes
Cooking time: 10 minutes
Difficulty: very easy
Accompaniment

DEEP-FRIED EGGPLANT WITH MISO SAUCE

Nasu no Miso Kake

p. 147

Preparation: 15 minutes + 20 minutes soaking time
Cooking time: 5 minutes
Difficulty: easy
Accompaniment

FRIED EGGPLANTS WITH GARLIC AND BASIL

p. 148

Preparation: 10 minutes
Cooking time: 20 minutes
Difficulty: very easy
Accompaniment

PEPPERS AU GRATIN

p. 148

Preparation: 40 minutes + 1 hour chilling time
Cooking time: 55 minutes
Difficulty: easy
Accompaniment

STIR-FRIED PEPPERS IN SHARP SAUCE

p. 149

Preparation: 15 minutes
Cooking time: 20 minutes
Difficulty: very easy
Accompaniment

TSATSIKI

p. 149

Preparation: 15 minutes
Difficulty: very easy
Accompaniment

STIR-FRIED ZUCCHINI WITH PARSLEY, MINT, AND GARLIC

p. 149

Preparation: 15 minutes
Cooking time: 10 minutes
Difficulty: easy
Accompaniment

ZUCCHINI PURÉE

p. 150

Preparation: 10 minutes
Cooking time: 20 minutes
Difficulty: very easy
Accompaniment

ZUCCHINI RUSSIAN STYLE

p. 150

Preparation: 15 minutes
Cooking time: 15 minutes
Difficulty: easy
Accompaniment

AVOCADO CREAM DESSERT

p. 150

Preparation: 15 minutes + 1 hour chilling time
Difficulty: very easy
Dessert

AVOCADO ICE CREAM

p. 151

Preparation: 40 minutes + 2 hours chilling time
Difficulty: very easy
Dessert

CANDIED PUMPKIN

p. 151

Preparation: 30 minutes + 24 hours soaking time + 1 week standing time
Cooking time: 40 minutes
Difficulty: very easy
Dessert

PUMPKIN AND AMARETTI CAKE

p. 151

Preparation: 20 minutes
Cooking time: 50 minutes
Difficulty: very easy
Dessert

TURKISH PUMPKIN DESSERT

Kabak Tatlisi

p. 152

Preparation: 20 minutes + 24 hours standing time
Cooking time: 45 minutes
Difficulty: very easy
Dessert

GLOBE ARTICHOKES IN OIL

20 baby globe artichokes (or very small side buds)

juice of 1 lemon

olive oil

4½ cups vegetable stock (see page 38) or court-bouillon *(see page 37)*

Half fill a very large bowl with cold water and the lemon juice to acidulate. Prepare the artichokes (see page 18), stripping them down to the heart and slicing off the leaves about three-quarters of the way down to the base. Cut lengthwise into quarters. As each artichoke is prepared, drop it into the bowl of acidulated water to prevent discoloration.

Bring the aromatic cooking liquid or *court-bouillon* slowly to a boil over low heat. When it boils, drain the artichokes and boil gently in it until they are tender but still fairly crisp (about 12–15 minutes). Drain the artichokes upside down, then pack in layers into large, sterilized glass preserving jars. Pour in sufficient olive oil to completely cover the top layer, tapping the jar on the work surface to release any trapped air bubbles. Seal tightly. Store in a cupboard at cool room temperature and wait at least 2 weeks before sampling.

——— • ———

SUN-DRIED TOMATOES IN OIL WITH BREAD AND CHEESE

1 lb sun-dried tomatoes

white wine vinegar

bay leaves

olive oil

garlic

canned anchovy fillets, drained

small, firm black olives, pitted

dried chili peppers

oregano

capers

Serve with:

mild cheese

crusty white or whole wheat bread

Place the tomatoes in a large, nonmetallic bowl, pour in sufficient vinegar to completely cover them, and leave to stand for 2 hours. Drain off the vinegar and spread the tomatoes out on paper towels; cover with more paper towels and blot up excess vinegar. Layer the tomatoes with all the other ingredients in sterilized canning jars, starting with a small bay leaf in the bottom of the jar, and covering it with a layer of tomatoes. Sprinkle these with some oil, then place a small, peeled clove of garlic on top, one or two pieces of anchovy fillet, a couple of olives, a piece of chili pepper, a pinch of oregano, and 2 or 3 capers. Position the small bay leaves against the sides of the jar as you continue layering until the jar is almost full. Make sure the olive oil completely covers the top layer and tap the jar gently to release any air bubbles that have been trapped in the layering process.

Seal the jars tightly and store in a dark cupboard at cool room temperature. Wait at least 2 weeks before using the tomatoes. Drain shortly before serving. These tomatoes are delicious with a fairly mild cheese and crusty bread.

——— • ———

BEAN SPROUT, CUCUMBER, AND CRAB SALAD
Kani to Momiyashi

¾ lb fresh bean sprouts (momiyashi)

1 small or ½ large cucumber

1 cup white crab meat

For the sambai zu dressing:
3 tbsp Japanese light soy sauce
3 tbsp Japanese rice vinegar or 2 tbsp cider vinegar mixed with 1 tbsp water and 1½ tbsp granulated sugar
½ cup sunflower oil
salt and freshly ground white or black pepper

Bring 9 cups water to a boil in a large saucepan, adding 2 tbsp salt. Rinse the bean sprouts; remove any green seed cases and any wilted sprouts. Add to the boiling water and blanch for only 2–3 minutes. Drain quickly and leave to cool in the colander.

Rinse and dry the cucumber, cut off both ends, then cut in half. Use a very sharp vegetable knife to remove the skin in strips with about ¼ in of the underlying flesh attached to it. This will bring you close to the central, seed-bearing section, which is not used for this recipe. Shred the "peelings" and place them in a large bowl of iced water. Chop the crab meat coarsely and mix with the bean sprouts and well drained cucumber strips in a large salad bowl. Mix the ingredients with the tips of your fingers.

Cover and chill for 1–1½ hours; transfer to individual plates and make the dressing just before serving: mix the vinegar with 2 pinches of salt and the soy sauce in a bowl. Using a small whisk, beat in the oil, adding a little at a time so that it forms an emulsion. Add pepper to taste and serve.

— • —

ASPARAGUS WITH BÉARNAISE SAUCE

1¾ lb large asparagus spears
For the béarnaise sauce:
¼ cup white wine vinegar
¼ cup dry vermouth or dry white wine
1 medium-sized shallot, finely chopped
½ tbsp dried tarragon
3 egg yolks
¾ cup butter
3 tbsp fresh French tarragon or parsley, finely chopped
½ cup heavy cream
salt and black peppercorns
Garnish with:
carrot bundles (see page 28)
chives

Clean and prepare the asparagus (see page 17), place in a steamer, sprinkle with a small pinch of salt, and steam for 10–12 minutes or until tender.

Make the sauce: heat the vinegar, vermouth or wine, shallot, dried tarragon, a generous pinch of salt, and 4 black peppercorns in a small, heavy-bottomed saucepan. Simmer, uncovered, over moderate heat until the liquid has reduced to about 3 tbsp. Pour through a very fine sieve and set aside. This stage of the sauce can be prepared before you cook the asparagus.

Have the remaining ingredients of the béarnaise sauce ready to use before you start cooking the asparagus; set aside 2 tbsp of well chilled butter, cut into pieces about the size of a filbert. Once the asparagus is steaming, melt the rest of the butter very gently. Beat the egg yolks until creamy in a small, heatproof bowl or in the top of a double boiler; gradually beat in the reduced vinegar mixture, adding a little at a time; beat in half the pieces of chilled butter. Set over gently simmering water and continue beating until the mixture starts to thicken. Immediately remove the double boiler from the heat and beat in the remaining chilled butter, beating in each piece before adding the next. Keep beating as you add the melted butter, starting with a few drops at a time and gradually increasing this to a very thin continuous trickle as the sauce thickens. Stir in the chopped fresh tarragon or parsley.

Beat the cream until stiff and fold into the sauce. Place the asparagus on plates, garnish with the carrot bundles tied with chives, and serve with the sauce.

STEAMED ASPARAGUS WITH MOUSSELINE SAUCE

1 ¾–2 lb large, fat white or green asparagus spears

For the mousseline sauce:

1 ¾ cups hollandaise sauce (see page 85)

½ cup whipping cream

Garnish with:

tomato rosebuds (see page 27)

sprigs of parsley or basil

Prepare the garnishes, if used. Trim and wash the asparagus (see page 17), place in a steamer, sprinkle with a very little salt, and steam for about 12 minutes or until tender.

Make the hollandaise sauce while the asparagus is cooking, beat the cream until fairly stiff and fold into the hollandaise sauce.

Transfer the asparagus to heated plates; garnish each serving with a tomato rosebud and a sprig of parsley or basil. Serve with the sauce.

—— • ——

JAPANESE ASPARAGUS SALAD

1 ¾ lb fresh asparagus spears

For the Japanese dressing:

5 tbsp Japanese light soy sauce

2 tsp sugar

2 tsp sesame oil

Clean and prepare the asparagus (see page 17). For this dish only the tenderest tips are used, the remainder can be kept for soups. Rinse and drain these tips and cut each one diagonally in half; turn one half upside down, so that the cut sides are still next to one another but with the top end of one half dovetailed against the lower end of its other half. Steam for 4–5 minutes or until tender but firm. Drain off any moisture and arrange on small, individual plates, sprinkling each serving with a little of the Japanese dressing. Make this by first mixing the soy sauce and sugar and then beating in the oil with a small whisk.

—— • ——

EGGPLANT TURKISH STYLE
Imam Bayildi

2 long, thin eggplants

4 medium-sized onions

approx. 1 ¼ lb large ripe tomatoes

4 small green peppers

½ fresh green chili pepper

4 large cloves garlic, peeled

6 tbsp extra-virgin olive oil

2 ½ tsp sugar

2 ½ tsp chopped parsley

salt and pepper

Cut off both ends of the eggplants and rinse; slice lengthwise in half. Use a potato peeler to peel two strips along each half, leaving a strip of skin nearest the cut edges and along the middle. Make 3 or 4 deep lengthwise cuts along the inner, cut side of each half where the flesh is at its thickest. Sprinkle liberally with salt and leave at an angle in a colander to allow the bitter juices to drain away from the flesh.

Slice the onions very thinly and place in a sieve resting in a bowl; sprinkle with 3 tbsp salt and mix well. Add sufficient cold water to just cover the onions. Place a plate that fits easily inside the sieve on top of them, press down, and place a weight on top. Leave for 30 minutes.

Blanch, peel, and seed the tomatoes. Chop the flesh into very small pieces. Remove the stalks, seeds,

107

and pale inner membrane from the peppers and cut them into thin rings.

Rinse the eggplants thoroughly under running cold water and then dry thoroughly. Rinse the onions by pouring a kettle of near-boiling water over them, drain, and squeeze out excess moisture. Mix the onions in a large bowl with the tomatoes and peppers; season with a little salt and plenty of freshly ground black pepper.

Use 1-1/2 tbsp olive oil to grease a wide, shallow flameproof casserole dish large enough to accommodate the eggplants in one layer, cut side uppermost. Use your fingers to open out the deep incisions and push some of the onion and tomato mixture down into them; spread the rest over the exposed cut surfaces. Place a slightly minced whole garlic clove on top of each eggplant and sprinkle the remaining oil over the topping and garlic cloves. Sprinkle with the sugar. Pour in 1 cup cold water through a gap between the vegetables, bring to a boil, cover, and reduce the heat to very low. Simmer very gently for 1-1/4 hours. Remove the lid and allow to cool. Sprinkle with chopped parsley just before serving straight from the cooking dish.

— • —

EGGPLANT MOCK PIZZAS

2 large, round eggplants
2 medium-sized cans Italian chopped tomatoes
¾ cup freshly grated Parmesan cheese
½ lb mozzarella cheese
oregano
extra-virgin olive oil
butter
salt and pepper

Preheat the oven to 400°F. Oil two cookie sheets or trays. Cut off both ends of the eggplants, rinse, and dry. Cut into round, even slices about ¼ in thick.

Place these on the oiled trays, brush the exposed cut surfaces with olive oil, and season with salt and freshly ground pepper. Bake in the oven for 6 minutes. Drain off excess liquid from the chopped tomatoes. Take the trays of eggplant slices out of the oven and place a few pieces of tomato on top, sprinkle with a small pinch each of salt, pepper, dried oregano, Parmesan, and about ½ tsp butter, then cover with a slice of mozzarella cheese. The "pizzas" can be prepared up to this point a few hours in advance if desired. Return to the oven for 15 minutes, then serve piping hot while the mozzarella cheese is still very soft.

— • —

TURKISH EGGPLANT DIP
Patlican Ezmesi

2 lb eggplants
1 small clove garlic, peeled
6 tbsp natural yoghurt
approx. ¼ cup fresh lime or lemon juice
4 tbsp extra-virgin olive oil
salt and pepper
Garnish with:
1 sprig parsley
1 black olive
Serve with:
pita bread

Preheat the oven to 400°F. Rinse the eggplants and place them whole on the center shelf of the oven to bake for 30 minutes, turning them after the first 15 minutes. Remove from the oven and allow to cool. Cut off the stalk ends, peel, and cut lengthwise into quarters. If they have seeds inside, scoop these out neatly and discard. Cut the flesh into small pieces and reduce to a very smooth purée in the food processor at high speed with the garlic, yoghurt, lime or lemon juice, oil, salt, and freshly ground pepper.

Transfer the dip to a serving bowl. Decorate the top with a sprig of parsley and an olive. If prepared in advance, cover with plastic wrap and refrigerate. Serve at room temperature, with hot pita bread.

———— • ————

EGGPLANT JAPANESE STYLE
Yaki Nasubi

---

4 round, medium-sized eggplants

wasabi powder (Japanese horseradish)

6 tbsp grated igname (Chinese yam or yamaimo)

3 tbsp white sesame seeds

Japanese light soy sauce

Preheat the oven to 400°F. Rinse the eggplants and place them whole on the central shelf of the oven to bake for about 30 minutes or until tender, turning them after 15 minutes. Remove from the oven and allow to cool. Slice off the ends, peel, and cut lengthwise in quarters. Neatly remove any seed-bearing sections, if visible, then cut the flesh into small, rectangular "brick" shapes and heap these up on 4 small plates.

Mix 1½ tbsp wasabi powder with just enough hot water to moisten and shape into 4 tiny balls; place one on each plate.

Peel a piece of igname (Chinese *yam* or *yamaimo*) and grate finely; you will need about 6 tbsp of its rather slimy grated flesh. Place a quarter of this neatly in a little heap beside the wasabi ball on each plate. Heat a nonstick skillet and roast the sesame seeds, stirring with a wooden spoon until they start to jump about. Remove from the heat immediately and crush slightly with the back of the spoon. Sprinkle over the heaps of eggplant pieces.

Serve this dish in the Japanese way: hand round tiny bowls of light soy sauce, mix some of the wasabi with a little soy sauce, then stir in a little of the grated igname. Take hold of a piece of eggplant with your chopsticks or fork and dip in the resulting highly flavored relish before eating.

DEEP-FRIED EGGPLANT TURNOVERS
Melítzano Burakakia

½ lb feta cheese

½ cup béchamel sauce (see page 36)

5 eggs

light olive oil

2 large eggplants

fine dry breadcrumbs

salt and white peppercorns

Use a fork to crumble the feta cheese finely in a bowl; stir in the béchamel sauce and 3 lightly beaten eggs. Season with salt and freshly ground white pepper.

Heat plenty of light olive oil in a wok or deep-fryer to 350°F. Cut off the ends of the eggplants, rinse, dry, and slice into rounds just under ¼ in thick. Do not use the tapering ends: the slices need to be large and flat. Deep-fry in batches for 1 minute or until pale golden brown but not at all crisp or dry. Drain well when removing from the oil; spread out on paper towels to finish draining. Beat the 2 remaining eggs lightly in a bowl with a little salt and pepper; set aside for dipping the turnovers. Place about 1 tbsp of the cheese and egg mixture slightly right of center on each slice; fold the left side of the slice over it, making the edges meet to form a pasty or turnover. Press gently but firmly to make the surfaces adhere; use a little beaten egg as a paste if necessary. Dip each of the turnovers twice in the beaten egg before coating with the breadcrumbs; as each one is coated, place it on a very large platter or chopping board.

Deep-fry in batches in the oil for 1 minute, or until the breadcrumb coating is golden brown. Keep hot on paper towels to finish draining while the later batches are fried; serve at once while hot.

———— • ————

GREEK EGGPLANT DIP
Melitzanosaláta

2¼ lb eggplants

1 medium-sized onion

2 small green peppers

1 green chili pepper (optional)

1 lemon

½ clove garlic

⅓ cup natural yoghurt

pinch granulated sugar

6 tbsp extra-virgin olive oil

salt and pepper

Garnish with:

black olives

Serve with:

pita bread

Preheat the oven to 400°F. Slice off the ends of the eggplants, peel them, and cut in half lengthwise. Cover a cookie sheet with foil and place the eggplants on it, cut sides uppermost. Bake in the middle of the oven for 30 minutes or until tender. Meanwhile, peel and chop the onion very finely. Remove the stalk, seeds, and inner membrane of the green peppers and the chili pepper if used. Chop these finely also. Fry the onion and peppers slowly in 3 tbsp of the oil for about 10 minutes, stirring frequently, until tender but not at all browned. Take the eggplants out of the oven when they are done and allow to cool for a few minutes before scooping out and discarding any seed-bearing sections with a spoon. Cut the eggplants into small pieces, place in the food processor with the lemon juice, the peeled garlic clove, yoghurt, a pinch of sugar, the remaining oil, and ½ tsp salt; process until very smooth. Transfer this thick purée into a serving bowl, stir in the onion and pepper mixture, and add a little more salt and freshly ground pepper to taste. Chill for about 30 minutes. Decorate with black olives. Serve with pita bread.

DEEP-FRIED EGGPLANT WITH YOGHURT
Patlican Kizartmasi

1½ lb eggplants

1 small clove garlic, peeled

1¼ cups natural yoghurt

light olive oil

salt and pepper

Garnish with:

parsley

Peel the eggplants and cut into round slices about ¼ in thick. Spread these out on large plates, sprinkle liberally with salt, and leave for 30 minutes on the drainer of your sink, having first tipped up the plates to enable the liquid to drain off. Rinse the slices under running cold water and blot dry thoroughly between clean cloths or paper towels. Heat plenty of oil to 350°F in the deep-fryer and fry the slices a few at a time until golden brown. If preferred, shallow-fry in hot oil, turning once. Drain well, spread out on paper towels, and sprinkle lightly with salt and freshly ground pepper. Transfer to a serving platter or individual plates.

Grate the garlic finely and mix with the yoghurt (low fat or standard natural yoghurt or the Greek version) and a pinch of salt. Spoon the yoghurt over the eggplants, decorate with parsley, and serve warm or cold. Eggplants absorb a great deal of oil when fried; for a version of this dish that is lower in fat, broil the eggplant slices, brushing them lightly with oil.

———— • ————

GREEK SALAD

4 very large ripe tomatoes

2 large or 4 medium-sized scallions

1 cucumber, sliced

20 black olives

4 thick slices feta cheese

oregano

extra-virgin olive oil

salt and pepper

Cut the tomatoes into eighths then cut these wedges across, in half. Place an equal number in 4 fairly deep individual plates. Slice the scallions. Arrange the scallions, cucumber, and the olives in the plates and place a slice of feta cheese in the center of each. Sprinkle oregano over the salads, season, and drizzle some olive oil onto each serving.

———•———

STUFFED BAKED PEPPER AND ANCHOVY ROLLS

4 large yellow or red peppers

16 canned anchovy fillets

savory breadcrumb stuffing (see page 114)

10 canned tomatoes, drained, seeds removed

extra-virgin olive oil

salt and pepper

Preheat the oven to 350°F. Rinse the peppers and cut lengthwise into quarters; remove the stalk, seeds, and white membrane. Place on a chopping board or work surface in a single layer, inside facing upward. Drain the anchovy fillets and lay one out flat along the center of each pepper piece; cover with 1½–3 tbsp of the bread stuffing mixture, smoothing it neatly, and top with a few strips of tomato flesh. Season with a very little salt and some freshly ground black pepper; sprinkle with olive oil. Arrange the stuffed pepper quarters in a single layer

in a lightly oiled shallow ovenproof dish, carefully pour in some boiling water into one corner of the dish, adding enough to barely cover the bottom and bake in the oven for 35–40 minutes. When the peppers are very tender, remove from the oven. As soon as they are no longer too hot to handle, roll them up loosely from one end to the other.

———•———

AVOCADO, TOMATO, AND GRAPEFRUIT SALAD WITH CITRONETTE DRESSING

1 pink grapefruit

1 large beefsteak tomato

2 ripe firm avocados

½ cup citronette dressing (see page 35)

Garnish with:

mint or basil leaves

Peel the grapefruit and slice into segments. Discard the seeds. Cut the flesh into dice and place in a large bowl. Peel the tomato, remove the seeds, and dice the flesh. Mix with the grapefruit.

Cut the avocados in half and remove the pit. Fill the hollow with the tomato and grapefruit mixture and sprinkle liberally with the citronette dressing. Place a small sprig of mint or basil on top and serve at once.

———•———

AVOCADO AND SHRIMP IN TUNA SAUCE

1 cup tuna sauce (see page 125)

1 cup thawed peeled, cooked small shrimp
1½–2 tbsp lemon juice
extra-virgin oil
2 firm ripe avocados
salt and pepper
Garnish with:
1 tbsp capers or chopped dill

Make the tuna sauce and chill in the refrigerator. Drain the shrimp and blot dry with paper towels. (If raw, steam for 2 minutes and allow to cool before using.) Mix with the lemon juice, 3 tbsp olive oil, a pinch of salt, and freshly ground white or black pepper. Cut the avocados lengthwise in half, remove the pit, and fill the hollow with the shrimp. Cover with a thin layer of tuna sauce. Garnish with the capers or dill and serve the remaining tuna sauce separately.

———— • ————

JAPANESE CUCUMBER AND HAM SALAD
Sunomono

1 cucumber
2 slices lean cooked ham, just under ¼ in thick
½ cup sambai zu dressing (see page 106)
1 tbsp finely grated fresh ginger
salt
Garnish with
cucumber fans (see page 29)

Wash the cucumber and cut off the ends. Take a handful of salt and rub it hard into the skin all over; this will preserve and enhance the green color. Wipe the skin with a clean cloth. Cut the cucumber lengthwise into quarters. Use a curved vegetable knife or tip of a metal spoon to remove the seed-bearing section and a thin layer of flesh beneath it.

Cut the remaining flesh and skin into even-sized matchstick strips. Shred the ham into strips of the same size. Using your fingertips, mix together in a bowl as briefly as possible to avoid breaking up the strips. Cover the bowl with plastic wrap and chill in the refrigerator.

Make the sambai zu dressing and stir in the grated ginger. Cover and refrigerate for at least 1 hour.

Just before serving, transfer the salad to small bowls and decorate each portion with a cucumber fan. Serve the dressing separately.

———— • ————

CUCUMBER WITH SALMON ROE AND HARD-BOILED EGGS
Ikura to Kyuri

4 hard-boiled eggs (see method)
1 medium-sized cucumber
1½ tbsp black sesame seeds
1 small can salmon roe or red lumpfish roe
½ lemon

Place the eggs in a small saucepan, add enough cold water to amply cover them, and bring slowly to a boil over low heat. As soon as the water starts to boil, turn off the heat, cover the saucepan, and leave for 20 minutes before draining and shelling the eggs. When completely cold, cut lengthwise in half: the yolks will not be completely hard-boiled and will be bright yellow and creamy in the center. Slice a small piece off the white curved side of the egg halves so they are stable when placed yolk uppermost on small, individual plates.

Rinse and dry the cucumbers; slice off the ends and cut lengthwise in half; cut each half in 2. Scoop out the seed-bearing section neatly. Carefully cut off a thin slice all down the curved side so that they will remain stable. Cut a fairly thick layer off the top but stop about ½ in short of the end, so that you can bend and fold this long piece over and secure with a cocktail stick, to represent a sail at the end to which it is still attached. If the removal of the seed section means that you cut 2 parallel strips, this will look just as attractive. Place one of these "boats" on each plate, between the two egg halves.

Roast the sesame seeds in a nonstick skillet over moderate heat for 1–2 minutes, stirring with a wooden spoon; as soon as they begin to jump about in the skillet, remove from the heat. Spoon over the egg yolks.

Spread the salmon or lumpfish roe along the center of the upper surface of the cucumber "boats," squeeze lemon juice over them, and serve.

———•———

GUACAMOLE

2 ripe avocados

1 juicy lime

1½ tbsp extra-virgin olive oil

cayenne pepper or chili powder

1 large scallion, finely chopped

1 large ripe tomato

1 green chili pepper

3 tbsp chopped coriander leaves

salt and pepper

Peel the avocados, remove the pit, and slice the flesh into a large mixing bowl. Immediately add the lime juice (or use 3 tbsp lemon juice) and 1½ tbsp olive oil and season to taste with salt, freshly ground pepper, and a pinch of cayenne pepper or chili powder. Mix well, mincing with a fork to a fairly smooth, thick consistency. Stir in the scallion.

Peel the tomato and remove the seeds and any tough parts; cut the flesh into small dice. Remove the stalk and seeds from the fresh chili and chop very finely. Stir the tomato, chili, and chopped coriander into the avocado. Cover and chill. Guacamole is best eaten within 2 hours of being prepared. Serve as a dip.

PILI-PILI TOMATO SALAD

1¼ lb firm, ripe tomatoes

3 tbsp finely chopped parsley

½ cup olive oil

salt

For the pili-pili hot relish:

1 cup small fresh red chili peppers

3 cloves garlic

1 large red pepper

1 medium-sized Bermuda onion

½ cup extra-virgin olive oil

1½ tsp salt

You can buy pili-pili relish from good delicatessens; in some versions tiny chilis are used and left whole. If you prefer to make your own relish, store it in very small batches in the freezer and thaw on the day you plan to use it as it does not keep well once thawed.

Prepare the relish: rinse the chilis, remove their stalks and the surrounding tough parts, place in the food processor with the peeled garlic cloves, and process until very smooth and creamy. Rinse the red pepper and cut lengthwise into quarters; remove the stalk, seeds and white membrane. Chop coarsely. Chop the onion very finely. Add the peppers and the salt to the puréed chilis in the food processor or blender and process at high speed, adding a thin trickle of olive oil through the hole in the lid. Stop adding the oil once the mixture is very smooth and thick, rather like a bright red mayonnaise. Add the onion and process for a 2–3 seconds only, just long enough to distribute the pieces evenly.

Transfer to very small, nontransparent freezer containers with tightly sealed lids and freeze until required.

When you want to use the pili-pili, thaw if you have used very tiny containers, or use the tip of a strong knife to scrape off the required quantity from a still-frozen portion. Mix this with a little olive oil and a pinch of salt and use as a very hot dip for crudités or as an added flavoring and seasoning for a wide var-

113

iety of dishes or sauces; do not mix with oil and salt before adding to the latter. Pili-pili is very good in minute quantities with vegetable omelets, and with fresh or matured cheeses.

Wash the tomatoes (use the full-flavored, oval variety if you can buy or grow them), neatly remove the stalk and all the tough, fibrous parts (see page 14), cut lengthwise in half, and discard all the seeds. Place cut side uppermost on a large serving platter and sprinkle with the parsley. Place 1–2 tsp pili-pili in a bowl with a pinch of salt and gradually beat in 1/2 cup olive oil; sprinkle over the tomatoes.

———•———

STUFFED TOMATOES AU GRATIN

8 medium-sized ripe tomatoes
For the savory breadcrumb stuffing:
1 large onion, finely minced
¾ cup fine fresh breadcrumbs
¾ cup finely grated hard cheese
3 tbsp finely chopped parsley
1½ tbsp chopped chives
olive oil
salt and pepper
Serve with:

Stuffed Baked Pepper and Anchovy Rolls (see page 111), Eggplant Mock Pizzas (see page 108), Curried Fennel (see page 91), or other stuffed, baked or braised vegetables

Preheat the oven to 400°F. Make the bread stuffing: fry the onion very gently in ⅛ cup olive oil for 10 minutes, stirring at intervals. When tender and very pale golden brown, add the breadcrumbs and continue cooking over low heat for a few seconds while stirring. Remove from the heat and leave to cool before stirring in the grated cheese, parsley, and chives. Season to taste with salt and freshly ground pepper. Rinse and dry the tomatoes. Cut off the tops about ¼ of the way down the tomato and scoop out all the

seeds and some of the flesh. Sprinkle with a little salt and then fill with the bread mixture, smoothing with a knife to form a dome. Lightly oil a wide, shallow ovenproof dish and arrange the tomatoes in it in a single layer. Bake in the oven for 10 minutes. Serve hot or warm on their own as an appetizer, as an accompaniment or as part of an assortment of other stuffed, baked vegetables.

———•———

ZUCCHINI ESCABECHE

8 zucchini
1 sprig mint
4 cloves garlic, peeled
light olive oil
red or white wine vinegar
salt

Trim off the ends of the zucchini, rinse them, and dry. Cut diagonally into ¼-in slices: spread these out in a single layer on a clean cloth or paper towels and blot dry with another cloth or paper towels.

Heat plenty of oil in a deep-fryer to 350°F and fry the zucchini slices a few at a time until they are pale golden brown and slightly crisp on the surface. Finish draining on paper towels while you fry the remaining batches.

Separate the mint leaves, rinse and dry them. Chop the garlic coarsely.

Spread half the fried zucchini slices out in a wide, fairly deep nonmetallic dish. Sprinkle with a little salt and with half the chopped garlic. Moisten with plenty of vinegar. Place half the mint leaves, laid flat, on top of this layer. Repeat with the remaining half of the ingredients. There should be enough vinegar to soak the zucchini but they should not be swimming in it. Cover with plastic wrap and chill for several hours, or overnight.

DEEP-FRIED OKRA

1 cup chickpea flour

½ tsp ground roasted cumin seeds

1 tsp ground roasted coriander seeds

pinch chili powder

1¼ cups okra

sunflower oil

salt

Garnish with:

coriander leaves

Mix the chickpea flour in a large mixing bowl with the spices, salt, and chili powder. Remove the stalks from the okra, rinse and dry them; cut lengthwise into thin strips. Spread out on a very large chopping board or on the work surface and sprinkle with the spiced flour, using your hands to lift and turn them so that they are completely coated. Heat plenty of sunflower oil in a deep-fryer to 350°F and fry the okra in small batches for 2–3 minutes or until crisp and golden brown. Remove from the oil and keep warm in a low oven on paper towels to finish draining until the remaining batches have been fried. Serve without delay, garnished with coriander leaves.

———•———

SPRING VEGETABLE SOUP

3 very small globe artichokes

¼ lb young tender spinach

¼ lb escarole

1 medium-sized onion

2–3 slices smoked bacon

½ cup dry red wine

1¼ lb very young, fava beans, shelled and skinned

1¾ cups fresh or frozen young, tender peas or petits pois

6 tbsp extra-virgin olive oil

salt and black peppercorns

Serve with:

thick slices of hot French, or coarse white, bread

Trim and wash the vegetables. Remove the outer leaves of the artichokes and cut the remaining, inner parts lengthwise into thin slices. Unless you can buy very fresh, baby artichokes, preferably the more tender varieties, it is best to use only the bottoms; otherwise leave them out of this soup. Do not use canned ones; it is better to substitute tender green asparagus tips. Shred the spinach and the escarole. Chop the onion finely and cut the bacon into small dice.

Traditionally this soup is made in a flameproof earthenware cooking pot, but an enameled cast-iron casserole or a heavy-bottomed stainless steel saucepan are also fine. Fry the onion and bacon in 3 tbsp of the oil over low heat and stir for a few minutes. Pour in the wine when they start to brown. Cook, uncovered, until the wine has evaporated. Add the beans and artichokes, stir briefly, then pour in 6½ cups cold water; and add 1–1½ tsp salt and 5 black peppercorns.

Bring to a boil over moderate heat, then reduce the heat, cover, and simmer gently for 45 minutes. Add the peas and leaf vegetable *chiffonade* and simmer for another 15–20 minutes. Remove from the heat for 5 minutes, then serve in individual bowls, adding a little olive oil to each at the last minute. This is best enjoyed warm or fairly hot rather than boiling hot. Serves 6.

———•———

CREAM OF ASPARAGUS SOUP

2 lb young, tender green asparagus

4½ cups chicken stock (see page 37)

2¼ cups milk

2½ tbsp butter

115

¼ cup all-purpose flour
2 egg yolks
½ cup light or heavy cream
salt and white pepper
Serve with:
croutons

Heat the milk and the stock separately. Trim and prepare the asparagus (see page 17); only the green, tender part is used for this soup. Wash very thoroughly, drain, and cut into pieces.

Melt the butter over low heat in a large, heavy-bottomed saucepan; add the flour and keep stirring as you cook it gently for 1 minute; it should not color. Draw aside from the heat and pour in the hot milk while beating with a whisk to prevent lumps forming. Gradually stir in the hot stock.

Return the saucepan to a moderate heat and keep stirring as it heats to boiling; add the pieces of asparagus. Cover and simmer over low heat for 40 minutes. Remove from the heat and beat with a hand-held electric beater until creamy (or put through a potato ricer). Reheat for a few minutes until very hot but do not allow the soup to boil again. Place the egg yolks in a small bowl and mix well with the cream. Take the soup off the heat and keep beating continuously as you gradually add the egg and cream thickening liaison. Add salt and pepper to taste. Serve with croutons.

——— • ———

CASTILIAN CREAM OF ARTICHOKE SOUP

4 very large or 8 medium-sized globe artichokes
¾ cups chicken stock (see page 37)
juice of 1 lemon

¼ cup butter
½ cup all-purpose flour
1 cup milk
¼ cup heavy cream
2 egg yolks
salt and freshly ground white pepper

Heat the stock and prepare the artichokes by stripping off all the leaves and removing the choke. You will only need the bottoms for this recipe. As soon as you have prepared each artichoke bottom, drop it into a bowl of cold water acidulated with two-thirds of the lemon juice to prevent discoloration. When all the artichoke bottoms are ready, drain and pat dry; place them flat on a chopping board and slice downward into thin pieces.

Place these slices in a wide nonstick skillet, add generous ½ cup of the stock, cover with a lid or foil, and poach for 10 minutes.

Melt the butter in a large, heavy-bottomed saucepan; stir in the flour and cook over low heat stirring continuously for 1–2 minutes. It should not color. Remove from the heat and keep beating energetically with a whisk or hand-held electric beater as you gradually pour in the hot stock.

Return to a gentle heat, add the artichoke slices and their cooking juices, cover, and simmer for 30 minutes. When the artichokes are very tender, beat with a hand-held electric beater or put through a potato ricer. Stir the milk into this purée and reheat to boiling point. Season the soup with salt and pepper. Beat the egg yolks and cream together briefly in a small bowl with a pinch of salt and a little freshly ground white pepper. Gradually beat in 3 tbsp of the hot soup. Remove the saucepan with the boiling hot soup from the heat; pour the slightly thinned egg and cream liaison into the soup in a thin stream while beating continuously.

Return the saucepan to a very low heat and keep stirring as the soup thickens slightly. It must not boil. Remove from the heat and stir in some of the remaining lemon juice.

——— • ———

CHILLED TOMATO SOUP WITH SCAMPI
Crème de tomates Bec Fin

2¾ lb very ripe tomatoes

1 clove garlic

1 tsp fresh marjoram leaves

pinch oregano

1 large sprig basil

1 red chili pepper

¼ cup extra-virgin olive oil

3–4 tbsp lemon juice

3–4 tbsp fine, fresh white breadcrumbs

2 large shallots, finely chopped

1 tbsp butter

1½ tbsp light olive oil

¼ cup dry vermouth or dry white wine

12–16 raw, unpeeled scampi or jumbo shrimp

salt and pepper

Garnish with:

4 small sprigs basil

Use a full-flavored variety of tomato for this soup. Blanch if necessary before peeling them, cut lengthwise in half, remove the seeds and any tough parts, and place the flesh in the food processor. Process at high speed with the peeled clove of garlic, the marjoram, oregano, basil, chili pepper (remove its stalk and seeds first), the extra-virgin olive oil, lemon juice, breadcrumbs, and a little salt and freshly ground pepper. When very smooth, taste and add more salt or pepper if needed. Process for a few seconds more. Chill for 30 minutes in the refrigerator before serving; do not chill for longer or the flavor will be blunted.

While the soup is chilling, fry the shallot very gently in the butter and 1½ tbsp light olive oil for 5 minutes. Add ¼ cup dry vermouth or dry white wine and the scampi, then cover, and cook gently for 5 minutes. Take them out of the pan and peel, leaving the end flippers attached. Pour the tomato soup into individual bowls and arrange 3 or 4 scampi radiating out from the center like the spokes of a wheel. Place a small sprig of basil in the center and serve.

———•———

CREAM OF TOMATO SOUP

1½ lb ripe full-flavored tomatoes (or 1 large can Italian tomatoes)

4½ cups chicken stock (see page 37)

1 medium-sized onion, finely chopped

¼ cup butter

generous ¼ cup all-purpose flour

juice of ½ lemon

salt and white peppercorns

Bring the stock to a boil. Blanch and peel the tomatoes and remove any tough parts and all the seeds. Reserve any juice for other uses. Chop the flesh coarsely and place in a bowl.

Melt 3 tbsp of the butter in a large, heavy-bottomed saucepan and fry the onion very gently for 10 minutes, stirring frequently, until very tender but not at all browned. Add the flour and cook gently, stirring for 1–2 minutes. Remove from the heat and pour in one-third of the hot stock all at once, beating vigorously with a whisk until it has blended evenly with the roux mixture without forming any lumps. Pour in the remaining stock, mix well, and return to a gentle heat. Add the chopped tomato and stir. Simmer for 30 minutes, covered, stirring occasionally.

Remove from the heat and beat with a hand-held electric beater until all the solid ingredients have blended into a creamy liquid. Add salt, freshly ground white pepper, and lemon juice to taste. Add the remaining butter in one piece and stir.

———•———

117

PUMPKIN SOUP WITH AMARETTI

1 lb prepared pumpkin flesh

2¾ cups chicken stock (see page 37)

3 tbsp butter

4 large shallots, peeled and chopped

1¼ cups light cream

superfine sugar

10 amaretti di Saronno cookies

salt and white pepper

Heat the stock. Dice the pumpkin flesh. Melt 2 tbsp of the butter in a large heavy-bottomed cooking pot or saucepan and sweat the shallots for 5 minutes, stirring frequently. Add the pumpkin and cook over moderate heat for 5 minutes. Pour in the hot stock, bring to a boil then cover, reduce the heat, and simmer gently for 20 minutes, or until the pumpkin is tender.

Remove from the heat and allow to cool a little. Process in batches in the blender until smooth; return to the saucepan. Stir in the cream, a generous pinch each of sugar and salt, and some freshly ground white pepper. Reheat to boiling point. Turn off the heat. Taste, adding a little more salt if necessary. Add the remaining butter and stir until it has melted. Crush the amaretti coarsely. Ladle the soup into heated soup bowls and sprinkle the amaretti on top. Serve at once.

—— • ——

ARGENTINE ICED PUMPKIN SOUP
Fonda del Sol

1 ¾ lb pumpkin or winter squash

4 fresh or canned tomatoes

1 medium-sized onion

2 leeks

2 vegetable or chicken bouillon cubes

1 cup light cream

salt and pepper

Garnish with:

¼ cup chopped chives

Peel the pumpkin or winter squash, cut out all the seed-bearing sections, and weigh out 1¾ lb of flesh. Cut this into small pieces. Blanch and peel the tomatoes, if fresh; remove any tough parts and discard the seeds. Trim and peel the onion, slice lengthwise into quarters, and then cut across into small pieces. Slice the leek.

Place all the vegetables in a large saucepan or pot; add 3½ cups water and the 2 bouillon cubes. Bring to a boil, cover, and reduce the heat to low so that the liquid remains at a gentle boil. After 1 hour the pumpkin should be very tender. Remove from the heat and put through a blender.

Allow to cool completely. Then stir in the cream, season to taste with salt and pepper, cover, and chill. Serve cold, with a sprinkling of chopped chives.

—— • ——

BAMBOO SHOOT AND ASPARAGUS SOUP

¼ cup dried bamboo shoots

5 cups chicken stock (see page 37)

1 lb fresh, tender asparagus spears

juice of 1 lemon

4 canned bamboo shoots, drained

1 thin piece fresh ginger, peeled

½ lb tofu

2 tbsp sunflower oil

3 tbsp Chinese rice wine, dry sherry or dry vermouth

1½ tsp sesame oil

salt and pepper

Soak the dried bamboo shoots in plenty of very hot water for 30 minutes. While they are soaking, heat the chicken stock. Trim and prepare the asparagus, washing very thoroughly: only the tender, upper halves of the spears are used for this recipe; slice them diagonally into lengths and place in a bowl of cold water mixed with half the the lemon juice. Cut the canned bamboo shoots into 1-1/4-in thick slices, unless they are already sliced, then cut these to approximately the same size as the pieces of asparagus. Chop the fresh ginger very finely. Drain and squeeze the soaked, dried bamboo shoots, pat dry in paper towels, then shred. Cut the tofu into small cubes.

Heat the sunflower oil in a large, heavy-bottomed saucepan and fry the ginger and shredded bamboo shoots briefly, until the ginger releases its aroma. Then stir in the pieces of canned bamboo shoots and fry gently for a few seconds. Add the chicken stock. Bring to a boil, reduce the heat, cover, and simmer gently for 30 minutes. Add the asparagus and the tofu and simmer for another 10 minutes, until the asparagus is tender but still a little crisp.

Remove from the heat, add the rice wine or substitute and the sesame oil; season to taste with salt and pepper and serve very hot. This soup is traditionally served at the end of a Chinese meal.

———— • ————

AVOCADO SOUP
Sopa de Aguacate

3 ripe avocados
4½ cups chicken stock (see page 37)
1 cup light cream
¼ cup dry sherry
salt and white peppercorns
Serve with:
tortillas

Bring the stock to a gentle boil over moderate heat. While it is heating, cut the avocados lengthwise in half, discard the pit, and peel. Cut all but one of the peeled avocado halves into pieces and blend at high speed with ½ cup of the hot stock. In a few seconds it will have turned into a smooth, thick purée. Add the cream and blend at high speed for 30 seconds more. When the stock reaches a gentle boil, reduce the heat to very low and gradually beat in the avocado cream mixture with a whisk or hand-held electric beater, adding a little at a time to blend smoothly. Do not allow to boil. Remove from the heat as soon as you have added all the avocado mixture and stir in the sherry, salt, and freshly ground white pepper to taste. Ladle into very hot soup bowls. Cut the remaining avocado half lengthwise into thin slices and place 2 or 3 of these in each bowl of soup. Serve with hot tortillas.

———— • ————

CHINESE CUCUMBER, PORK, AND CHICKEN SOUP

3½ cups chicken stock (see page 37)
1 thick piece fresh ginger, peeled
1 cucumber
6 thin slices pork tenderloin
2½ tsp cornstarch
salt and pepper

Bring the chicken stock slowly to a boil with the slice of ginger in it. Wash the cucumber and cut off the ends; run the potato peeler lightly over the skin to remove just the outermost tough layer. Cut lengthwise into quarters and then into matchstick strips. Shred the pork into very thin strips, place in a bowl, and dust with the cornstarch mixed with 1 tsp salt, rubbing it gently into the pieces of meat. When the chicken stock comes to a boil, add the pork and simmer gently for 4 minutes; add the cucumber and simmer for 3 minutes more. Taste and correct the seasoning. Remove and discard the piece of ginger. Serve very hot in heated soup bowls.

119

ZUCCHINI SOUP WITH BASIL AND LEMON

1 lb baby zucchini

¼ cup fresh basil leaves

3 tbsp chopped parsley

1 lemon

2 eggs

¼ cup grated hard cheese

extra-virgin olive oil

salt and pepper

Serve with:

slices of coarse white bread crisped in the oven

Bring 5 cups water to a boil. Wash the zucchini, trim off their ends, and cut them into small dice.

Heat 6 tbsp olive oil in a large, heavy-bottomed saucepan over moderate heat; add the zucchini and brown lightly, stirring frequently. Add the boiling water and a little salt and freshly ground pepper. When the liquid comes to a boil, reduce the heat, cover the saucepan, and simmer for 40 minutes.

Bake the thick slices of bread in the oven until golden brown or dry-fry in a nonstick skillet. Chop the basil and parsley, and finely grate the rind of half the lemon. Beat the eggs lightly in a bowl and add the basil, parsley, cheese, grated lemon rind, and a small pinch of salt.

When the zucchini have cooked for 40 minutes, remove from the heat and beat vigorously with a whisk or hand-held electric beater as you add the egg mixture in a very thin stream; it will thicken the soup slightly.

Taste and add more salt and some freshly ground pepper before serving.

---•---

GAZPACHO

2¼ lb large, ripe tomatoes

3 cloves garlic, peeled

¼ cup white wine vinegar

⅓ cup extra-virgin olive oil

1 red pepper

1 small red chili pepper, seeds and stalk removed

¼ cup fine fresh breadcrumbs

1 small or ½ a large cucumber, chilled

5 thin slices white bread

salt

Blanch, peel, and strain the tomatoes; process them in the blender at high speed with 3 cloves garlic, a pinch of salt, the vinegar, ¼ cup of the oil, the red pepper, the chilli pepper, and the breadcrumbs until very well blended. Add a little more salt than you would normally allow as chilling the soup reduces the taste of the salt. Pour into a soup tureen, cover, and chill in the refrigerator for 2–6 hours. Shortly before serving the soup; wash, dry, peel, and cut the chilled cucumber lengthwise in quarters; remove the seed-bearing section and peel if wished; cut the flesh into small dice, place in a bowl, and chill.

Cut the crusts off the bread slices and cut into small cubes. Heat the remaining olive oil in a nonstick skillet and fry the cubes over moderate heat, turning frequently until crisp and golden brown. Sprinkle with a little salt and set aside in a bowl.

Serve the gazpacho in chilled soup bowls, passing the bowls of cucumber and croutons at the table.

---•---

RISOTTO WITH ZUCCHINI FLOWERS

5 cups chicken stock (see page 37)

4 large shallots

24 freshly picked zucchini flowers
¼ cup unsalted butter
1½ tbsp olive oil
2 cups risotto rice (e.g. arborio)
½ cup dry white wine
1 sprig thyme
4–6 fresh basil leaves
¼ cup light or heavy cream
½ cup freshly grated Parmesan cheese
salt and white peppercorns

Bring the stock slowly to a boil. While it is heating, peel and finely chop 3 of the shallots. Rinse the zucchini flowers, set 8 of them aside and slice the rest into rings about ¼ in thick.

Heat 1½ tbsp of the butter together with 1½ tbsp olive oil in a large, fairly deep, heavy-bottomed skillet. Fry the chopped shallots very gently for 10 minutes, stirring now and then; do not allow to color. Add the sliced zucchini flowers and fry over low heat for 30 seconds, stirring gently. Add the rice, turn up the heat, and cook, stirring continuously, for 1–2 minutes. Add the wine and continue cooking until it has evaporated. Add 1 cup boiling hot stock, stir, and continue cooking over moderately high heat so that the liquid boils; keep adding more hot stock as the liquid in the skillet reduces and is absorbed by the rice. Stir occasionally. After about 14 minutes, the rice should be tender but still have a little bite left in it. The risotto should be very moist and creamy.

While the rice is cooking, peel and chop the remaining shallot and fry with the thyme in 1 tbsp of the remaining butter in a very small saucepan. Add the remaining, whole, zucchini flowers and the finely chopped basil leaves; cook very gently, turning once or twice, for 1 minute. Stir in the cream, a pinch of salt, and freshly grated white pepper; cook gently for another 30 seconds. The whole flowers should be still slightly crisp.

When the rice is done, remove from the heat, add the remaining butter and the Parmesan cheese; stir gently as the butter melts. Correct the seasoning.

Remove and discard the thyme sprig.

Serve the risotto in bowls or soup plates and place a few of the whole zucchini flowers with some of the creamy sauce in the center of each serving.

———— • ————

ARTICHOKE RISOTTO

4 tender globe artichokes (see method)
5 cups chicken stock (see page 37)
1 small onion
¼ cup unsalted butter
1½ tbsp olive oil
1¾ cups risotto rice (e.g. arborio)
¼ cup dry vermouth
½ cup grated Parmesan cheese
½ clove garlic, peeled and grated (optional)
salt and pepper

Slowly bring the stock to a boil. While it is heating prepare the artichokes. You will need only the heart, the inner leaves, and base; if you cannot buy very young, tender varieties of artichoke, use 12 artichoke bases or bottoms (*fonds d'artichauts*). If using artichoke hearts, cut them lengthwise in quarters, remove and discard the choke, if there is one, from the center, and slice thinly.

Peel and finely chop the onion and sweat gently in 2 tbsp of the butter and 1½ tbsp oil in a large, heavy-bottomed skillet. Add the artichokes and fry gently for 5 minutes, stirring and turning them. Add the rice, turn up the heat to moderately high, and cook, stirring for 1–2 minutes. Pour in the vermouth and continue stirring until it has evaporated. Pour in about 1 cup of the boiling hot stock and stir. Cook briskly, adding more hot stock when necessary and stirring occasionally. The rice should be tender, with just a little bite left in it after 14 minutes.

Remove from the heat, add the remaining butter and the Parmesan cheese; if desired, peel ½ clove garlic and grate over the risotto.

121

SPAGHETTI WITH TOMATOES AND CHILI

¾ lb spaghetti

4 large ripe tomatoes

¼ cup extra-virgin olive oil

4 cloves garlic, peeled, minced but still whole

1 dried chili pepper (optional)

1 large sprig basil

grated Parmesan cheese

salt and black peppercorns

Bring a large pot of salted water to a boil to cook the spaghetti. Blanch and peel the tomatoes; cut them in quarters and remove the seeds. If you cannot buy ripe, flavorsome tomatoes, use 8–10 canned, drained Italian tomatoes. Heat the olive oil in a large, heavy-bottomed saucepan and fry the garlic cloves gently until they are pale golden brown. Crumble the chili pepper into the saucepan (discard the seeds if you do not want a very peppery dish), stir and then add the tomatoes, a pinch of salt, and 2 basil leaves. Cook, uncovered, over high heat for 5 minutes, stirring occasionally. Remove from the heat. Tear the rest of the basil into small pieces.
Cook the spaghetti in the boiling salted water until just tender but still firm. Drain, add to the tomatoes and cook for 1 minute, stirring. Turn off the heat, stir in the Parmesan cheese, plenty of freshly ground pepper, and the basil. Serve at once.

———•———

SPAGHETTI WITH MEDITERRANEAN SAUCE

¾ lb spaghetti

For the sauce:

2 large ripe tomatoes

3 tbsp chopped parsley

2 large sprigs fresh basil

4 large cloves garlic

generous pinch oregano

2 tbsp capers

16 black olives, pitted and chopped

1 dried chili pepper

⅓ cup extra-virgin olive oil

salt and pepper

Bring a large pot of salted water to a boil to cook the spaghetti. Peel the tomatoes and chop; save all their juice and transfer both juice and tomatoes to a bowl; use 1 cup canned chopped Italian tomatoes if preferred. Add the chopped parsley, the shredded basil leaves, the peeled and minced garlic cloves, the oregano, capers, chopped olives, crumbled chili pepper, a pinch of salt, and some freshly ground pepper. Stir in the olive oil and set aside for a few minutes. When the water comes to a boil, add the spaghetti and cook until tender but still firm. Drain, add the sauce, and stir. This sauce is also delicious with steamed brown rice.

———•———

SPAGHETTI WITH EGGPLANT

¾ lb eggplants

1 lb ripe tomatoes or 1 large can tomatoes

½ fresh green chili pepper

2 large cloves garlic, peeled and minced

½ cup dry white wine

pinch oregano

¾ lb spaghetti

¼ cup grated Pecorino cheese

few fresh basil leaves

3 tbsp extra-virgin olive oil

oil for frying
salt and black peppercorns
Garnish with:
basil leaves

Bring plenty of salted water to a boil in a large pot, for the spaghetti. Slice the ends off the eggplants, peel them, and cut lengthwise into slices about ¼ in thick. Cut these slices into rectangles measuring approximately 1½ × 1 in. Deep- or shallow-fry briefly in batches in very hot oil until pale golden brown; take out of the oil with a slotted spoon or ladle and finish draining on paper towels. Sprinkle with a little salt and freshly ground black pepper.

Blanch, peel, and seed the tomatoes; cut the flesh into small pieces. Remove the stalk and seeds from the chili pepper and chop finely; fry the chili pepper gently with the minced garlic in the extra-virgin olive oil for a few seconds. Add the tomatoes and a pinch of salt and cook briskly, uncovered, for 7 minutes, stirring frequently; the sauce should thicken as the water content evaporates. Add the wine and stir while continuing to cook over moderately high heat until the sauce has thickened again. Stir in a good pinch of oregano and remove from the heat.

Cook the spaghetti until tender but still firm. Drain and return to the pot; add the tomato sauce and ¼ cup grated Pecorino cheese. Season with more freshly ground black pepper. Serve without delay in hot, deep plates. Spread the eggplant slices on top and garnish each serving, with a sprig of basil.

—— • ——

SPAGHETTI WITH ZUCCHINI FLOWERS, CREAM, AND SAFFRON

24 freshly picked zucchini flowers
¾ lb whole wheat spaghetti
4 large shallots
3 tbsp butter

1 cup light or heavy cream
1 sachet or generous pinch saffron threads
salt and pepper

Heat a large pot of salted water for the spaghetti. While it is heating, prepare the zucchini flowers: cut off their stalks, then, by inserting your longest, middle finger into each flower, carefully break off the yellow, fleshy pistil and discard it. Rinse the flowers gently in cold water. Drain and spread out to dry on a clean cloth or paper towels. Peel and finely chop the shallots; sweat gently over low heat in the butter in a skillet. Do not allow to color.

When the water reaches a fast boil, add the spaghetti and cook until just tender. When the spaghetti has been cooking for 5 minutes, slice across the flowers to form rings, add to the shallot, and cook gently for 30 seconds. Add the cream and saffron and simmer, uncovered, over slightly higher heat for 1 minute. Add salt and freshly ground white pepper to taste. As soon as the spaghetti is cooked, drain and gently stir in the sauce.

—— • ——

PASTA CAPRI

¾ lb pasta shapes (e.g. quills)
2 cloves garlic, peeled
1½ tbsp extra-virgin olive oil
1 dried red chili pepper
1 lb fresh or canned tomatoes, chopped
1 large sprig fresh basil
¾ lb eggplants
vegetable or sunflower oil for deep-frying
½ lb mozzarella cheese
salt and black peppercorns

Heat plenty of salted water in a large saucepan for the pasta. Smash the garlic cloves with the flat of a heavy knife; fry gently in the extra-virgin olive oil in a large, heavy-bottomed saucepan until pale golden brown. Add the crumbled chili pepper (discard the seeds if you want the dish to be less peppery); stir in the tomatoes and 2 basil leaves. Season lightly with salt and freshly ground black pepper. Bring to a gentle boil and then simmer, uncovered, for 15–20 minutes to allow the sauce to reduce and thicken considerably. Stir occasionally. While the sauce is simmering, slice the ends off the eggplants, peel, and cut them lengthwise into slices about 1/4 in thick; cut these in turn into rectangles about 1-1/4 × 3/4 in. Heat the oil in the deep-fryer to 350°F and fry the eggplant slices in batches until pale golden brown; alternatively, shallow-fry, turning once. Drain well, spread out on paper towels, and sprinkle with a little salt.

Cut the mozzarella cheese into small cubes or dice and tear all but a few of the remaining basil leaves into thin strips. Cook the pasta until just tender; drain, add to the tomato mixture, and stir over moderately high heat for a few seconds. Add the eggplant pieces and the diced mozzarella and stir over slightly lower heat for a few seconds more, until the mozzarella melts. Remove from the heat, stir in the shredded basil leaves and serve in deep, heated plates, garnished with the reserved whole basil leaves and sprinkled with a little more freshly ground black pepper.

———•———

TURKISH STUFFED PEPPERS
Zeytinyagli Biber Dolmasi

4 medium-sized green or red peppers
1 lb onions
2 tbsp chopped parsley
2 tbsp chopped fresh dill
1 cup long-grain rice
scant ¼ cup pine nuts
1 tsp ground cinnamon
1½ tbsp dried mint
3 tbsp lemon juice
⅓ cup olive oil
salt and pepper

Preheat the oven to 350°F. Wash and dry the peppers; cut off a lid at the stalk end, reserving these lids. Remove the seeds and inner membrane. Turn the peppers upside down on a chopping board.

Peel the onions and chop finely. Chop the parsley together with the dill (use 2–3 pinches dried dill if fresh is unavailable). Rinse the rice in a sieve under running cold water until the water comes out completely clear. Drain.

Heat 1½ tbsp olive oil in a large, heavy-bottomed skillet and cook the pine nuts over moderate heat until they are golden brown. Remove with a slotted spoon and set aside. Add the rest of the olive oil to that remaining in the skillet and fry the onions gently in it for 10 minutes, stirring frequently, until tender and pale golden brown. Add the rice and pine nuts. Cook for 1 minute, stirring, then add 1 cup boiling water and 1½ tsp salt. Continue stirring for a few minutes, until the water has almost completely disappeared but the rice is still very moist. Remove from the heat and allow to cool slightly; stir in the cinnamon, the crumbled dried mint, the parsley, dill, 3 tbsp lemon juice, and plenty of freshly ground black pepper. Add a little more salt if desired. Stuff the peppers with this mixture and replace their lids. Lightly oil a wide, shallow ovenproof dish, just large enough to accommodate all the upright peppers packed closely together in a single layer. Pour 1 cup boiling water into the dish between the peppers and bake uncovered in the oven for 30 minutes. Take the dish out of the oven, baste the peppers with the cooking liquid, cover with foil, and return to the oven for another 15 minutes. Serve warm or cold.

———•———

PALM HEART, SHRIMP, AND AVOCADO SALAD

½ lb thawed frozen peeled small shrimp

1 large can palm hearts (1 lb)

2 avocados

12 cherry tomatoes

For the curry dressing:

2 tsp mild curry powder

1 tsp chili powder

¼ cup fresh lime juice

⅓ cup sunflower oil

½ cup light cream

3 tbsp chopped fresh coriander leaves

salt and pepper

Garnish with:

fresh coriander leaves

If the shrimp are raw, steam them for 2 minutes and allow to cool. Drain the palm hearts and slice them into rings.

Cut the avocados in half lengthwise, remove the pit, and peel them; slice lengthwise, then cut each of these slices in half. Rinse and dry the cherry tomatoes.

Arrange the vegetables in sections, the tomatoes to one side, the palm hearts in the middle, and the avocado slices to the right. On the far side of each plate, heap up some of the shrimp. Make the dressing: mix the curry powder and chili powder with the lime juice, 1 tsp salt, and some freshly ground pepper. Gradually beat in the sunflower oil, adding a little at a time so that it emulsifies evenly with the other ingredients. Beat in the cream. Stir in the chopped coriander leaves.

Sprinkle this dressing over each portion of shrimp, garnish with a sprig of coriander and serve at once, before the avocado has time to discolor.

———— • ————

CHICKEN, TOMATO, AND AVOCADO SALAD WITH TUNA SAUCE

2 large boneless chicken breasts

1 lemon

½ cup chicken stock (see page 37)

¼ cup dry white wine

¼ lb tender shoots asparagus chicory or escarole

4 ripe tomatoes

2 ripe avocados

For the tuna sauce:

1 egg

½ tsp mustard

1 cup light olive oil

3 tbsp lemon juice

3–4 drops Worcester sauce

3 tbsp white wine vinegar

6 tbsp dry white wine

1 small can tuna

1 tbsp pickled capers, finely chopped

Garnish with:

chopped capers

fresh basil leaves or parsley

Season the chicken breasts with salt and freshly ground pepper and sprinkle lemon juice on both sides. Marinate for 10 minutes, then arrange in a single layer in a nonstick skillet, add the stock, the first quantity of dry white wine, and cover. Barely simmer over very low heat for 3 minutes, turn them, and simmer for 3 minutes more. Test whether they are done by inserting the point of a sharp knife deep into the thickest part of the breast; if the juice comes out at all pink, cook for another minute on each side. Remove from the heat, transfer the breasts to a chopping board, and allow to cool.

Make the sauce: break the whole egg into the blender or processor and add the mustard and a pinch of salt. Process at high speed, gradually drizzling in the olive oil through the hole in the blender or processor lid. Within about 2 minutes the mayonnaise will be thick and pale. Add 3 tbsp lemon juice, the Worcester sauce, 3 tbsp wine vinegar, the white wine, and the drained tuna. Process until a very smooth, much thinner sauce is formed. Add a little salt if necessary and plenty of freshly ground pepper to taste. Pour the sauce into a sauceboat or bowl. Stir in the finely chopped capers, reserving 1 tsp for decoration. (This sauce can be prepared in advance and chilled in the refrigerator until needed.)

Shortly before serving the salad, cut the chicken breasts into thin, diagonal slices. Rinse and drain the asparagus chicory; cut into short lengths. If using escarole, shred finely. Wash and dry the tomatoes, cut lengthwise in half and then slice each half. Fan out some of these slices on the side of each plate furthest away from you, slightly overlapping. Cut the avocados lengthwise in half, remove the pit, peel, and slice the halves lengthwise; fan out these slices in turn, allowing them to overlap both the tomato slices and each other slightly. Finally, arrange the chicken slices and the asparagus chicory or escarole in the same way, slightly overlapping the avocado slices, on the side of the plate nearest you. Spoon plenty of tuna sauce over the chicken. Garnish with the remaining chopped capers and basil.

———— • ————

BEAN SPROUTS WITH PORK AND STEAMED RICE
Momyashi to Gyuniku

4 thin scallions
1 piece fresh ginger, about ½ in thick
6 thin slices pork tenderloin
6 tbsp sunflower oil
1 clove garlic, peeled
1 dried chili pepper

1 lb bean sprouts
salt
For the sauce:
¼ cup mirin, dry sherry or sweet white vermouth
3 tbsp soy sauce
6 tbsp beef stock (see page 37)
1 tbsp concentrated tomato paste
½ tsp sugar
¼ beef bouillon cube, mixed with ½ cup boiling water
freshly ground pepper
½ tsp cornstarch or potato flour
For the egg roll garnish:
2 eggs
1½ tbsp sunflower oil
pinch sugar
salt
Serve with:
1 cup long-grain rice (see method)

Cook the rice. Use 1¼ cups water for every cup of rice. Place the rice in a sieve and rinse under running cold water until the water draining out of the rice runs clear. Leave the rice to soak in a bowl of cold water for 20 minutes, then drain it, transfer to a heavy-bottomed saucepan with a tight-fitting lid and add your measured quantity of cold water. Bring to a boil, reduce the heat to very low, cover tightly, and simmer gently for 18 minutes without lifting the lid. Check to see if the rice is done; all the water should have been absorbed and there should be little depressions in the surface of the rice. Cover and cook for another 2 minutes if necessary. Remove from the heat and leave to stand, with the lid on, for 10 minutes before serving. The rice will keep hot for about 30 minutes. Do not add salt. Good-quality rice is needed for this absorption method to work well.

Chop the scallions. Peel the ginger and chop finely. Shred the pork into thin strips. Mix the cornstarch with 1½ tbsp cold water then combine in a bowl with all the sauce ingredients.

Make the egg roll garnish: beat the eggs lightly with

a pinch each of salt and sugar and cook in a large, lightly oiled nonstick skillet; the omelet must be very thin. Do not allow to brown at all. Spread out flat on a chopping board; roll up when cool and cut this roll into thin strips. Heat 3 tbsp oil in the skillet and when very hot, stir-fry the pork for 1-1/2 minutes. Stir the sauce well again before pouring all over the meat. Reduce the heat and keep stirring while simmering for 30 seconds as the sauce thickens and coats all the meat. Remove the saucepan from the heat and set aside.

Heat 3 tbsp oil in a large wok or very large skillet, add the peeled, minced garlic clove, the scallions, ginger, and crumbled chili pepper. Stir-fry over high heat for 30 seconds; add the bean sprouts and 1 tsp salt and stir-fry for 2 minutes. Add the meat and sauce, reduce the heat a little, and continue stir-frying for 1 minute more. The bean sprouts should still be quite crisp. Serve immediately with the rice, garnished with the egg roll. Serves 4 as part of an Oriental meal; 2 as a meal in itself.

———•———

WHITE ASPARAGUS WITH SPICY SAUCE

2¾ lb white asparagus spears

For the sauce:

4 eggs, at room temperature

1½ tbsp white wine vinegar

½ cup light olive oil

pili-pili hot relish (see page 113), optional

salt and pepper

Place the eggs in a saucepan, add cold water, and bring to a boil over low heat. Turn off the heat, put the lid on the saucepan, and set aside for 30 minutes. When they are completely cold, shell them. Prepare the asparagus (see page 17), wash thoroughly, and

tie into 4 equal bundles. Boil or steam for 12–14 minutes or until tender.

While the asparagus is cooking, chop the eggs finely and mix in a bowl with the vinegar, oil, and a little salt and pepper. If you like a hot, peppery sauce, mix ½–1 tsp pili-pili with 1 tbsp oil and add to the sauce.

Transfer the asparagus bundles to a hot plate, remove the strings, and serve with the sauce.

———•———

ASPARAGUS WITH EGGS AND PARMESAN

2¼ lb asparagus

½ cup butter

8 eggs

¾–1 cup freshly grated Parmesan cheese

salt and pepper

This Milanese recipe makes an excellent lunch or supper dish. Prepare the asparagus as called for in the preceding recipe but boil or steam for only 10–12 minutes, until they are just tender and still a little crisp. Keep hot while you cook the eggs.

Place 1 tbsp of the butter in each of 2 nonstick skillets then break 4 eggs into each as soon as the butter has melted, before it starts to foam. Fry gently over low heat. Sprinkle with a little salt if desired. Heat the remaining butter in a small saucepan and turn off the heat as soon as it has stopped foaming. Transfer the asparagus bundles to hot plates, leaving room for the eggs. Remove the strings, and sprinkle each serving with plenty of Parmesan cheese. Drizzle the melted butter over the tips and place 2 fried eggs beside each mound of asparagus. Serve at once and have a pepper mill on the table, for each person to add freshly ground pepper.

———•———

ASPARAGUS MOLDS

1 scallion, chopped

¼ cup + 2 tbsp butter

½ cup coarsely chopped asparagus tips

2 tbsp vegetable stock (see page 38)

4 eggs

salt and pepper

nutmeg

½ cup light cream

½ cup milk

¼ cup Parmesan cheese

Serve with:

Mousseline Sauce (see page 107)

Preheat the oven to 350°F. Sweat the scallion in ¼ cup butter for 10 minutes. Add the asparagus tips, cook gently for 5 minutes then add the stock, cover and cook for a further 5 minutes. Reserve 3 tbsp and purée the rest in the food processor. Use the remaining butter to grease 4 ¾ cup molds or ramekins. Beat 2 whole eggs with 2 yolks, a pinch of salt, pepper, and nutmeg. Beat in the cream, milk, the asparagus purée, the reserved chopped asparagus, and the Parmesan. Correct the seasoning and divide the mixture between the molds. Place these in a roasting pan and pour in enough boiling water to come halfway up the sides of the molds. Cook for 30 minutes. Allow to stand briefly before unmolding. Serve with Mousseline Sauce.

— • —

GLOBE ARTICHOKES ROMAN STYLE

4 very fresh young artichokes

juice of ½ lemon

1 clove garlic

3 tbsp finely chopped parsley

½ cup grated mature hard cheese

best-quality olive oil

½ chicken bouillon cube

salt and black pepper

You can use large green artichokes or medium-sized ones with violet-colored leaves; make sure they still have 2–3 in of their stalks attached. They must be young and freshly picked or they will not be tender enough for this dish. Rinse thoroughly under running cold water. Cut the stalks flush with the base of the artichoke; use a potato peeler or sharp knife to peel away the tough section round the base, and to strip off the outer layer of the stalks. It is best to prepare the artichokes one at a time; as soon as you have finished each one, drop it into a very large bowl containing cold water mixed with the juice of a lemon. Chop the stalks together with the garlic and parsley. Transfer to a bowl, stir in the cheese, and gradually work in the oil to make a thick paste. Season with salt and pepper.

Strip off the lower, outer leaves and cut off the top half of the artichoke with a sharp knife. Push the inner leaves apart to reach the feathery choke inside the artichoke and remove this, leaving only the heart and the remaining leaves. Snip off their pointed tips. Drain all the prepared artichokes well, then place them upside down on a chopping board and press down hard while turning the artichoke from side to side to make the leaves open out wide like a flower. Pack the space inside the artichokes with the prepared stuffing.

Place the artichokes right side up in a straight-sided saucepan or fireproof dish large enough to contain them all snugly in one layer. Pour in sufficient water near the edge of the dish to come one-third of the way up the side of the artichokes. Sprinkle plenty of olive oil over them. Crumble the bouillon cube finely and sprinkle it into the water wherever there is a gap between the artichokes. Heat until the water begins to boil, cover, and simmer for 20 minutes.

— • —

ARTICHOKE AND EGG PIE

1 lb olive oil pastry (see recipe for Italian Easter Pie on page 50) or ½ lb thawed frozen puff pastry

12 tender young globe artichokes

juice of 1 lemon

1 medium-sized onion (4½ oz), finely minced

1 bay leaf

9 eggs

½–¾ cup grated hard cheese

1 small clove garlic, peeled and finely minced

1 tbsp finely chopped fresh marjoram leaves

1 cup ricotta cheese

olive oil

butter

salt and pepper

Make the pastry as described in the recipe for Italian Easter Pie on page 50 or use ready-made puff pastry.

Have a bowl of cold water acidulated with the lemon juice standing ready. Prepare the artichokes by stripping off the leaves, removing the choke, and leaving only the saucer-shaped bottoms (see page 17). Cut these horizontally into thin round slices. Trim the more tender varieties down to the heart; cut the hearts lengthwise in quarters then cut each quarter lengthwise into thin slices. Place the artichokes in the acidulated water.

Sweat the onion with the bay leaf over low heat in 2 tbsp butter and 3 tbsp olive oil until tender. Drain the artichokes and blot dry with paper towels. Add to the onion and fry gently over slightly higher heat for 10 minutes, turning frequently, until they have browned lightly. Season. Add 3 tbsp water, cover, and cook for another 5 minutes or until tender. Cook, uncovered, stirring for a few minutes if necessary to allow any liquid to evaporate. Remove from the heat and allow to cool slightly.

Beat 5 of the eggs in a large bowl with a little salt and pepper. Stir in the grated cheese, garlic, marjoram,

ricotta, and the artichokes. Preheat the oven to 475°F.

Roll out the pastry and assemble the pie (see page 50). Once you have the third layer of pastry in place, spread out the filling evenly on top of it and make 4 well spaced depressions with the back of a tablespoon, each just large enough to take a raw egg broken into it. Season these with salt and pepper. If you use puff pastry, you will only need to line the pan and cover it with a lid. Glaze with beaten egg. Complete as for Italian Easter Pie and bake for 30 minutes.

—— • ——

ARTICHOKE FRICASSÉE

12 tender young globe artichokes or 16 artichoke bottoms

1½ quarts vegetable stock (see page 38)

5 eggs

1 cup crème fraîche *or heavy cream*

1 tbsp lemon juice

1½ tbsp olive oil

3 tbsp butter

1½ tbsp chopped parsley

3 tbsp chopped fresh basil

nutmeg

salt and pepper

Serve with:

thick slices of French bread, toasted

green salad

Prepare the vegetable stock. Prepare the artichokes (see preceding recipe and page 18); cook until tender but still fairly crisp (about 10 minutes, less for the artichoke bottoms) in the prepared liquid. Drain in a colander and refresh under running cold water. Drain well again.

Beat the eggs just enough to blend with the cream,

1 tbsp lemon juice, salt, freshly ground pepper, and a pinch of grated nutmeg. Heat the oil and butter in a wide, nonstick skillet and fry the artichokes over moderate heat for 2–3 minutes, turning frequently, until lightly browned. Reduce the heat a little, add the beaten egg mixture, and cook for only 20–30 seconds, just long enough for the eggs to start setting and form a thick cream halfway in consistency to scrambled eggs. Remove immediately from the heat, sprinkle with the parsley and basil, and serve without delay on hot plates.

Serve with slices of toasted French bread. A crisp green salad goes well with this dish.

———•———

BAKED VEGETABLE OMELET

2 young tender globe artichokes or 4–6 artichoke bottoms
2 young tender zucchini
½ cup ceps (boletus edulis) or cultivated closed cap mushrooms
½ lb spinach or Swiss chard leaves
1 lettuce
1 red pepper
1 small clove garlic, peeled
3 tbsp light olive oil
2 tbsp butter
6 eggs
½ cup ricotta or cream cheese
6 tbsp grated hard cheese
1 tbsp chopped fresh marjoram
¾ cup fine dry breadcrumbs
salt and pepper

Preheat the oven to 300°F. Prepare all the vegetables, cutting the artichokes into quarters, then into lengthwise slices. See page 17 and recipe for Artichoke and Egg Pie on page 129 for method of preparation. Slice the zucchini into thin rounds. Slice the mushrooms about ¼ in thick.

Shred the spinach or Swiss chard and the lettuce. Shred the pepper into very thin strips. Heat the butter and oil gently in a large, deep, nonstick skillet and cook the artichokes, zucchini, spinach or Swiss chard, lettuce, and garlic over low heat for 10 minutes, stirring occasionally. Add the peppers and the mushrooms and continue cooking gently for another 5 minutes. Season with a little salt and some freshly ground pepper. Beat the eggs lightly in a large bowl with a pinch of salt and pepper; beat in the ricotta or cream cheese, the hard cheese, and the marjoram. Lightly oil a large, round pie plate, or grease with butter; sprinkle the plate with half the breadcrumbs, tipping out any excess. Stir the tender, crisp vegetables into the egg and cheese mixture and transfer to the pie plate. Sprinkle the surface with more breadcrumbs and bake in the oven for 20 minutes, or until set.

———•———

DEEP-FRIED EGGPLANT AND CHEESE SANDWICHES

2 large round eggplants
1 lb mozzarella cheese
3 eggs
approx. 1 cup fine dry breadcrumbs
8 large basil leaves
sunflower oil
salt and pepper
Serve with:
tomato salad

Preheat the oven to 400°F. Lightly oil 2 cookie sheets. Slice off both ends of the eggplants where they begin to taper; rinse and dry, then slice into rounds of even thickness (approximately ¼ in thick). Lay these slices out flat in a single layer on the oiled cookie trays and brush their exposed surfaces lightly with olive oil; season lightly with salt and pepper. Bake in the oven for 6 minutes, then take out and allow to cool. These first stages of preparation can be carried out several hours in advance. Slice the mozzarella cheese into 1/4-in thick rounds when you are ready to start cooking the "sandwiches." Beat 3 eggs lightly in a small, deep bowl with a pinch of salt and pepper. Spread the breadcrumbs out on a plate.

Place a slice of cheese on top of half the eggplant slices, press a basil leaf on top, and cover with the remaining eggplant slices. Press together gently, then dip in the beaten egg to coat all over. Cover with the breadcrumbs, pressing lightly to make them adhere.

When all the sandwiches have been dipped in egg and coated in breadcrumbs, pour olive oil into a nonstick skillet; it should be about 1/2 in deep. When it is hot, but not smoking hot, add the sandwiches and fry over moderately high heat, turning once, until they are golden brown on both sides. Take out of the oil with a slotted spoon and drain briefly on paper towels on a very hot plate. Serve without delay on hot plates, while they are piping hot and the mozzarella is still creamy. A tomato salad makes the perfect accompaniment to this dish.

———— • ————

STUFFED EGGPLANTS

2 large long eggplants
wine vinegar
1 clove garlic
1 medium-sized ripe tomato
9 firm black olives, pitted
½ cup grated hard cheese
2 tbsp capers
3 tbsp finely chopped parsley
fine dry breadcrumbs
olive oil
salt and pepper
Serve with:
broiled chicken
green salad

Preheat the oven to 400°F. Bring a large pot, three-quarters full of salted water, acidulated with 1½ tbsp wine or cider vinegar, to a boil. Wash and dry the eggplants, slice off their ends, and cut lengthwise in half. Scoop out most of the flesh, leaving behind an even layer next to the skin about ¼ in thick. Blanch these in the boiling water for 1 minute; take out with a slotted ladle and leave to drain upside down on a clean cloth. Chop the flesh together with the garlic. Heat ¼ cup olive oil in a large, nonstick skillet and gently fry the chopped mixture, stirring, for 2–3 minutes. Stir in ¼ cup breadcrumbs and season with salt and pepper. Continue cooking over low heat for another minute, then remove from the heat.

Blanch, peel, seed, and chop the tomato (use drained canned Italian tomatoes if desired). Finely chop the olives. Stir the tomato, olives, grated hard cheese, capers, and parsley into the fried mixture, adding 1½ tbsp more oil. Add a little more salt if needed and plenty of freshly ground black pepper. Lightly oil the inside of a wide, shallow, ovenproof dish or small roasting pan; fill the eggplant skins with the mixture, sprinkle with more breadcrumbs, and place carefully in the pan in a single layer. Bake in the oven for about 15 minutes and brown under a hot broiler for the last 3 minutes. Serve warm or cold.

———— • ————

EGGPLANT, TOMATO, AND CHEESE BAKE

4 cloves garlic, peeled
1 dried chili pepper
1¾ lb canned tomatoes
generous pinch oregano
2 large eggplants

1 lb mozzarella cheese
1 cup freshly grated hard cheese
1 large sprig basil
light olive oil
salt and pepper

Make the tomato sauce: sauté the garlic in 3 tbsp olive oil until pale golden brown; crumble in the chili pepper, discarding the seeds if you prefer a less peppery dish. Remove the seeds from the tomatoes and chop them coarsely, then add to the garlic and chili together with a pinch of salt. Stir. Cover and simmer for 20 minutes, stirring occasionally. Add a generous pinch of oregano.

While the tomatoes are simmering, prepare and fry the eggplants. Trim off the ends, peel, and slice lengthwise just under ¼ in thick. Heat plenty of oil in a deep-fryer to 350°F and fry a few slices at a time; alternatively, shallow-fry them. They should be pale golden brown but not crisp or dry. Drain well, spread out on paper towels to finish draining, and sprinkle with a little salt. Thinly slice the mozzarella. Preheat the oven to 400°F. Lightly oil the inside of a wide, shallow ovenproof dish or small roasting pan, cover the bottom with a layer of eggplant slices, sprinkle with a little of the sauce and with some of the grated hard cheese. Top with a sprinkling of some of the basil leaves, torn into small pieces, and with some mozzarella slices. Keep layering in this way until you have used all the ingredients, finishing with a mozzarella layer.

Place in the oven and cook for 15–20 minutes; finish by browning the top lightly under a very hot broiler.

——— • ———

Neapolitan Stuffed Eggplants

2 long large eggplants
3 tbsp wine vinegar

2 cloves garlic
9 black olives
2 tbsp capers
1 tbsp finely chopped parsley
2–2½ tbsp finely chopped basil
3 small cans whole tomatoes, drained
¾ cup fine dry breadcrumbs
olive oil
salt and pepper
Serve with:
broiled meat, poultry or fish steaks

Preheat the oven to 400°F. Bring a large pot three-quarters full of salted water acidulated with the wine vinegar to a boil. Rinse and dry the eggplants, cut off the ends, and slice lengthwise in half. Scoop out the flesh, leaving a layer about ¼ in next to the skin. Blanch the hollowed skins in the saucepan of boiling water for 1 minute, drain, and place upside down on a clean cloth.

Cut the flesh into small, thick slices or dice. Heat ¼ cup olive oil in a wide, nonstick skillet over moderately high heat and fry 2 minced garlic cloves together with the flesh for 2 minutes, stirring with a nonstick spoon. Season with salt and freshly ground pepper, cover, and reduce the heat to low; cook gently for 15 minutes, adding 3 tbsp of hot water if the mixture shows signs of becoming too dry.

During this time, pit and chop the olives. Remove the eggplant mixture from the heat and stir in the olives, capers, parsley, and basil, adding a little more salt and pepper if desired. Fill the drained skins with this stuffing and cover with chopped tomato flesh. Sprinkle with a very little salt and a little olive oil and top with a light covering of breadcrumbs.

Place the eggplants in an oiled ovenproof dish or small roasting pan and bake for 15 minutes. Brown the topping under a very hot broiler for 2–3 minutes. Serve hot or warm.

——— • ———

STUFFED EGGPLANTS GREEK STYLE
Papuzakia

2 large long eggplants

wine vinegar

1 medium-sized onion, finely minced

1 cup ground lean lamb

pinch chili powder (optional)

1 small garlic clove, finely minced

3 tbsp finely chopped parsley

1 cup tomato sauce (see pages 74–75)

6 tbsp grated mature hard cheese

olive oil

salt and pepper

For the sauce:

1¾ cups béchamel sauce (see page 36)

2 egg yolks

3 tbsp grated mature hard cheese

Preheat the oven to 350°F. Heat plenty of salted water in a large pot. Trim off the ends of the eggplants, wash and dry them, and cut lengthwise in half. Scoop out the flesh, leaving a layer about ¼ in thick next to the skin.

Add 2 tbsp wine vinegar to the boiling salted water and blanch the hollowed eggplants for 5 minutes. Drain and arrange upside down on a clean cloth or tilted chopping board to continue draining.

Heat ¼ cup olive oil in a large nonstick skillet and fry the onion gently for about 10 minutes, stirring frequently. When it is very tender and has started to brown, add the meat and fry for 1 minute, using a wooden spoon to prevent it from forming balls or lumps as it cooks. Add the chopped flesh scooped out of the eggplants. Cook over low heat, stirring for 2–3 minutes only. Season with salt and freshly ground pepper, adding a pinch of chili powder if desired.

Mix the garlic and parsley and stir into the meat mixture; add the tomato sauce. Keep stirring over low heat for 2 minutes to blend and thicken the mixture. Remove from the heat and stir in the grated cheese. Taste and correct the seasoning if necessary.

Lightly oil a shallow ovenproof dish. Fill the eggplant skins with the meat and vegetable mixture, smoothing the surface. Bake in the oven for 30–35 minutes. Make the sauce. When the stuffed eggplants have finished cooking and it is time to serve them, stir the egg yolks and remaining grated cheese into the sauce off the heat. Return to a very low heat to cook, stirring gently, for a few seconds; do not allow to boil. Pour over the stuffed eggplants and serve at once.

EGGPLANT CUTLETS WITH TOMATO RELISH

1 very large firm eggplant

2 eggs

¾–1 cup fine dry breadcrumbs

light olive oil

salt and pepper

For the relish:

2 ripe tomatoes

pinch oregano

4 basil leaves, shredded

1 clove garlic, peeled and minced

1½ tbsp extra-virgin olive oil

salt and pepper

Make the relish: blanch the tomatoes for 10 minutes and peel (or use drained, canned tomatoes). Remove the seeds and chop the flesh coarsely. Mix in a bowl with a little salt, freshly ground pepper, the oregano, the shredded basil, the garlic, and the extra-virgin olive oil. Stir well and set aside.

133

Cut off the ends of the eggplants, peel them, and cut lengthwise into thick slices. Beat the eggs lightly with a pinch of salt and pepper, then dip the eggplant slices in the beaten egg and coat with breadcrumbs.

Heat 1 cup light olive oil in a large skillet and when very hot, fry the eggplant slices for 1 minute on each side or until the coating is golden brown. Remove from the oil, draining well and place in a single layer on paper towels to finish draining. Do not cover. Transfer to a large dish or platter and serve with the fresh tomato relish mixture

————— • —————

EGGPLANT TIMBALES WITH FRESH MINT SAUCE

1 firm fresh eggplant
2 cloves garlic, peeled
1 sprig thyme
butter
4 eggs
½ cup light cream
½ cup milk
½ cup freshly grated hard cheese
¼ cup olive oil
salt and pepper
For the mint sauce:
¼ cup mint leaves
½ cup white wine vinegar
1½ tbsp sugar
salt and pepper
Serve with:
Macedoine of Vegetables with Thyme (see page 196)

These quantities yield enough to fill four 1-cup or six ½-cup dariole molds or ramekins. Preheat the oven to 350°F. Trim, wash, dry, and peel the eggplant and dice the flesh. Pour ¼ cup olive oil into a large non-

stick skillet, add the garlic cloves, still whole but lightly minced with the flat of a knife blade, and the sprig of thyme, and heat. When hot, add the diced eggplant and fry over moderate heat, stirring, for about 3 minutes. Season with salt and pepper, add 1 cup water, cover, and reduce the heat. Simmer over low heat for 20 minutes, stirring occasionally, until the eggplant is very tender. Cook, uncovered, for a further 10 minutes, stirring, to allow excess moisture to evaporate completely. Remove and discard the garlic and the remains of the thyme sprig; crush the eggplant flesh to a pulp with a potato masher to turn it into a very thick purée, firm enough to hold its shape easily. If necessary, cook for another 2 minutes over low heat, stirring, to thicken further. Add salt to taste.

Grease the molds generously with butter and have some boiling water ready in the kettle to pour into a roasting pan for a *bain-marie*.

Beat 2 egg yolks with 2 whole eggs in a bowl with a pinch of salt and pepper; beat in the cream, milk, the eggplant, and the finely grated cheese. Taste and add more seasoning if necessary. Fill the molds with this mixture, tapping their bases on the work surface to make the mixture settle evenly. Place the molds in the roasting pan, add sufficient boiling water to come three-quarters of the way up the sides of the molds, and cook in the oven for 30 minutes.

Make the mint sauce. Chop the mint leaves and place in a bowl. Add the sugar, vinegar, salt and pepper, and 3 tbsp cold water.

Let the molds stand for a few minutes after removing them from the oven before unmolding them on to individual plates; run a pointed knife round the inside of each mold or dish to loosen. Spoon a little mint sauce to one side of the plate and garnish each timbale with a very small mint sprig.

————— • —————

EGGPLANT PIE

2¼–2½ lb eggplants
½ cup ricotta cheese

1 cup coarsely grated smoked semihard cheese
⅓ cup grated Parmesan cheese
¼ cup fine fresh or dry breadcrumbs
1 small clove garlic, peeled and finely grated
generous pinch oregano
3–4 fresh basil leaves or small pinch dried basil
4 eggs
nutmeg
olive oil
salt and pepper

Preheat the oven to 400°F. Rinse the eggplants, place them whole on the central shelf of the oven, and bake for 30 minutes, turning after 15 minutes. Take them out of the oven and reduce the temperature to 350°F. Trim off their ends and peel; cut lengthwise into quarters and remove the seeds if there are any. Cut the flesh into small pieces and transfer to a large bowl. Reduce to a smooth pulp with a fork or potato masher; blend in the ricotta, the coarsely grated smoked cheese, the Parmesan, 3 tbsp of the breadcrumbs, garlic, oregano, and the basil leaves torn into small pieces.

Beat the eggs briefly with 1 tsp salt, plenty of freshly ground pepper, and a pinch of nutmeg. Mix well with the eggplant and cheese mixture.

Lightly oil a 10-in quiche dish or pie plate about 1½ in deep. Spread the mixture out in it evenly and sprinkle the surface with a thin layer of breadcrumbs. Bake for 35–40 minutes.

———•———

CAPONATA

2 medium-sized eggplants
1 large green celery stalk
½ cup olive oil
1 large onion, finely minced
6 large firm black olives, pitted
6 green olives, pitted
2–2½ tbsp capers
2 large ripe tomatoes
pinch oregano
3 tbsp wine vinegar
1 tbsp sugar
2 cloves garlic
1 sprig basil
salt and pepper
Serve with:
cold meat
crusty white bread

Trim and peel the eggplants, cut the flesh into small, rectangular pieces, and place them in a large bowl. Sprinkle with plenty of salt, stir well, and leave to stand for 1 hour. Trim and wash the celery and cut into small pieces. Fry the onion gently in 3 tbsp of the olive oil for 10 minutes, stirring frequently; add the celery, olives, and capers and continue cooking over low heat while stirring for 2 minutes. Add the coarsely chopped tomatoes, a pinch of salt, some freshly ground pepper, a pinch of oregano, the vinegar, and the sugar. Cover and cook over low heat for 20 minutes.

Drain off all the liquid that the salt will have drawn out of the eggplants, rinse them briefly, and dry them thoroughly with paper towels. Heat a nonstick skillet over moderately high heat for a few minutes; when it is very hot, add the eggplant cubes and dry-fry while stirring until they form a dry "skin," which will prevent them from soaking up as much oil as usual. Take them out of the skillet and pour in ¼ cup olive oil. Mince the garlic cloves and add them to the skillet together with the eggplant. Fry, stirring, for 3 minutes. Season with salt and pepper. Add ½ cup water, cover, and cook for 8 minutes or until the cubes are tender but still firm.

Mix with the tomato sauce and simmer over low heat for a minute or two. Sprinkle with the basil leaves torn into small pieces and serve. Caponata may also be served chilled.

PROVENÇAL STUFFED VEGETABLES
Petits Farcis

3 small round or oval eggplants

3 young zucchini

1 small red pepper

1 small yellow pepper

6 small ripe tomatoes

3 small onions

6 large zucchini flowers (optional)

5 thin slices lean veal

1 cup button mushrooms, sliced

1 sprig thyme

4 slices ham

1 medium-sized garlic clove

2 tsp chopped fresh marjoram leaves

2–3 tbsp velouté sauce (see page 35)

1 egg

¾ cup grated hard cheese

generous pinch oregano

olive oil

2 tbsp butter

¾–1 cup fine dry breadcrumbs

3 tbsp wine vinegar

salt and pepper

Serve with:

French bread

Bring a large pot of salted water to a boil; add the vinegar. Prepare and rinse all the vegetables. Cut the eggplants and zucchini lengthwise in half; pare off a thin piece from the most rounded part so they will be stable. Cut the peppers lengthwise in quarters, removing all the seeds, the stalk, and the inner white membrane. Cut "lids" off the tomatoes, taking a slice off about ¼ of the way down from the top. Slice the peeled onions horizontally in half.

Rinse the freshly picked zucchini or pumpkin flow-

ers, remove the pistils, and drain on paper towels or a clean cloth.

Preheat the oven to 320°F. Sauté the veal slices and button mushrooms in a little hot butter with the sprig of thyme for 2–3 minutes, turning once. Transfer to a chopping board and chop fairly finely, together with the ham.

Scoop out the flesh from the eggplants and zucchini, leaving a fairly thick layer next to the skin. Reserve the flesh. Carefully remove the inner part of the onion, using a grapefruit knife or curved, small vegetable knife. Blanch these hollowed vegetables in the acidulated water, allowing 10 minutes for the onions, 5 minutes for the peppers and the zucchini, and 3 minutes for the eggplant. Drain all these blanched vegetables upside down. Remove the seeds and discard them; scoop out some of the flesh from the tomatoes and reserve; sprinkle the inside with a little salt.

Chop the reserved flesh from the vegetables, heat 1 tbsp olive oil in a nonstick skillet and sauté the chopped flesh for about 5 minutes. Season with salt and pepper, cover the skillet and simmer gently for 10 minutes.

Chop the marjoram leaves with the garlic and combine in a large mixing bowl with the chopped veal and mushrooms, the velouté sauce, the lightly beaten egg, grated cheese, cooked vegetable pulp, a pinch of oregano, and a little salt and pepper. If the mixture is too stiff, add a little more velouté sauce.

Lightly oil 2 cookie sheets. Fill all the vegetables with the prepared mixture. If you use summer squash flowers, use only 1 tsp for each and insert carefully as they tear very easily; pinch the ends of the flowers gently to enclose the filling.

Transfer the stuffed vegetables to the cookie sheet, sprinkle with a topping of breadcrumbs and place a piece of butter on each. Bake for about 30 minutes, placing each sheet in turn under a very hot broiler for the final 2–3 minutes to brown the surface.

Leave to stand for 5 minutes before serving; they are at their best warm rather than piping hot. Serve with crusty French bread warmed in the oven. Serves 6.

— • —

RATATOUILLE

2 medium-sized eggplants
2 large ripe tomatoes
4 zucchini
3 cloves garlic, peeled
1 large onion, finely minced
4 green peppers, finely chopped
1 fresh chili pepper, finely chopped (optional)
½ lb small new potatoes, scrubbed or peeled
oregano
3 tbsp finely chopped parsley
1 cup olive oil
salt and pepper
Serve with:
a selection of French goat's cheeses dressed with extra-virgin olive oil and freshly ground black pepper
French bread

Peel the eggplants, cut the flesh into small cubes, and place these in a large bowl; sprinkle them with salt, stir well, and leave to stand for 1 hour. Peel the tomatoes, remove their seeds, and chop the flesh coarsely.

Trim, wash, and dry the zucchini, cut them lengthwise in quarters, and then cut these long pieces into 1¼-in lengths. Heat 6 tbsp olive oil in a very large, fairly deep nonstick skillet and fry the zucchini with 1 minced garlic clove for 2–3 minutes, while stirring. Sprinkle with a pinch of salt, cover, reduce the heat and simmer gently for 10 minutes or until the zucchini are tender but still a little crisp. Set aside.

Fry the onion very gently with the chili pepper in 6 tbsp olive oil for 5 minutes in a very large, heavy-bottomed flameproof casserole dish, stirring frequently. Add the small new potatoes, turn up the heat, and fry for 5 minutes, stirring frequently; add the tomato flesh, 1 cup hot water, a pinch of oregano, a little salt, and some freshly ground pepper. Cover, reduce the heat, and simmer for 15–20 minutes, by which time the potatoes should be tender.

Stir from time to time and add a little more hot water at intervals when needed.

While these vegetables are simmering drain and squeeze the eggplants to eliminate the liquid the salt has drawn out; dry thoroughly on paper towels or in a clean cloth. Heat a large nonstick skillet without any oil in it and when it is very hot, add the eggplants and dry-fry them for 2–3 minutes over high heat, stirring the cubes continuously. This will seal them and prevent them from soaking up too much oil. Transfer them to a bowl, heat ¼ cup oil in the skillet, and return the eggplant to it with a minced garlic clove. Fry for 3 minutes while stirring and turning. Season with a very little salt and some freshly ground pepper. Add 3 tbsp water, reduce the heat, cover and simmer for 5 minutes or until the eggplant is tender but not at all mushy, then transfer to the casserole dish. Add the zucchini and stir over a low heat for about 2 minutes, taste, and add more salt or freshly ground pepper if needed.

Remove from the heat, sprinkle with the parsley and the last garlic clove, finely minced. The flavors are at their best when ratatouille is served warm, not piping hot. Small, fresh goat's cheeses served cold or broiled, each sprinkled with a little olive oil and some freshly ground black pepper, and hot French bread will make a perfect foil.

———•———

PEPERONATA

2¾ lb large firm red, yellow and green peppers
3 large onions
1 fresh green chili pepper (optional)
1 lb ripe fresh tomatoes (or drained canned)
¼ cup wine vinegar
¼ cup green olives, pitted
6 tbsp olive oil
salt and pepper
Serve with:
broiled meat
crusty white, or whole wheat bread

Rinse and dry the peppers, cut them lengthwise into quarters, remove and discard the stalk, seeds, and white membrane, then slice each quarter lengthwise into 3 broad strips. Peel the onions, slice off the ends, and cut lengthwise into quarters; slice each quarter lengthwise very thinly. Discard the stalk and seeds from the chili pepper and chop very finely. Rinse the tomatoes, cut in quarters, remove any tough parts and all the seeds.

Heat 6 tbsp olive oil in a wide, heavy-bottomed flameproof casserole dish and gently fry the onions and chili pepper together over low heat, stirring frequently, for 15 minutes. Add the peppers and the tomatoes and stir. Cook over slightly higher heat for 5 minutes, still stirring frequently. Season with salt and pepper and sprinkle with 1/4 cup vinegar. Reduce the heat to moderately low and cook for another 10 minutes, stirring occasionally.

Add the olives and continue simmering until the peppers are tender, but still fairly firm, and the liquid has thickened. Taste and add a little more salt if necessary. Turn off the heat and set aside for 10 minutes before serving.

— • —

STUFFED PEPPERS

4 large red or yellow peppers
3 slices 2–3-day old firm white bread
10 firm, fleshy black olives
3 anchovy fillets
4 basil leaves
1–1½ tbsp capers, chopped
1 clove garlic, peeled and finely grated
pinch oregano
fine dry breadcrumbs
6 tbsp light olive oil
1½ tbsp extra-virgin olive oil
salt and pepper

138 | Preheat the oven to 400°F. Line a cookie tray or a

wide shallow ovenproof dish with foil. Rinse and dry the peppers and place on their sides on the foil. Bake in the oven for about 30 minutes, turning them half way through, until their skin has browned and lifted, making them very easy to peel. Remove from the oven and allow to cool completely. Turn down the oven temperature to 325°F.

Peel the peppers, then cut neatly round the stalk, making an opening large enough to enable you to remove the seeds and the pale membrane sections attached to them from the inside with two fingers, holding each pepper upside down in your other hand. Make the filling: slice off the crusts from the bread, cut it into dice, and fry these in 4 tbsp hot olive oil in a nonstick skillet, stirring frequently, until they are crisp and golden brown all over. Drain on paper towels, sprinkle with a pinch of salt, and place in a large mixing bowl.

Cut the pitted olives into small pieces. Chop the anchovies. Tear the basil leaves into small pieces (use a small pinch of dried basil if fresh basil is unavailable). Add all these ingredients with the capers, garlic, and oregano to the fried bread dice and mix; use a teaspoon to fill the peppers loosely with this mixture and then place them upright in a lightly oiled shallow ovenproof dish. Sprinkle the top of each pepper with a small pinch of salt, some fine dry breadcrumbs, and a little extra-virgin olive oil.

Bake in the oven for 35 minutes and then place under a very hot broiler to brown and crisp the breadcrumb topping. Serve the peppers hot, sprinkling them with any juices left in the cooking dish.

— • —

MEXICAN PEPPERS WITH POMEGRANATE SAUCE
Chiles en nogada

4 large green peppers
1 medium-sized onion, finely minced
½ clove garlic, very finely minced
½–1 fresh green chili pepper, finely chopped

1¼ cups lean ground beef

2 ripe, medium-sized tomatoes, peeled and seeded

1½ tbsp white wine or cider vinegar

¼ cup seedless white raisins

1 tsp light brown sugar

pinch ground cloves

½ tsp ground cinnamon

1½ tbsp slivered or flaked almonds

1½ tbsp butter

sunflower oil

salt and pepper

For the cream and pomegranate sauce:

4 walnut halves

¼ cup whole almonds, blanched and peeled

1 cup light cream

2–2½ tbsp chopped fresh coriander leaves

pinch sugar

½ tsp ground cinnamon

fleshy seeds from half a pomegranate

few whole, fresh coriander leaves

1 juicy lime

salt

Preheat the oven to 400°F. Rinse and dry the peppers and bake in the oven for 20–25 minutes, turning halfway through to loosen their skins (see previous recipe). Peel the whole peppers carefully, remove the stalk neatly, cutting away just sufficient flesh surrounding it to make it easy to remove the seeds and pale membrane using two fingers.

Make the filling while the peppers are in the oven. Heat the butter with 1½ tbsp oil in a wide nonstick skillet over low heat and sweat the onion, garlic, and chili pepper for 10 minutes, stirring occasionally. Increase the heat, add the meat and a pinch of salt and pepper, and cook for 2 minutes, stirring continuously. Add the finely chopped tomatoes, the wine or vinegar, seedless white raisins, sugar, cloves, and cinnamon. Stir well, turn down the heat and simmer, uncovered, over very low heat for 15 minutes or

until all excess moisture has evaporated, stirring at frequent intervals. Stir in the almonds, cook for another 2–3 minutes, and then remove from the heat.

Make the sauce: put the walnuts and almonds in the blender with the cream, coriander leaves, cinnamon, and a pinch each of salt and sugar; process briefly.

Fill the peeled peppers with the hot stuffing mixture, pressing it down into them gently but firmly. Place each stuffed pepper on a plate, pour a little of the cream sauce over each serving and sprinkle with a few pomegranate seeds and a couple of coriander leaves (the colors echo those of the Mexican flag). Place a wedge of lime on each plate for each person to squeeze over the stuffed pepper.

Serve at once. If you prefer to serve the peppers very hot, after stuffing them return to a moderate oven (320°F) for 15 minutes in an ovenproof dish to heat through before covering with the sauce and garnishing.

———— • ————

PEPPER, ZUCCHINI, AND TOMATO FRICASÉE
Pisto Manchego

3 large green peppers

½ green chili pepper

3 medium-sized zucchini

3 medium-sized onions

10 medium-sized ripe tomatoes

pinch dried oregano

pinch dried basil

6 eggs, 2 of which hard-boiled

olive oil

salt and pepper

Serve with:

rice pilaf

139

The Spanish equivalent of the French dish *pipérade*, this dish comes from the La Mancha region. Rinse, dry, and prepare the peppers and the chili pepper and chop them coarsely. Dice the zucchini. Chop the onions coarsely. Blanch, peel, and seed the tomatoes and chop coarsely. Use drained, canned tomatoes if necessary.

Heat 1/2 cup olive oil in a large, heavy-bottomed flameproof casserole dish and when it is very hot, add the peppers and chili peppers, zucchini, and onions. Stir well, sprinkle with a pinch of salt and some freshly ground black pepper. Cover, and reduce the heat to very low; simmer gently for 25 minutes, stirring occasionally.

Prepare the garnish. Cut the 2 hard-boiled eggs lengthwise in half. Remove the yolks and push them through a sieve; cut the whites into strips or very small dice. Mix the chopped tomato flesh with a pinch of salt, 1 tbsp oil, the oregano, and a small pinch of dried basil in a small saucepan; bring to a boil, stirring. Simmer, uncovered, for about 10 minutes, stirring frequently as the sauce reduces and thickens considerably. When it is very thick, stir into the pepper and onion mixture and cook over low heat, stirring, for 1–2 minutes to reduce the liquid further. The vegetable mixture should be very thick, with no visible liquid. Beat 4 eggs in a bowl with a pinch of salt and pepper, stir into the vegetables and cook, stirring, over very low heat. When the eggs have started to set and thicken the mixture, which will take about 30 seconds, remove from the heat and serve this thick, creamy mixture immediately on hot plates, decorating each serving with a little of the egg garnish.

—— • ——

LOBSTER CATALAN

1 large yellow pepper, roasted
1 large red pepper, roasted
2 frozen lobsters or crawfish
4 large scallions
4 medium-sized ripe tomatoes
10 black olives, pitted
10 green olives, pitted
generous pinch oregano
2 tbsp capers
1 lemon
2 sprigs fresh basil
6 tbsp light olive oil
3 tbsp extra-virgin olive oil
1 red chili pepper, finely chopped (optional)
salt and black peppercorns

The day before you plan to serve this dish, roast the peppers as described on page 16 and when cold, peel them, and cut into thin strips (discard the stalk, seeds, and inner white membrane). Cut these strips into small pieces or dice shortly before using.

Thaw the lobsters or crawfish very slowly, leaving them on the bottom shelf of the refrigerator for 30 hours or thaw more quickly at room temperature (about 3 hours). When completely thawed, poach in gently simmering salted water or in a *court-bouillon* (see page 37) for 5 minutes; drain and cool. Place each one on a chopping board, the right way up and place the tip of a heavy kitchen knife in the center of the lobster's head where there is usually a cross-shaped indentation. Push firmly downward or use a mallet to tap the handle of the knife sharply to pierce the shell, cut the lobster lengthwise in half, and remove all the white meat. Reserve the soft meat in the head for other uses. Cut the white, tail meat into slices about ¼ in thick and season with a little salt and freshly ground pepper; set aside.

Trim off the roots and leaves of the scallions, remove the outermost layer, and slice the bulb into thin rings; blanch these for 30 seconds in simmering salted water; remove with a slotted spoon and drain. Blanch, peel, and seed the tomatoes; cut them into small pieces. Heat 3 tbsp of the light olive oil in a large nonstick skillet and sauté the scallions, chili pepper if used, tomatoes, peppers, and all the olives. Add the oregano, capers, a pinch of salt, and some freshly ground black pepper. Reduce the heat and stir while simmering for about 30 seconds. Turn off the heat. Heat 3 tbsp of light olive oil in another nonstick skillet and sauté the pieces of lobster over moderately high heat for about 30 seconds on each side. Spoon some of the vegetable mixture onto heated individual plates, place an equal number of lobster slices on top of each serving, and squeeze a little lemon juice over them. Sprinkle with 1 tbsp extra-virgin olive oil and serve.

———•———

AVOCADO AND CHICKEN SALAD WITH BLUE CHEESE DRESSING

1 chicken, poached in aromatic stock (see page 72)

2¼ cups mayonnaise (see page 35)

1 small bunch lamb's lettuce

1 head chicory

1 bunch arugula

1 Belgian endive

1 head radicchio

4 young, tender zucchini

¼ cup olive oil

2 cloves garlic, peeled and minced

3 tbsp finely chopped parsley

1 cup light cream

2 oz blue cheese (e.g. Gorgonzola, Roquefort or Stilton)

1 lemon

3 ripe avocados

salt and white peppercorns

Cook the chicken as described on page 72. While it is cooking, make the mayonnaise and set aside. Trim, wash, and dry all the salad leaves and shred. Mix well and place on individual plates. Rinse, dry, and trim the ends from the zucchini and slice them into thin rounds. Heat ¼ cup olive oil in a wide non-stick skillet and sauté the zucchini slices with the 2 whole garlic cloves, lightly minced with the flat of a large knife blade. Season with salt and pepper, and add most of the parsley, reserving a little for a final

garnish. When the slices are tender but still crisp, remove from the skillet, leaving the garlic behind, and place on top of the shredded salad bed.

Stir the cream into the mayonnaise; add the blue cheese, rubbed through a fine sieve. Season with salt and freshly ground white pepper to taste and stir in a little lemon juice.

Slice the cooked, cooled chicken into fairly small pieces. Peel the avocados, cut lengthwise in half, and remove the pit; cut the flesh lengthwise into thin slices and fan these out on top of the zucchini and salad. Top each serving with some of the chicken and cover with a few spoonfuls of the blue cheese dressing. Decorate with sprigs of parsley and serve the remaining dressing in a sauceboat or bowl.

———•———

TEMPURA WITH TENTSUYU SAUCE

1 egg yolk

1 cup all-purpose flour

1 lb pumpkin, peeled

8 baby zucchini

16 thin scallions

16 large pumpkin or any winter or summer squash flowers (optional)

sunflower oil

salt

For the Tentsuyu sauce:

1 tsp crumbled instant dashi cube

½ cup Japanese light soy sauce

½ cup mirin (sweet Japanese rice wine) or dry sherry sweetened with 2 tsp sugar

one-third daikon root and 1 or 2 chili peppers (momiji oroshi)

Serve with:

steamed or plain boiled rice

Make the batter: beat the egg yolk well until creamy in a large bowl with ½ tsp salt; gradually beat in 1 cup

141

iced water, adding a little at a time. Sift in the flour and stir only just enough to mix; do not worry if there are a few small lumps (if stirred too much, this batter becomes gluey). Cover and chill in the coldest part of the refrigerator for at least 1 hour.

Prepare the vegetables, trimming, rinsing, and drying them. Leave about 1 in of stalk attached to the flowers; remove the pistil from inside the flower carefully and then rinse the flowers; spread out on a clean cloth to dry. If these flowers are unavailable, the vegetables will do very well on their own. Cut the peeled pumpkin or winter squash flesh into pieces just under 1/4 in thick and about 2 in long. Cut the zucchini lengthwise into slices just under 1/4 in thick and then cut each of these long slices across, in half. Trim the roots and the ends of the leaves off the scallions, remove the outermost layer of the bulb and leaves, and cut the inner, tender part of both lengthwise in half. Make the sauce: pour 1 cup water into a small saucepan, add the crumbled instant dashi cube, the soy sauce, and the mirin; stir well as you heat to just below boiling point. Turn off the heat; this will be reheated, without letting it boil, just before serving. Have the momiji oroshi ready to be added to the liquid just before serving: peel one-third of a daikon root, cut into round slices and process in the food processor with 1 or 2 chili peppers (seeds and stalks removed) until it forms a thick purée. Place this purée in a fine mesh plastic sieve and leave to drain over a bowl.

Cook the rice. When the rice is ready, reheat the sauce and pour into tiny bowls, one for each person; stir about 1 tbsp of the momiji oroshi into each bowl. Serve the rice in one large bowl or in small, individual bowls. Fry the vegetables just before serving. Heat plenty of oil in a deep-fryer to 350°F. Divide the prepared vegetables into 4 batches; you will serve each batch as soon as it is fried. Stir the chilled batter very briefly and dip each piece of vegetable in it before adding to the hot oil, working quickly once the first batter-coated piece has been added to the oil. At this temperature the battered vegetables will fry gently and will take a few minutes to become lightly crisp and pale golden brown; as soon as they do so, take out of the oil, drain briefly on paper towels, and serve at once. While this first batch is being eaten, fry the next and so on.

To eat, pick up a piece of fried vegetable with chopsticks and dip it in the hot sauce. The rice makes a suitable foil.

———•———

ZUCCHINI WITH DILL SAUCE
Gefüllter Kurbiss mit Dillsauce

6 large zucchini
1 medium-sized onion, finely minced
1¼ cups ground lean veal
½ cup ground ham
1 egg
¼ cup grated Swiss cheese
¼ cup long grain rice (uncooked weight), boiled or steamed
6 tbsp chopped fresh dill leaves
1 lemon
1¼ cups light cream
1½ tbsp all-purpose flour
1½ tbsp sunflower oil
¼ cup butter
white wine vinegar
nutmeg
salt and pepper
Serve with:

Sweet-sour Baby Onions (see page 192)

Preheat the oven to 350°F. Bring a large pot two-thirds full of salted water to a boil; add 3 tbsp vinegar. Rinse the zucchini and cut off about ¾ in from each end. Use a potato peeler to remove a very thin layer of skin from them and hollow out each section with an apple corer; if they are very long, cut them in half. Blanch in the boiling acidulated water for 5 minutes, drain, and leave to cool. Fry the onion very gently for 10 minutes in 2 tbsp butter and 1-1/2 tbsp oil, stirring frequently. When tender but not browned, stir in the veal, season with a little salt and pepper, and stir over higher heat for 2 minutes. Transfer to a bowl and mix with the ground ham, the egg, cheese, and boiled rice, adding a little more salt and pepper to taste and a pinch of nutmeg. Stuff the hollowed zucchini with this mixture. Oil a wide, shallow ovenproof dish and arrange the stuffed zucchini in it in a single layer.

Melt 2 tbsp butter in a small saucepan and add the dill; fry gently for 1 minute. Add 1/2 cup water and the finely grated rind of the lemon. Simmer, uncovered, for 2 minutes. Mix the cream with the flour and stir into the contents of the saucepan, continuing to stir and cook until the sauce has thickened. Stir in the juice of the lemon and season with salt and pepper. Pour this sauce all over the stuffed vegetables. Bake in the oven for 15–20 minutes, then serve without delay. Serves 6.

———•———

until tender. Drain and reserve.

Sweat the onion in the butter for 10 minutes over very low heat, stirring frequently. Turn up the heat and add the meat. Sprinkle with a generous pinch of cinnamon, and cook for 2 minutes, stirring continuously over moderately high heat. Season with salt and pepper and leave to cool. Mix with the peas, adding a little more salt, pepper, and cinnamon to taste.

Stuff the hollowed zucchini with this mixture, arrange in a single layer in a greased wide, shallow flameproof dish. Add 2¼ cups water, cover with a lid or with foil, and simmer very gently for 25–30 minutes or until the vegetables are tender. Transfer to heated individual plates and serve with a bowl of natural yoghurt mixed with a pinch of salt. Serves 6.

———•———

PERSIAN STUFFED ZUCCHINI
Dolmeh Kadu

6 large zucchini
1 cup dried split green peas
1 large onion, finely minced
2 cups ground lean lamb or beef
generous pinch ground cinnamon
¼ cup butter
salt and pepper
Serve with:
natural yoghurt, lightly salted to taste

Rinse the zucchini and prepare as for the previous recipe, cutting off their ends, hollowing out the centers and blanching them.

Rinse the dried peas well in a sieve under running cold water; boil in salted water for 30 minutes or

ZUCCHINI FRITTERS
Mücver

2¼–2½ lb zucchini
1 medium-sized onion
1¾ cups all-purpose flour
1 cup crumbled feta cheese
2 egg yolks
3 tbsp chopped fresh dill leaves
3 tbsp chopped parsley
olive oil
nutmeg
salt and pepper
Garnish with:
sprigs of parsley or dill
Serve with:
Tsatsiki (see page 149)

143

Trim off the ends of the zucchini, peel them with a potato peeler, and grate finely into a large mixing bowl. Do the same with the onions. Stir in the sifted flour, the finely crumbled feta cheese, and the lightly beaten eggs. Season with a very little salt if wished, with plenty of freshly ground pepper, and a pinch of nutmeg. Stir in the dill and parsley, blending the mixture very thoroughly.

Heat enough oil for shallow-frying in a wide skillet or wok. When hot, drop tablespoonfuls of the mixture into the hot oil, taking care to space them out so that they do not stick to one another. Fry until crisp and golden brown, turning once. Remove with a slotted spoon and finish draining on paper towels. Serve at once, while still piping hot, garnished with sprigs of parsley or dill, accompanied by tsatsiki.

BRAISED ARTICHOKES

8 very young, tender globe artichokes
juice of 1 lemon
6 tbsp extra-virgin olive oil
¼ chicken bouillon cube
1 large clove garlic, peeled
1½ tbsp chopped parsley
salt and pepper

Prepare the artichokes, removing the outer leaves and chopping the top third off the remaining, tender heart. Trim the stems to about 2 in in length and use a potato peeler to peel off the tough outer layer (see page 18). The artichokes most suitable for use are the very tender varieties; if you grow your own Breton variety, use side buds trimmed off to let the main bud develop as they will not be old enough to have developed a choke. Rinse them inside and out. As each one is cut, peeled, and trimmed, drop it into a large bowl of water that has been acidulated with the lemon juice.

Drain the artichokes, place them in one layer, turned upside down in a saucepan that is just wide enough to accommodate them all. Add 1¼ cups water, the oil, the crumbled bouillon cube, garlic, parsley, and the pieces of artichoke stalk you have trimmed off and peeled. Add a very small pinch of salt and some freshly ground pepper.

Bring to a boil over moderate heat, cover tightly, and turn down the heat to low; simmer gently for 15 minutes or until the artichokes are tender and the liquid has reduced considerably.

ARTICHOKE BOTTOMS IN CHEESE SAUCE

6 large globe artichokes
vegetable stock (see page 38)
1 lemon
2 tbsp grated Swiss cheese
butter
For the cheese sauce:
1¾ cup béchamel sauce (see page 36) or velouté sauce (see page 35)
2 egg yolks
¼ cup grated Gruyère cheese
Garnish with:
tomato rosebuds (see page 27)
Serve with:
zucchini purée

Gently boil the artichoke bottoms in vegetable stock. Have a large bowl of water acidulated with the lemon juice standing ready. Prepare the artichokes as described on page 18, stripping each one right down to its fleshy base; remove the choke and cut off the stalk. Use a small sharp knife to pare away any rough edges and drop immediately into the acidulated water. When they are all ready, poach in the cooking stock for about 20 minutes or until they are tender. Preheat the oven to 350°F.

Make the cheese sauce: prepare the béchamel or velouté sauce; it should be of pouring consistency and not too thick for this recipe. As soon as it has thickened, take the saucepan off the heat and beat with a whisk as you add the egg yolks one at a time, followed by the grated cheese.

Drain the cooked artichoke bottoms and cut each one into 3 or 4 thin sections. Cover the bottom of the dish with half the sauce, place the sliced artichokes on top, cover with the remaining sauce, and sprinkle with the grated cheese. Dot little flakes of butter here and there on the surface. Bake in the oven for 15 minutes, to form a golden brown layer on top. Serve immediately, garnished with tomato rosebuds. Zucchini purée makes a good complement to the artichokes, both in taste and appearance.

— · —

CRISPY GLOBE ARTICHOKES

8 young tender globe artichokes
juice of 1 lemon
light olive oil
salt

If you grow your own artichokes, the small side buds can be used for this recipe, although they will not look as spectacular as the large yet tender ones grown in Mediterranean climates. Large Breton type artichokes are too tough. Half fill a very large pot with salted water, add the lemon juice, and bring to a boil. Prepare the artichokes, trimming off the stems about 2 in from their bases; use a sharp, curved vegetable knife to scrape off the outer layer from the stem. Remove the lower, outer leaves which are too tough for this recipe, even when tender varieties are used. Remove the feathery choke enclosed by the leaves. Rinse thoroughly, drain well, and place upside down on a chopping board. Take hold of the stems one at a time and rotate the artichokes while pressing down hard against the board; this will make them open out like flowers. Blanch the artichokes for a full 5 minutes in the fast

boiling acidulated water, drain well, and dry as best you can with clean cloths. Replace them upside down on the chopping board to dry off completely (this will take up to an hour in a warm room). Pour sufficient olive oil into a very wide, heavy-bottomed saucepan to a depth of about 2 in. When it is very hot, but not smoking, place the artichokes upside down in it and fry over moderately high heat until they are tender and the leaves are crisp and lightly browned. If using a deep-fryer, the oil should be kept at an even 350°F throughout the frying time. Remove the artichokes from the saucepan and finish draining briefly on paper towels. Serve hot.

— · —

ARTICHOKE BOTTOMS VENETIAN STYLE

12 large globe artichokes
1 lemon
1 cup chicken stock (see page 37)
6 tbsp olive oil
2 cloves garlic, peeled
3 tbsp chopped parsley
salt and white peppercorns

Prepare the artichokes in the same way as for Artichoke Bottoms in Cheese Sauce on page 144. Place them in a bowl of acidulated water to prevent discoloration.

Choose a very large skillet or saucepan to accommodate all the drained artichoke bottoms in a single layer (or use 2 saucepans). Pour the stock and the oil all over them. Add the whole but lightly minced garlic cloves and sprinkle with the parsley. Season with a small pinch of salt and some freshly ground pepper. Cover tightly and simmer for about 20 minutes, or until the artichokes are tender, adding a little more hot stock when necessary. Serve hot, moistened with their cooking juices.

145

BAKED TOMATOES WITH MINT

4 beefsteak tomatoes
extra-virgin olive oil
2 sprigs mint
salt and pepper

Preheat the oven to 350°F. Rinse the tomatoes and cut them horizontally in half. Lightly oil 2 wide roasting pans and place the halved tomatoes, cut side uppermost in a single layer in them. Season with salt and freshly ground pepper and sprinkle with a little olive oil.

Tear the mint leaves into small pieces and sprinkle these over the cut surfaces of the tomatoes. Pour in just enough water, between the tomatoes, to cover the bottom of the pans and come about one-quarter of the way up the sides of the tomatoes. Bake in the oven for about 50 minutes.

— • —

TOMATOES AU GRATIN

2 tbsp chopped parsley
1 small clove garlic, peeled
1½ tbsp capers
½ cup fine breadcrumbs
oregano
4 beefsteak tomatoes
extra-virgin olive oil
salt and pepper

Preheat the oven to 320°F. Chop the parsley very finely with the garlic and transfer to a bowl. Chop the capers coarsely and place in the bowl; add ¼ cup of the breadcrumbs and mix well. Stir in 1½ tbsp oil and a generous pinch each oregano and freshly ground black pepper.

Rinse the tomatoes and cut horizontally in half.

Place cut side uppermost on an oiled cookie sheet. Sprinkle with a little salt; cover each cut surface with some of the prepared mixture. Sprinkle the top of each tomato with a little oil and cover with a layer of plain breadcrumbs. Bake in the oven for 25–30 minutes, placing the tomatoes under a very hot broiler for the last 3 minutes to brown the topping. Serve hot or warm.

— • —

BAKED EGGPLANTS

4 medium-sized long eggplants
2 cloves garlic, peeled
extra-virgin olive oil
salt and pepper

Preheat the oven to 350°F. Trim off the eggplant stalks, cut them lengthwise in half, and make deep intersecting cuts in the exposed flesh to form a lattice design. Season with salt and freshly ground pepper. Sprinkle with the finely minced garlic, pressing it down into the deep cuts in the flesh.

Oil a roasting pan and arrange the eggplants, cut sides uppermost, in a single layer. Sprinkle the cut surfaces with olive oil and bake for about 35 minutes or until the eggplants are tender and are lightly browned on the surface. Serve hot.

— • —

NEAPOLITAN EGGPLANT SALAD

4 long firm eggplants
2–3 cloves garlic, peeled and thinly sliced
1 fresh red or dried chili pepper
generous pinch oregano
¼ cup wine vinegar
extra-virgin olive oil
salt and pepper

Serve with:

fresh sliceable cheese

crusty white bread or whole wheat bread

Bring a large pot of salted water to a boil. While it is heating, chop off the ends of the eggplants, rinse and dry them, then cut lengthwise in quarters. Scoop out any central seed-bearing sections.

Slice the flesh lengthwise into strips, cutting these in half if they are very long. Soak in a large bowl of lightly salted water for about 10 minutes or until the water turns a brownish-gray color. Drain the strips and blanch in the pot of boiling water for 1½ minutes. Drain again and spread out in a single layer on a clean cloth; cover with another cloth and blot dry.

Place the eggplant strips in a wide, shallow serving dish and sprinkle with the very thinly sliced garlic, the finely chopped or crumbled chili pepper, oregano, and a little salt and pepper. Sprinkle with the vinegar and about 6 tbsp olive oil. Stir and leave to marinate for at least 30 minutes before serving.

———— • ————

EGGPLANT AND MINT RAITA

2 white or purple small, round or oval eggplants

1 large scallion

2½ cups plain yoghurt

1 large sprig mint

salt and pepper

Garnish with:

small mint leaves

Serve with:

curry

Set up a steamer ready for cooking the vegetables. Trim and peel the eggplants, cut the flesh into ¾-in cubes, and steam for 10 minutes. Remove the root end and outermost layer of the scallion; cut the inner, tender parts of the bulb and leaves into very thin rings. Blanch these for 5 seconds in boiling salted water and drain.

Mix the scallion with the yoghurt in a large bowl, adding 1 tsp salt and plenty of freshly ground pepper. Mix very thoroughly. Stir in 1½ tbsp chopped mint.

Place the cooked eggplant cubes in a large bowl, mash them coarsely with a fork or potato masher, and allow to cool. Mix them with the yoghurt, add a little more salt and pepper if needed, and transfer to a serving dish. Decorate with small mint leaves. This cooling raita provides a refreshing foil to curried dishes.

———— • ————

DEEP FRIED EGGPLANT WITH MISO SAUCE
Nasu no Miso Kake

4 very small, round or oval white or purple eggplants

sunflower oil

salt

For the miso sauce:

3 tbsp shiro miso (sweet white bean paste)

6 tbsp saké, dry sherry or dry vermouth

2 tsp sugar

½ tsp cornstarch or potato flour

1 generous pinch monosodium glutamate, optional

Garnish with:

radish flowers (see page 28)

parsley sprigs

Slice off the ends of the eggplants; rinse and dry them, then cut lengthwise into quarters. Cut across the middle of each quarter to obtain pieces about 2½ in long. Use a small, sharp vegetable knife to "turn" these pieces, cutting off the sharp edges. Make 2 or 3 parallel lengthwise cuts, about ⅛ in deep in the skin of each piece and place in a large bowl of cold, salted water; set aside. The water should cover all the pieces. Drain after 20 minutes and blot dry in clean cloths.

Make the miso sauce: warm the sweet white bean paste gently with the saké and sugar, stirring with a wooden spoon. Mix the cornstarch or potato flour with 6 tbsp cold water and stir into the mixture; keep stirring as the sauce comes to a boil and thickens. Remove from the heat and stir in the monosodium glutamate, if used. Do not add salt.

Heat plenty of sunflower oil in a deep-fryer to 320°F (or use a wok); fry the pieces of eggplant in batches until they are pale golden brown but not crisp; each batch will take about 5–7 minutes. Remove with a slotted spoon and finish draining on paper towels while keeping hot. When the last batch is cooked, serve without delay on individual plates, coated with the sauce. Garnish with radish flowers and sprigs of parsley.

———— • ————

FRIED EGGPLANTS WITH GARLIC AND BASIL

2 large firm eggplants
2 large cloves garlic, peeled
6 tbsp olive oil
salt and pepper
Garnish with:
fresh basil leaves

Trim off the ends of the eggplants; peel and quarter them lengthwise. Scoop out the central section if there are seeds in it. Chop into dice. Heat the oil in a very large, nonstick skillet and stir-fry the eggplant dice with the finely sliced garlic cloves over fairly high heat. As soon as they have browned lightly, cover the skillet, reduce the heat to very low, and cook gently for 15 minutes; they will produce some moisture but you may need to add 2–3 tbsp water af-

ter a while. Season with salt and pepper and serve, garnished with fresh basil leaves.

———— • ————

PEPPERS AU GRATIN

5 large peppers (red, yellow, green and purple)
few basil leaves
5 cloves garlic, peeled
2 canned anchovy fillets, drained
12 fleshy black olives
2½–3 tbsp capers
generous pinch oregano
fine dry breadcrumbs
extra-virgin olive oil
salt and pepper

Preheat the oven to 400°F. Rinse and dry the peppers, wrap each one tightly in foil, enclosing it completely, and bake in the oven for 40 minutes, turning halfway through this time. Take the parcels out of the oven and allow to cool completely (chill in the refrigerator for 1 hour if you wish) before unwrapping them. Nearly all the skin will come away with the wrapping; what little is left will peel off easily. Remove and discard the stalk and seeds, then cut the flesh into thin strips.

Oil two wide, shallow ovenproof dishes and sprinkle the basil leaves, torn into very small pieces, over the bottom. Use parsley instead if preferred. Spread half the pepper strips out in a single layer in each dish, mixing the colors. Season with salt and pepper. Chop the garlic and the anchovies fairly coarsely and sprinkle over the peppers. Pit the olives, cut them in quarters, and sprinkle evenly over the peppers. Finish with the capers, a pinch of oregano for each dish, and a thick covering of breadcrumbs. Drizzle a little olive oil over this topping and bake in the oven for 15 minutes to form a crunchy, golden brown layer on the surface.

STIR-FRIED PEPPERS IN SHARP SAUCE

2¼ lb peppers

6 tbsp extra-virgin olive oil

4 cloves garlic, peeled

1 fresh or dried chili pepper

½ cup wine vinegar

½ cup fine breadcrumbs

2–3 tbsp capers

oregano

salt

Prepare the peppers: remove the stalks, seeds, and white inner membrane. Cut them into fairly thin strips. Heat the oil in a very large skillet or wok and stir-fry the garlic, chili pepper, and pepper strips for 2–3 minutes over high heat. Reduce the heat and sprinkle with about one-third of the vinegar and 3 tbsp of the breadcrumbs. Add a pinch of salt. Keep stirring the peppers as you gradually add the remaining vinegar and breadcrumbs. Stir in the capers. Cover and cook over low heat for about 15 minutes, or until the peppers are tender, moistening with a little water when necessary. Add a little more salt if needed, a pinch of oregano, stir once more and remove from the heat.

This mixture also makes a delicious stuffing for vegetables and poultry.

— • —

TSATSIKI

1 medium-sized firm, fresh cucumber

1 large clove garlic, peeled and finely grated

3 tbsp chopped fresh dill leaves

2¼ cups natural yoghurt

few fleshy black olives

extra-virgin olive oil

salt

Serve with:

Zucchini Fritters (see page 143)

Wash and dry the cucumber, slice off the ends, and peel. Grate into a large bowl and mix in the garlic, ½ tsp salt, the dill, and the yoghurt. Drizzle a little oil over the surface, garnish with a few black olives, and serve lightly chilled as a refreshing side dish item to a Greek or Turkish meal.

— • —

STIR-FRIED ZUCCHINI WITH PARSLEY, MINT, AND GARLIC

3 tbsp chopped parsley

1 sprig mint

1 lb baby zucchini

6 tbsp olive oil

2 cloves garlic, peeled

wine vinegar

salt and pepper

Chop the parsley and tear the mint leaves into small pieces by hand. Rinse and dry the zucchini, slice off the ends and cut them lengthwise into ¼-in thick slices. Cut these into 1½-in lengths.

Heat the oil in a large skillet and stir-fry the pieces of zucchini with the lightly minced but whole garlic clove over fairly high heat for 5 minutes; they should be just tender but still crisp. Remove the garlic clove.

Sprinkle the zucchini with ¼ cup vinegar, a pinch of salt, and some freshly ground pepper. Stir-fry for a few seconds, reduce the heat, then sprinkle with the parsley and mint. Stir briefly, then remove from the heat.

Drain off any excess oil and serve the zucchini at once. This dish goes well with broiled meat, poultry or fish. It is also excellent served cold with cured meat, sausages, and smoked cheese.

——— • ———

ZUCCHINI PURÉE

2¼–2½ lb zucchini

1 ¾ cups vegetable stock (see page 38)

1 clove garlic, peeled

few basil leaves

4 egg yolks

6 tbsp finely grated Parmesan cheese

3 tbsp unsalted butter

nutmeg

salt and white peppercorns

Garnish with:

sprigs of fresh basil

Serve with:

hot meat dishes

Rinse and dry the zucchini, cut off their ends and slice them into thick rounds. Place these in a pot with the stock and the garlic clove, partially crushed with the flat of a heavy knife blade. Cover and bring to a gentle boil, then simmer for about 20 minutes or until the zucchini are tender but still firm. Drain, discard the garlic clove, and reduce to a thick purée, together with a few basil leaves torn into small pieces, by beating with a hand-held electric beater or processing in batches in a food processor.

Beat the egg yolks in a large bowl with the grated cheese, a pinch of salt, some freshly ground white pepper, and a pinch of nutmeg. Heat the purée, add the butter, and beat as it melts; keep beating as you gradually add the egg yolk and cheese mixture. Do

not allow the purée to boil. Stir over low heat until it is very hot.

Serve, garnished with sprigs of fresh basil. This purée goes well with meat dishes.

——— • ———

ZUCCHINI RUSSIAN STYLE

1 ¾ lb baby zucchini

all-purpose flour

¼ cup butter

3 tbsp sunflower oil

½ cup light cream

1½ tbsp chopped parsley

1½ tbsp chopped mint

salt and pepper

Trim off the ends of the zucchini. Wash and dry them and cut into thick rounds. Place these in a bowl that has been dusted with flour, and coat all the pieces, turning them over and mixing by hand. Tap the bottom of the bowl on the work surface to make any excess flour fall to the bottom. Heat the butter with the oil in a large skillet and when very hot add the zucchini pieces, leaving the unwanted flour behind in the bowl. Fry over moderately high heat for 2–3 minutes, stirring continuously. Reduce the heat to low, cover and cook gently for about 10 minutes, stirring occasionally, until the zucchini are tender but still crisp. Season with a little salt and pepper. Stir in the cream, parsley, and mint and continue cooking while stirring gently over low heat for 1–2 minutes. Serve very hot.

——— • ———

AVOCADO CREAM DESSERT

2 large ripe avocados

1 large lime

⅓ cup confectioner's sugar, sifted

Decorate with:

small mint leaves

thin lime slices

Cut the avocados lengthwise in half, remove the pits, and peel. Cut the pulp into fairly small pieces and purée in a food processor. Immediately mix this purée with part or all of the lime juice to prevent discoloration and sweeten with the sifted sugar. Stir well.

Spoon the avocado cream into ice cream coupes or crystal dishes, cover with plastic wrap, and chill. The dessert can be made up to 1 hour in advance; it is best served chilled but not ice cold. Decorate with mint leaves and lime slices.

———•———

AVOCADO ICE CREAM

¼ cup corn syrup or clear honey

3¼ cups superfine sugar

2¼ lb avocado flesh

juice of 2 limes

Decorate with:

clusters of redcurrants, or wild strawberries

Combine the corn syrup or honey and the sugar with 1 quart cold water in a saucepan. Bring to a boil, stirring frequently. When the sugar has completely dissolved, remove from the heat and allow to cool. Wait until the syrup is cold before you start preparing the avocados (their flesh discolors very quickly once exposed to the air). Cut them in half, take out the pit, and scoop all the flesh out of the skin. Working quickly, measure the required weight of flesh, place in the blender with the lime juice and ½ cup of the syrup; blend until very smooth. The lime juice prevents the flesh from discoloring.

Add the remaining syrup and blend. (If you do not have a large-capacity blender, you can process the mixture in two batches; stir the two batches thoroughly in a bowl before freezing). If you have an electric ice-cream maker, pour the mixture into it and process for 15 minutes. Transfer into suitable containers and place in the freezer for at least 2 hours. Alternatively, pour into ice trays and freeze these, stirring several times as the mixture starts to freeze and thicken.

Ten minutes before serving the ice cream, take it out of the freezer. Use an ice cream scoop and transfer to glass bowls or coupes. Decorate each individual serving with redcurrants or wild strawberries.

———•———

CANDIED PUMPKIN

1 lb pumpkin flesh

1 lb sugar

1 lemon

Once you have peeled the piece of pumpkin and removed the seeds and surrounding fibers, measure out the above weight. Cut into thick slices and then cut these into small cubes, place in a bowl, and cover with cold water. Leave to soak for 24 hours, then drain.

Melt the sugar in 2¼ cups water over low heat, stirring continuously. Cook over low heat until the syrup temperature reaches about 244°F. Use a candy thermometer to measure the temperature if you have one, or allow a large drop to fall into a saucer of iced water: if it forms a soft ball, it is ready. Add the pumpkin, the juice of the lemon, and its finely grated rind. Stir, cover the pot, and cook gently for about 30 minutes or until the pumpkin is tender.

Leave to stand for 1 week at room temperature, then take out the pumpkin pieces and spread out in one or more very large sieves. Place in the oven at about 225°F or even lower to dry out gradually; when dry,

151

store in tightly sealed glass jars.

Excellent eaten as candy or with liqueurs at the end of a meal.

———— • ————

PUMPKIN AND AMARETTI CAKE

1 lb pumpkin flesh

2¼ cups milk

5 slices 2-day old white bread, crusts removed

3 eggs

¼ cup caster sugar

½ cup seedless white raisins

½ cup pine nuts

9 double amaretti di Saronno (i.e. 18 small cookies)

softened butter

flour

coarse sea salt

Dice the pumpkin flesh and cook in the pressure cooker with 1 cup water and a pinch of coarse sea salt for 20 minutes.

Preheat the oven to 350°F. Prepare all the other ingredients: bring the milk to a boil in a saucepan, draw aside from the heat, and soak the bread in it. After 10 minutes, remove the bread from the milk with a slotted spoon. Do not squeeze the bread but beat it with a fork until smooth. The excess milk left behind in the saucepan can be poured away.

When the pumpkin is cooked, reduce it to a coarse purée by crushing it with a fork or a potato masher. Mix very thoroughly with the bread purée.

Beat the eggs lightly with the sugar and stir into the pumpkin mixture, adding the seedless white raisins, pine nuts, and all but 1 of the amaretti cookies (i.e. half the contents of one of the twists of tissue paper), coarsely crushed. Grease the inside of a 9-in diameter deep pan or shallow cake pan (about 1½ in deep) with butter, sprinkle with a little

flour, tipping out excess, and turn the mixture into this pan. Smooth the surface level with a spatula. Bake in the oven for a total of 30 minutes; after 20 minutes, open the oven door and quickly sprinkle the remaining amaretto, very coarsely crumbled, onto the center of the cake and close the oven door again for the last 10 minutes of the baking time. Turn off the oven and wait 15 minutes before taking out the cake. Serve warm.

———— • ————

TURKISH PUMPKIN DESSERT
Kabak Tatlisi

2¼ lb pumpkin flesh

¾ cup sugar

1 cup chopped walnuts

Cut the pumpkin flesh into thin, rectangular pieces measuring about 3 × 1½ in. Place these in a very wide, flameproof earthenware casserole dish, sprinkle with the sugar, and add just enough water to cover. Set aside for 24 hours.

Place the casserole dish on a moderate heat, bring to a boil, reduce the heat, and simmer gently for 45 minutes, stirring now and then, until the pumpkin is tender. Allow to cool at room temperature then serve in small bowls or dishes, sprinkled with the chopped walnuts.

·BULB AND ROOT VEGETABLES·

CARROT BREAD

Babka

p. 161

Preparation: 30 minutes + 1 hour 30 minutes rising time
Cooking time: 50 minutes
Difficulty: easy
Appetizer

ONION SALAD WITH COLD MEAT

p. 161

Preparation: 5 minutes
Difficulty: easy
Appetizer

ONIONS WITH SHARP DRESSING

p. 161

Preparation: 40 minutes
Cooking time: 20 minutes
Difficulty: easy
Appetizer

GRATED CARROT TURKISH STYLE

Yogurtlu Havuç Salatasi

p. 162

Preparation: 20 minutes
Difficulty: very easy
Appetizer

CRUDITÉS WITH SESAME DIP

p. 162

Preparation: 30 minutes
Difficulty: very easy
Appetizer

CARROT, CELERY, AND FENNEL SALAD WITH SCAMPI

p. 163

Preparation: 45 minutes
Cooking time: 5 minutes
Difficulty: easy
Appetizer

CHILI PEPPER AND DAIKON ROOT SALAD WITH SALMON ROE

Ikura to daikon no oroshi ae

p. 163

Preparation: 10 minutes
Difficulty: very easy
Appetizer

RADISH AND CRESS SALAD IN CREAM DRESSING

p. 163

Preparation: 15 minutes
Difficulty: very easy
Appetizer

CELERY ROOT SALAD

p. 164

Preparation: 25 minutes
Difficulty: easy
Appetizer

RUSSIAN BEET PÂTÉ

p. 164

Preparation: 25 minutes + chilling time
Cooking time: 27 minutes
Difficulty: very easy
Appetizer

BEET AND CELERY SALAD

p. 164

Preparation: 25 minutes
Cooking time: 1 hour 30 minutes–2 hours
Difficulty: very easy
Appetizer

BAKED SWEET POTATOES WITH SOUR CREAM AND CORIANDER

p. 165

Preparation: 15 minutes
Cooking time: 45 minutes
Difficulty: very easy
Appetizer

Pasta and potatoes

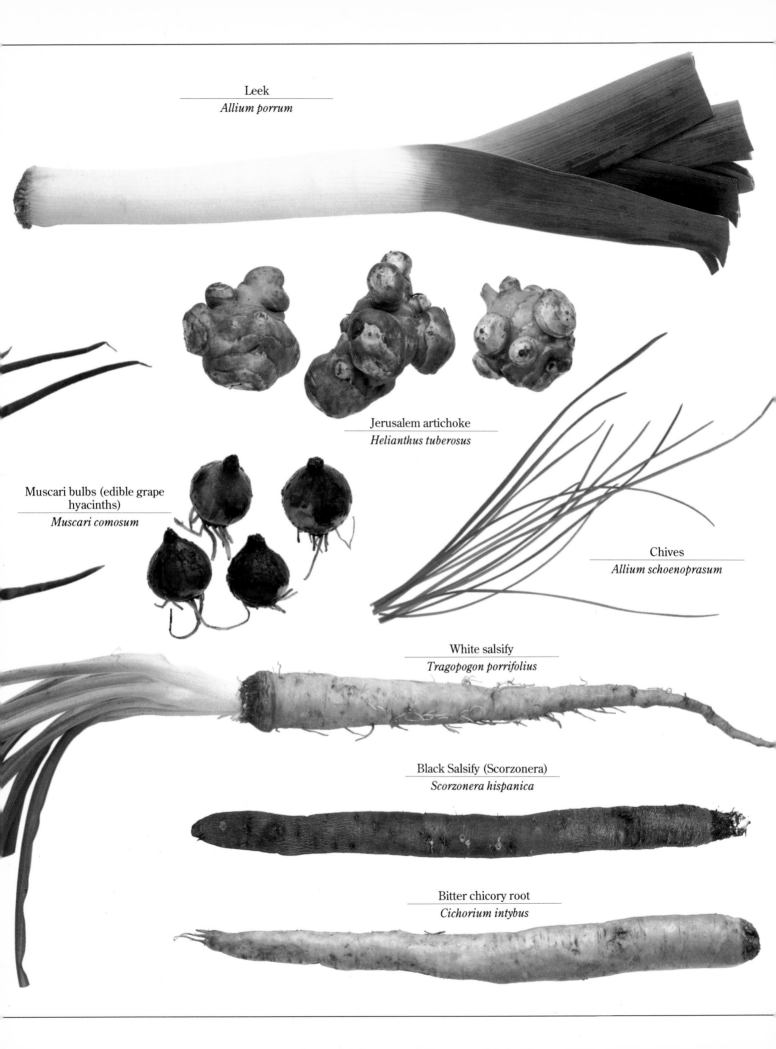

Leek
Allium porrum

Jerusalem artichoke
Helianthus tuberosus

Muscari bulbs (edible grape
hyacinths)
Muscari comosum

Chives
Allium schoenoprasum

White salsify
Tragopogon porrifolius

Black Salsify (Scorzonera)
Scorzonera hispanica

Bitter chicory root
Cichorium intybus

Yellow onion
Allium cepa

White onion
Allium cepa

Red onion (mild)
Allium cepa

Baby and pickling onions
Allium cepa

Garlic
Allium sativum

Scallion
Allium cepa

Radish
Raphanus sativus

Shallot
Allium ascalonicum

TURKISH POTATO SALAD

Patates Salatasi

p. 165

Preparation: 15 minutes
Cooking time: 30 minutes
Difficulty: very easy
Appetizer

BAKED POTATOES WITH CAVIAR AND SOUR CREAM

p. 166

Preparation: 10 minutes
Cooking time: 50 minutes
Difficulty: very easy
Appetizer

HUNGARIAN POTATO SALAD

p. 166

Preparation: 20 minutes
Cooking time: 20 minutes
Difficulty: very easy
Appetizer

JERUSALEM ARTICHOKE SALAD

p. 166

Preparation: 30 minutes
Difficulty: easy
Appetizer

RUSSIAN CARROT AND GREEN APPLE SALAD

p. 167

Preparation: 30 minutes
Difficulty: very easy
Appetizer or accompaniment

GRATED CARROT SALAD WITH INDIAN MUSTARD DRESSING

p. 167

Preparation: 25 minutes + 30 minutes soaking time
Difficulty: very easy
Appetizer or accompaniment

CARROT AND ONION SALAD INDIAN STYLE

p. 167

Preparation: 25 minutes
Difficulty: very easy
Appetizer or accompaniment

JAPANESE DAIKON ROOT AND CABBAGE SALAD

Namasu

p. 168

Preparation: 25 minutes
Difficulty: easy
Appetizer or accompaniment

JAPANESE MIXED SALAD

p. 168

Preparation: 20 minutes
Difficulty: easy
Appetizer or accompaniment

POTATO SALAD WITH SCALLIONS AND CHIVES

p. 169

Preparation: 20 minutes
Cooking time: 20 minutes
Difficulty: very easy
Appetizer or accompaniment

CHILLED LEEK AND POTATO SOUP

Vichyssoise

p. 169

Preparation: 15 minutes + 2 hours chilling time
Cooking time: 50 minutes
Difficulty: very easy
First course

CREAM OF CELERY ROOT SOUP

p. 169

Preparation: 20 minutes
Cooking time: 40 minutes
Difficulty: very easy
First course

CREAM OF PARSNIP SOUP WITH BACON AND GARLIC CROUTONS

p. 170

Preparation: 25 minutes
Cooking time: 20 minutes
Difficulty: easy
First course

PHILADELPHIA PEPPER-POT

p. 170

Preparation: 25 minutes
Cooking time: 1 hour 15 minutes
Difficulty: very easy
First course

RUSSIAN BEET AND CABBAGE SOUP

Borscht

p. 171

Preparation: 30 minutes
Cooking time: 2 hours 20 minutes
Difficulty: easy
First course

ICED BEET SOUP

p. 171

Preparation: 15 minutes + chilling time
Cooking time: 20 minutes
Difficulty: very easy
First course

CREAM OF POTATO AND LEEK SOUP

Potage Parmentier

p. 172

Preparation: 20 minutes
Cooking time: 50 minutes
Difficulty: very easy
First course

CREAM OF LEEK AND ZUCCHINI SOUP

p. 172

Preparation: 15 minutes
Cooking time: 50 minutes
Difficulty: very easy
First course

CREAM OF POTATO SOUP WITH CRISPY LEEKS

p. 173

Preparation: 30 minutes
Cooking time: 35 minutes
Difficulty: easy
First course

ONION SOUP

p. 173

Preparation: 30 minutes
Cooking time: 2 hours
Difficulty: easy
First course

ONION SOUP AU GRATIN

p. 174

Preparation: 35 minutes
Cooking time: 2 hours 30 minutes
Difficulty: easy
First course

PROVENÇAL GARLIC SOUP

Aïgo Bouïdo

p. 174

Preparation: 25 minutes
Cooking time: 35 minutes
Difficulty: easy
First course

BLACK SALSIFY AND FISH SOUP

p. 174

Preparation: 25 minutes
Cooking time: 40 minutes
Difficulty: easy
First course

ICED HAWAIIAN POTATO SOUP

p. 175

Preparation: 25 minutes
Cooking time: 35 minutes
Difficulty: very easy
First course

SPAGHETTI WITH SCALLION, TOMATO, AND HERB SAUCE

p. 175

Preparation: 35 minutes
Cooking time: 25 minutes
Difficulty: easy
First course

PASTA AND POTATOES

p. 176

Preparation: 20 minutes
Cooking time: 30 minutes
Difficulty: very easy
First course

PISSALADIÈRE

p. 176

Preparation: 20 minutes
Cooking time: 40 minutes
Difficulty: easy
First course

POTATO DUMPLINGS WITH BASIL SAUCE

Gnocchi al pesto

p. 177

Preparation: 45 minutes
Cooking time: 40 minutes
Difficulty: fairly easy
First course

GENOESE SEAFOOD AND VEGETABLE SALAD PLATTER

Cappon magro

p. 177

Preparation + cooking time: 2 hours 30 minutes
Difficulty: easy
Main course

JERUSALEM ARTICHOKE VOL-AU-VENTS RUSSIAN STYLE

p. 179

Preparation: 30 minutes
Cooking time: 20 minutes
Difficulty: easy
Main course or appetizer

SPANISH OMELET

Tortilla de papas

p. 179

Preparation: 20 minutes
Cooking time: 20 minutes
Difficulty: easy
Main course

ONION AND BACON PANCAKES

p. 180

Preparation: 30 minutes + 6 hours chilling time
Cooking time: 40 minutes
Difficulty: fairly easy
Main course

PROVENÇAL PANCAKES

p. 180

Preparation: 30 minutes + 2 hours chilling time
Cooking time: 30 minutes
Difficulty: fairly easy
Main course

SPICED VEGETABLES IN CREAMY SAUCE

Makhanwala

p. 181

Preparation: 45 minutes
Cooking time: 45 minutes
Difficulty: easy
Main course

SCRAMBLED EGGS WITH SCALLIONS INDIAN STYLE

p. 182

Preparation: 20 minutes
Cooking time: 15 minutes
Difficulty: easy
Main course

PERSIAN VEGETABLE AND EGG TIMBALE

Kukuye Sabzi

p. 182

Preparation: 45 minutes
Cooking time: 50 minutes
Difficulty: easy
Main course

LEEK AND CHEESE PIE

p. 183

Preparation: 15 minutes
Cooking time: 1 hour
Difficulty: very easy
Main course

STEAMED VEGETABLES WITH CURRY SAUCE

p. 183

Preparation: 30 minutes + 30 minutes standing time
Cooking time: 30 minutes
Difficulty: easy
Main course

RUSSIAN CARROT MOLD

p. 184

Preparation: 20 minutes
Cooking time: 45 minutes
Difficulty: easy
Main course

BLACK SALSIFY GREEK STYLE

p. 185

Preparation: 20 minutes
Cooking time: 25 minutes
Difficulty: easy
Main course

WHITE SALSIFY POLISH STYLE

p. 185

Preparation: 20 minutes
Cooking time: 15 minutes
Difficulty: easy
Main course

NEAPOLITAN POTATO CAKE

Gattò di patate

p. 186

Preparation: 35 minutes
Cooking time: 70 minutes
Difficulty: easy
Main course

PAN-FRIED POTATOES, PEPPERS, AND BEEF

p. 186

Preparation: 20 minutes
Cooking time: 20 minutes
Difficulty: easy
Main course

VEGETABLE AND CHEESE BAKE

p. 187

Preparation: 35 minutes
Cooking time: 1 hour 15 minutes
Main course

POTATO AND ARTICHOKE PIE

p. 187

Preparation: 30 minutes
Cooking time: 1 hour
Difficulty: very easy
Main course

POTATO SOUFFLÉ WITH ARTICHOKE SAUCE

p. 187

Preparation: 30 minutes
Cooking time: 1 hour
Difficulty: easy
Main course

SWEET POTATO PANCAKES WITH HORSERADISH AND APPLE SAUCE

p. 188

Preparation: 30 minutes
Cooking time: 20 minutes
Difficulty: easy
Main course

GREEK POTATO FRICADELLES

Patatakeftédes

p. 189

Preparation: 20 minutes + 1 hour chilling time
Cooking time: 50 minutes
Difficulty: easy
Main course

VEGETABLES WITH POTATO STUFFING

p. 189

Preparation: 30 minutes
Cooking time: 30 minutes
Difficulty: easy
Main course

STUFFED VEGETABLE SELECTION

p. 190

Preparation: 35 minutes + 2 hours chilling time
Cooking time: 1 hour
Difficulty: easy
Main course

BOILED OR STEAMED VEGETABLES WITH GUASACACA SAUCE

p. 191

Preparation: 35 minutes
Cooking time: 35 minutes
Difficulty: very easy
Main course

BRAISED ONIONS AND RICE

Soubise

p. 191

Preparation: 30 minutes
Cooking time: 70 minutes
Difficulty: easy
Accompaniment

SWEET-SOUR BABY ONIONS

p. 192

Preparation: 5 minutes
Cooking time: 20 minutes
Difficulty: very easy
Accompaniment

Sweet-sour baby onions

159

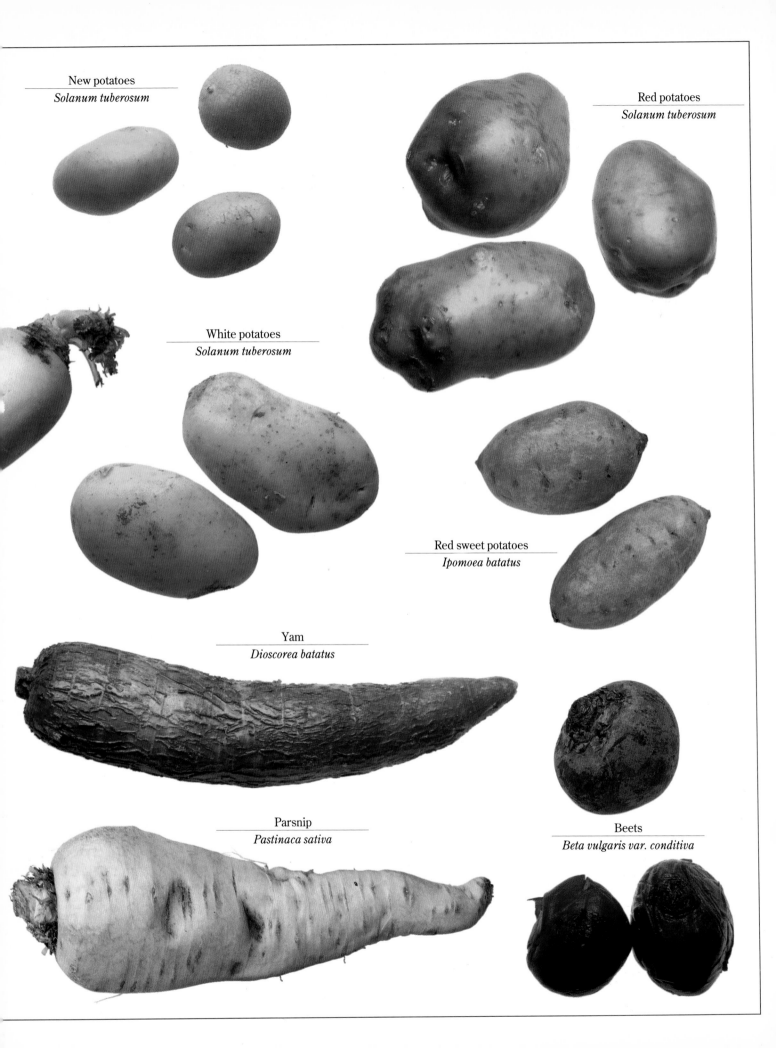

New potatoes
Solanum tuberosum

Red potatoes
Solanum tuberosum

White potatoes
Solanum tuberosum

Red sweet potatoes
Ipomoea batatus

Yam
Dioscorea batatus

Parsnip
Pastinaca sativa

Beets
Beta vulgaris var. conditiva

Horseradish
Armoracia rusticana

Navette French long turnip
Brassica rapa

Turnip
Brassica rapa

Daikon root or Japanese radish
Raphanus sativus var. major

Carrot
Daucus carota

Celery root
Apium graveolens var. rapaceum

Carrots with cream and herbs

TURKISH RICE WITH LEEKS

Zeytinyagli Pirasa

p. 192

Preparation: 25 minutes
Cooking time: 40 minutes
Difficulty: very easy
Accompaniment

CARROTS WITH CREAM AND HERBS

p. 192

Preparation: 30 minutes
Cooking time: 45 minutes
Difficulty: easy
Accompaniment

DEEP-FRIED BLACK SALSIFY

p. 193

Preparation: 20 minutes
Cooking time: 50 minutes
Difficulty: easy
Accompaniment

TURNIP AND RICE PURÉE

p. 193

Preparation: 20 minutes
Cooking time: 30 minutes
Difficulty: very easy
Accompaniment

BABY TURNIPS SAUTÉED WITH GARLIC AND PARSLEY

p. 194

Preparation: 30 minutes
Cooking time: 15 minutes
Difficulty: easy
Accompaniment

RADISHES WITH CREAM, CORIANDER, AND POMEGRANATE SAUCE

p. 194

Preparation: 30 minutes
Cooking time: 20 minutes
Difficulty: easy
Accompaniment

BEET, TOMATO, AND CUCUMBER SALAD

p. 195

Preparation: 25 minutes
Difficulty: very easy
Accompaniment

RÖSTI POTATOES

p. 195

Preparation: 10 minutes
Cooking time: 12 minutes
Difficulty: easy
Accompaniment

RÖSTI POTATOES WITH SHALLOTS

p. 196

Preparation: 15 minutes
Cooking time: 12 minutes
Difficulty: easy
Accompaniment

ENDIVE WITH RASPBERRY VINEGAR AND CREAM

p. 196

Preparation: 15 minutes
Cooking time: 10 minutes
Difficulty: easy
Accompaniment

MACEDOINE OF VEGETABLES WITH THYME

p. 196

Preparation: 30 minutes
Cooking time: 15 minutes
Difficulty: fairly easy
Accompaniment

DEEP-FRIED POTATOES

p. 197

Preparation: 25 minutes
Cooking time: 6 minutes
Difficulty: fairly easy
Accompaniment

POTATO MOUSSELINE

p. 197

Preparation: 15 minutes
Cooking time: 25 minutes
Difficulty: easy
Accompaniment

POTATO GRATIN

p. 197

Preparation: 20 minutes
Cooking time: 25 minutes
Difficulty: easy
Accompaniment

ALMOND CROQUETTE POTATOES

p. 198

Preparation: 25 minutes
Cooking time: 3 minutes
Difficulty: fairly easy
Accompaniment

JERUSALEM ARTICHOKES WITH CREAM SAUCE

p. 198

Preparation: 30 minutes
Cooking time: 30 minutes
Difficulty: easy
Accompaniment

CARROT CAKE

p. 199

Preparation: 20 minutes
Cooking time: 40 minutes
Difficulty: easy
Dessert

INDIAN CARROT DESSERT

Gajjar halva

p. 199

Preparation: 30 minutes
Cooking time: 1 hour 15 minutes
Difficulty: very easy
Dessert

SWEET POTATO CREAM INDIAN STYLE

p. 200

Preparation: 20 minutes + 2 hours chilling time
Cooking time: 20 minutes
Difficulty: easy
Dessert

CARNIVAL FRITTERS

p. 200

Preparation: 30 minutes
Cooking time: 45 minutes
Difficulty: easy
Dessert

CARROT BREAD
Babka

1 lb carrots

1 cake compressed yeast or ½ package active dried yeast

4½ cups all-purpose flour

3 eggs

½ cup butter, melted

½ cup sugar

½ cup flaked or slivered almonds

finely grated rind of 1 lemon

vanilla extract

salt

Serve with:

sour cream

Russian Beet Pâté (see page 164)

Trim, wash, and peel the carrots, then grate them very finely. Crumble the yeast into a small bowl containing ½ cup lukewarm water. Sift half the flour into a large mixing bowl; make a well in the center and break 1 egg into it (have the eggs at room temperature). Stir the yeast mixture gently, and work it into the flour and egg by hand. Add the grated carrot. Knead for 5–10 minutes until elastic, then shape into a ball. Leave to rise in the bowl in a warm place for 30–40 minutes, covered with a cloth that has been rinsed in hot water and wrung out.

When the dough has more than doubled in size, pour the melted butter over it, sift in the remaining flour, and add the remaining eggs, the sugar, almonds, grated lemon rind, and 2–3 drops of vanilla. Work the dough very thoroughly, mixing in all the ingredients evenly. Knead for 5–10 minutes, shape into a large ball, cover the bowl with a warm, damp cloth, and leave to rise once more in a warm place for 1 hour.

Preheat the oven to 350°F. Grease the inside of a ring tube mold (with an outer diameter of 10 in and an inner diameter of 4 in) with butter. Shape the dough into a fat ring and distribute it evenly inside the mold; smooth the surface level. Bake for 40 minutes or until the loaf is a deep russet color on top and a skewer inserted deep into it comes out clean and dry. Take out of the oven and leave to stand for 5 minutes before unmolding. Serve hot or warm with smetana and beet pâté.

———•———

ONION SALAD WITH COLD MEAT

1 large or 2 medium-sized mild Bermuda onions

1½ cups cold cooked ham or pork, shredded

wine vinegar or cider vinegar

2 tbsp finely chopped parsley

olive oil

salt and pepper

Serve with:

a carrot, lettuce, and chicory salad

hard-boiled eggs

Pressed, cooked meats are best for this dish; buy them thinly sliced and shred as neatly as possible. Trim and peel the onion; cut lengthwise into quarters and then shred each quarter thinly. Mix with the meat in a bowl, sprinkle with 3 tbsp good quality vinegar, and season with salt and freshly ground pepper. Stir and leave to stand for 5 minutes before sprinkling with approximately ¼ cup olive oil and the parsley. Serve at room temperature with a side salad of carrots, lettuce, and chicory, and quarters of hard-boiled egg.

———•———

ONIONS WITH SHARP DRESSING

1 lb very small onions

wine vinegar

extra-virgin olive oil

salt and pepper

Serve with:

vegetable omelets or sliced cold meats

Remove the skin and the outermost fleshy layer of the onions. Make a cross cut 1 in deep in the base of each, and place in a large bowl of cold water. Bring a large pot of salted water to a boil and cook the onions until tender. Drain and place in a serving dish and sprinkle with a little vinegar, salt, and freshly ground pepper. Set aside for 10 minutes, then sprinkle with a little best quality olive oil. Serve warm or cold.

———— • ————

GRATED CARROT TURKISH STYLE
Yogurtlu Havuç Salatasi

1¼ lb tender young carrots

For the dressing:

½ cup natural yoghurt

1 clove garlic, minced

½ cup olive oil

3 tbsp finely chopped parsley

salt and pepper

Peel the carrots and grate them finely. Transfer to a serving dish. Spoon the yoghurt into a bowl and stir in the garlic and a generous pinch of salt. Beat the yoghurt continuously with a small balloon whisk or hand-held electric beater as you gradually add the olive oil so that it forms an emulsion. Stir in the parsley and pour this dressing over the carrot, adding a little more salt if desired, and freshly ground pepper to taste.

———— • ————

CRUDITÉS WITH SESAME DIP

6 carrots

4 celery stalks

4 zucchini

4 scallions

For the sesame dip:

1 cup white sesame seeds

½ cup vegetable stock (see page 38)

6 tbsp white wine vinegar

2 tbsp sugar

3 tbsp mirin or sweet white vermouth or sweet sherry

5 tbsp Japanese light soy sauce

small pinch monosodium glutamate (optional)

salt

Begin by making the dip: spread the sesame seeds out in a large nonstick skillet and roast over moderate heat, stirring continuously to ensure that they cook evenly and do not burn. When they begin to jump about in the skillet, remove from the heat. Put them in the food processor and reduce to a smooth, creamy paste; transfer this to a bowl and stir in the stock, vinegar, sugar, mirin, soy sauce, a pinch of salt, and, if used, the monosodium glutamate. Taste and add a little more salt if you wish or a little more vinegar if you prefer a sharper taste. Keep at room temperature until ready to serve.

Clean and prepare the vegetables: cut lengthwise into fairly wide sections and then cut into large, thick matchstick shapes about 4 in long. Arrange the vegetables in alternating bundles on a large, round plate and place the bowl containing the sesame dip in the center. If you must prepare the vegetables a few hours in advance, keep them crisp in a bowl of iced water to which the juice of half a lemon has been added; take them out of the refrigerator and drain well shortly before serving.

———— • ————

Carrot, Celery, and Fennel Salad with Scampi

¾ lb peeled cooked scampi or jumbo shrimp

3 carrots

3 celery stalks

2 small fennel bulbs (outer layers removed)

extra-virgin olive oil

salt and pepper

If you bought raw crustaceans, steam them for 3 minutes and peel. Place in a bowl over barely simmering water, cover, and leave to keep hot while you trim, peel, rinse, and dry the vegetables. Use a mandoline slicer or similar utensil to slice the carrots into very thin rounds and the celery into very thin sections. Cut the fennel hearts lengthwise in half and then slice very thinly lengthwise.

Arrange the vegetables in layers on individual plates, starting with a fennel layer, followed by celery and topped with carrot. Cut the crustaceans lengthwise almost in half, open them out like a butterfly, and arrange on the vegetable bed. Sprinkle with a very little salt, some freshly ground pepper, and olive oil. Serve at once.

———•———

Chili Pepper and Daikon Root Salad with Salmon Roe
Ikura to daikon no oroshi ae

1–2 dried red chili peppers

½ daikon root

1 small jar salmon roe or red lumpfish roe

½ lemon

Garnish with:

sprigs of parsley

radish flowers (see page 28)

Remove the stalks and seeds from the dried chilis and place in the food processor. Peel and cut the daikon root into small pieces and add to the chilis. Process until you have a smooth, very pale pink creamy mixture. Place this in the center of a clean cloth, gather the edges together, and twist round tightly to squeeze out as much moisture as possible. Heap the mixture onto 4 plates and garnish with parsley and radishes cut into tulip shapes. Top each serving with a spoonful of the salmon or lumpfish roe, sprinkle with a little lemon juice, and serve immediately. This delicate and unusual appetizer should be prepared just before it is to be eaten as the daikon mixture (momiji oroshi) deteriorates rapidly once grated. This is an ideal beginning to a formal Japanese meal.

———•———

Radish and Cress Salad in Cream Dressing

1 small bunch radishes

1 bunch young watercress or bittercress (also known as American cress) or 3 boxes mustard and cress

2 small, tight Boston lettuce hearts

For the cream dressing:

3 tbsp white wine vinegar

pinch English dried mustard

¼ cup heavy cream

6 tbsp olive or sunflower oil

pinch cayenne pepper

pinch of salt

Trim the salad vegetables, rinse, drain, and pat dry with paper towels. Tear the larger lettuce leaves into smaller pieces and arrange on individual plates with

163

the smaller, inner leaves. Slice the radishes into thin rounds and spread over the lettuce; top with your chosen type of cress.

Make the dressing: mix all the ingredients except the oil in a bowl and gradually add the oil while beating continuously to form a smooth emulsion. Sprinkle over the individual salads just before serving.

———— • ————

CELERY ROOT SALAD

1¾ lb celery root

½ lemon

1¼ cups mayonnaise (see page 35)

½ cup whipping cream, stiffly beaten

salt and pepper

Served with:

sliced cold roast or cured meats and sausages

Peel the celery root, rinse well, and dry. Cut into quarters to make it easier to slice very thinly on a mandoline cutter; reassemble each quarter's slices on top of one another and slice downward into *julienne* strips (see page 24). Transfer the celery root strips into a salad bowl, season lightly with salt and freshly ground pepper, and sprinkle with the lemon juice. Stir in the mayonnaise, followed by the stiffly beaten cream. Add a little more salt and pepper if desired.

This is excellent as a salad vegetable in winter and is particularly good with a mixed platter of sliced cured meats and salami, and with cold roast meat or poultry.

———— • ————

RUSSIAN BEET PÂTÉ

1¼ lb raw beets

¼ cup butter

1 lemon

3 tbsp sugar

2 small cloves garlic, minced

3 tbsp finely chopped parsley

salt and black pepper

Serve with:

triangles of buttered toasted bread

Russian Carrot and Green Apple Salad (see page 167)

Scrub the beets under running cold water with a soft brush to remove all traces of dirt. Place them in a pressure cooker with a little water and cook for 20 minutes, or boil in a large pot of lightly salted water for about 2 hours, until tender. Leave to cool before peeling. Chop finely. Mix the beets in a bowl with the butter, softened at room temperature, the juice and finely grated rind of the lemon, a generous pinch of salt, and the sugar. Transfer this mixture to a nonstick saucepan and cook over moderate heat for about 7 minutes, stirring continuously; it should become thick and homogenous. Set aside. When cold, stir in the garlic and parsley, adding freshly ground pepper and more salt if needed. Press down firmly into 4 small terrines or ramekin dishes and chill in the refrigerator for a few hours or overnight. Serve as an appetizer with toasted bread triangles and Russian Carrot and Green Apple Salad.

———— • ————

BEET AND CELERY SALAD

2 medium-sized cooked beets

1¾ lb celery hearts

citronette dressing (see page 35)

3 tbsp finely chopped parsley

For the *blanc de cuisine*:
¼ cup all-purpose flour
juice of ½ lemon
2 tbsp butter
3 tbsp oil
salt

Make the *blanc de cuisine*: place the flour in a large mixing bowl and gradually stir in ¼ cup cold water, blending very thoroughly with a wooden spoon until no lumps remain. Gradually stir in another 4½ cups cold water. Pour this mixture through a fine sieve into a large, deep saucepan; add 2½ tsp salt, the lemon juice, the butter, and the oil. Bring to a boil, stirring continuously to prevent the flour from sinking to the bottom and catching. Once the liquid has reached a full boil, reduce the heat and maintain at a gentle boil without stirring. Trim and wash the celery; run the potato peeler down the outer surfaces to eliminate any strings and cut each stalk into 2½-in lengths. When they are all ready, drain and add to the cooking liquid. Cover and boil gently for 20 minutes or until tender. Drain and leave to cool.

Peel the cooked beets and slice them. Arrange the slices and the celery pieces in concentric circles on a serving platter, radiating out from the center. Sprinkle with the citronette dressing and the chopped parsley and serve.

———•———

BAKED SWEET POTATOES WITH SOUR CREAM AND CORIANDER

4 sweet potatoes, preferably red-skinned
1 large scallion, finely chopped
½ fresh green chili pepper, finely chopped
½ cup sour cream
1½ tbsp fresh coriander leaves, finely chopped
salt
Serve with:
1 lime, quartered

Preheat the oven to 400°F. Scrub the sweet potatoes under running cold water, dry them, and wrap each separately in foil. Bake for 45 minutes or until tender. Remove the outermost layer of the scallion and the seeds and stalk of the chili pepper before chopping. Combine with the sour cream and the coriander leaves. Season to taste with salt and freshly ground pepper.

Keep the sour cream mixture in the refrigerator until just before serving the sweet potatoes. Fold back the foil just enough to be able to make a deep crosswise incision across the top of each sweet potato. Place 1½ tbsp of the cream into those grooves and serve the rest separately. Garnish each serving with a wedge of lime for squeezing over the potato.

———•———

TURKISH POTATO SALAD
Patates Salatasi

6 medium-sized potatoes
For the dressing:
juice of 1½ lemons
½ cup olive oil
½ cup dry white wine vinegar
3–6 tbsp finely chopped parsley
salt and pepper
Serve with:
Greek Eggplant Salad (see page 110)
Zucchini Fritters (see page 143)

Scrub the potatoes under running cold water to eliminate any dirt; steam them until tender, leave to cool, then peel and cut lengthwise in half; cut each half into thin slices. Transfer to a deep serving dish or salad bowl.

Make the dressing: dissolve 1 tsp salt in the lemon juice in a bowl. Add the olive oil gradually, using a whisk or hand-held electric beater; keep beating as you then gradually add the wine vinegar. Season generously with freshly ground pepper and mix into the potatoes carefully, coating all the slices without breaking them up too much. Sprinkle with the parsley and serve at room temperature with other Turkish appetizers such as eggplant salad and zucchini fritters.

———— • ————

BAKED POTATOES WITH CAVIAR AND SOUR CREAM

4 medium-sized baking potatoes

1–1½ tbsp chopped fresh dill

1 cup sour cream or smetana (see page 59)

6 tbsp caviar or black lumpfish roe (Danish caviar)

salt

Garnish with:

4 lemon wedges

sprigs of fresh dill

Preheat the oven to 400°F. Scrub the potatoes completely clean under the running cold water, dry them, and wrap each one in foil. Bake in the oven for 50 minutes or until tender. Pull aside the foil just enough to be able to make a deep crosswise incision right across the top of each potato. Press the ends of each potato toward the middle to make the cooked potato emerge from the cut skin and top each with a pinch of salt, chopped dill, 3 tbsp sour cream, and the caviar or lumpfish roe. Serve at once, garnished with the lemon wedges and sprigs of dill.

HUNGARIAN POTATO SALAD

4 large potatoes

4 hard-boiled eggs

2 scallions

½ cup mayonnaise (see page 35)

¼ cup sour cream or smetana (see page 59)

salt and pepper

Serve with:

green salad

Peel the potatoes, cut them into small pieces, and place in a saucepan with 2¼ cups water and 1 tsp salt. Cover and cook over low to moderate heat. Bring to boil and simmer for about 15–20 minutes or until the potatoes are tender and have absorbed all the water. Chop the hard-boiled eggs coarsely; trim the scallions, removing the outer layer, and slice the bulb and leaves into rings. Use a fork to partially break up the hot, cooked potato pieces. Allow to cool at room temperature.

Make the mayonnaise, stir in ¼ cup sour cream, and mix with the potatoes, the scallions, and chopped eggs. Season with salt and freshly ground pepper to taste. Serve at room temperature, with green salad, as an appetizer or as part of a cold lunch.

———— • ————

GLOBE ARTICHOKE SALAD

juice of 2 lemons

1¼ lb Jerusalem artichokes

extra-virgin olive oil

salt and pepper

Garnish with:

1 tbsp finely chopped parsley

1 lemon, thinly sliced into rounds

Fill a bowl with cold water acidulated with the juice of 1 lemon. Peel the artichokes one at a time and immediately drop into the water to prevent discoloration. This stage of the preparation can be done several hours in advance.

Shortly before serving, slice the drained and dried raw artichokes very thinly on a mandoline cutter or in your food processor. Overlap the slices on individual plates; season with salt, freshly ground pepper, and sprinkle with olive oil, the remaining lemon juice, and parsley. Cut the lemon slices in half and arrange some on each plate, overlapping in a fan shape.

—•—

RUSSIAN CARROT AND GREEN APPLE SALAD

6 carrots

3 crisp green apples

juice of 2 lemons

1 egg

1 cup olive oil

1 tsp Dijon mustard

Worcester sauce

salt and pepper

Garnish with:

3 tbsp sour cream or smetana (see page 59)

4 sprigs fresh chervil

Peel the carrots and apples and grate them coarsely; mix together in a bowl with a pinch of salt and half the lemon juice. Make the mayonnaise as follows: break the whole egg into the blender, add the mustard, the juice of ½ lemon, 4–6 drops Worcester sauce, and some salt and pepper. Blend at high speed while gradually pouring a thin stream of the olive oil into the blender receptacle through the hole in its lid; the mayonnaise will become thick and

smooth. Add a little more salt, pepper, and lemon juice to taste and mix with the grated carrots and apples. Chill briefly before serving in individual bowls, topped with sour cream and a small sprig of fresh chervil.

—•—

GRATED CARROT SALAD WITH INDIAN MUSTARD DRESSING

3 tbsp seedless white raisins

10 medium-sized carrots

3 tbsp fresh lime juice

6 tbsp sunflower oil

1½ tbsp whole black mustard seeds

salt

Soak the seedless white raisins in warm water for 30 minutes, then drain well. Peel the carrots and grate coarsely. Mix with the raisins. Dissolve 1 tsp salt in the lime juice in a large mixing bowl. Heat the sunflower oil in a small saucepan and when very hot, add the mustard seeds; fry for just a few seconds then pour all over the grated carrot and raisins. Stir well and serve.

—•—

CARROT AND ONION SALAD INDIAN STYLE

1¼ lb carrots

1 medium-sized mild Bermuda onion

½ fresh green chili pepper, finely chopped

3 tbsp fresh lime juice

6 tbsp sunflower oil

1 tbsp peeled and finely grated fresh ginger

salt

Peel the carrots and grate coarsely. Peel the onion and cut into thin strips. Mix the carrot and onion with the chopped chili pepper in a large bowl. Dissolve a large pinch of salt in the lime juice in a small bowl, then gradually beat in the oil to form an emulsion like a vinaigrette. Stir in the grated ginger. Pour over the carrot and onion mixture, mix briefly, and leave to stand for 10–15 minutes. Serve at room temperature.

———— • ————

JAPANESE DAIKON ROOT AND CABBAGE SALAD
Namasu

---

1 medium-sized daikon root

¼ firm, tightly packed green or white cabbage

---

2 small, tender carrots

For the kimi zu dressing:

1½ tbsp cornstarch or potato flour

½ cup cold Japanese instant stock (dashinomoto)

---

4 egg yolks

---

6 tbsp Japanese rice vinegar or diluted cider vinegar (see method)

1½ tbsp sugar

---

salt

Trim, wash, and prepare the vegetables, drying them thoroughly before cutting them all into *julienne* strips (see page 24); grate the root vegetables if preferred. Mix together in a very large bowl, add sufficient iced water to cover and a few ice cubes, and leave in the refrigerator for 1 hour.

While the vegetables are chilling, make the dressing: heat some water in the bottom of a double boiler or saucepan. Place the cornstarch in a bowl and gradually stir in the cold stock, adding a little at a time to prevent lumps from forming. Using a balloon whisk or hand-held electric beater, beat the egg yolks continuously in the top of the double boiler or in a heatproof bowl over the simmering water, and

gradually add the cornstarch mixture. Add the sugar, 1 tsp salt, and continue beating as you gradually add the vinegar. The mixture will soon start to thicken; as soon as it does, remove from the heat and allow to cool. Cider vinegar is best diluted when substituted for Japanese rice vinegar: mix 5 tbsp of cider vinegar with 1½–2 tbsp cold water for this recipe and add a generous pinch extra sugar.

When the vegetables have crisped in the refrigerator, drain well and dry. Mix well with the dressing and serve.

———— • ————

JAPANESE MIXED SALAD

½ daikon root

---

2 young carrots

---

2 crisphead lettuce hearts

---

2 green celery stalks

---

2 thick slices ham

For the sambai zu dressing:

---

3 tbsp Japanese light soy sauce

---

3 tbsp Japanese rice vinegar or diluted cider vinegar (see method)

½ cup sunflower oil

---

salt and freshly ground pepper

Trim and peel the vegetables, wash, and drain well. Tear the larger lettuce leaves into small pieces. Use a mandoline cutter or food processor to slice the daikon root and carrot into wafer-thin rounds and the celery, across the stalk, into very thin pieces. Mix all these in a large bowl. Shred the ham.

Make the dressing: mix 2 pinches salt, a little freshly ground pepper, the soy sauce, and the vinegar in a bowl. If using cider vinegar, use 2 tbsp and add 1 tbsp of cold water and a generous pinch of sugar. Keep beating with a whisk as you gradually add the oil to form an emulsion. Add this dressing to the salad and mix in the shredded ham.

POTATO SALAD WITH SCALLIONS AND CHIVES

3 large potatoes
2 scallions
1¼ cups light or heavy cream
3 tbsp coarsely chopped chives
salt and freshly ground black pepper

Peel the potatoes, cut them into small pieces, and place in a saucepan with 2¼ cups water and 1 tsp salt. Bring to a boil, cover, and simmer over very low heat for 15–20 minutes or until the potatoes are tender and have absorbed all the water. Use a fork to break up the potatoes into very small pieces. Leave to cool but do not chill.

Pour the cream over the potatoes. Remove the outer layer of the scallions and slice the bulbs into thin rings. Add to the potatoes and cream together with plenty of freshly ground black pepper and a little more salt to taste. Mix well. Transfer to a serving bowl, sprinkle with the chives, and serve.

— • —

CHILLED LEEK AND POTATO SOUP
Vichyssoise

1 lb leeks, white part only
1 lb potatoes
5¼–5½ cups chicken stock (see page 37)
1 cup light cream
¼ cup chopped chives
salt and freshly ground white pepper

Cut the potatoes into small pieces and the white part of the leeks into very thin rings. Place both in a saucepan with the stock, bring slowly to a boil, cover, and simmer for about 50 minutes or until the vegetables are very tender. Use a hand-held electric beater to beat until smooth or put through a blender. Stir in the cream and add salt and pepper to taste.

When cool, chill in the refrigerator and serve very cold, garnishing the individual servings with the chopped chives.

— • —

CREAM OF CELERY ROOT SOUP

4½ cups chicken stock (see page 37)
1 lb celery root
2 tbsp butter
¼ cup rice flour
½ cup light or heavy cream
2 egg yolks
1 lemon
1–1½ tbsp finely chopped parsley
salt and freshly ground white pepper
Serve with:
croutons of white bread fried in butter

Heat the stock. Wash, dry, peel, and dice the celery root; place in a saucepan and add about half the boiling hot stock, enough to cover it. Cover and boil gently until the celery root is very tender; put through a vegetable mill to purée or process in batches in the blender. Heat the butter over low heat in a very large, heavy-bottomed saucepan; stir in the rice flour and cook for 1–2 minutes or until it begins to turn a pale golden brown. Remove from the heat and beat in the remaining hot stock. Return to the heat and bring very slowly to a boil, stirring continuously. Simmer for another 5 minutes before stirring in the celery root purée. Reduce the heat to very low.

Beat the egg yolks and cream together in a small bowl or jug until blended; gradually beat this thickening liaison into the soup. Continue to beat for a few minutes, until the soup is very hot, but do not

169

allow to boil. Draw aside from the heat, stir for 1–2 minutes, and season to taste with salt, pepper, a little lemon juice, and the parsley. Sprinkle with parsley and croutons, fried in butter until golden brown.

———•———

CREAM OF PARSNIP SOUP WITH BACON AND GARLIC CROUTONS

2 slices smoked bacon, ¾ in thick

1¼ lb parsnips, trimmed

¼ cup unsalted butter

2 shallots, peeled and finely chopped

½ cup fairly dry white wine

3¼–3½ cups chicken stock (see page 37)

3 egg yolks

1 cup light cream

nutmeg

salt and freshly ground white pepper

For the garlic croutons:

5 small slices white bread

3 tbsp olive oil

2 cloves garlic

salt

Make the garlic croutons: trim off and discard the crusts from the bread slices and cut the bread into small squares or dice. Heat the olive oil and the peeled, whole garlic cloves in a very wide, nonstick skillet, add the diced bread, and fry over moderate heat until crisp and golden brown all over, turning them occasionally. Sprinkle them with a little salt and set aside.

Use bacon with a good proportion of lean to fat and cut it into very small dice. Fry over moderate heat without any added oil, turning now and then, until the bacon is crisp and lightly browned. Remove the bacon and place on paper towels to drain. Set aside.

Dice the peeled parsnips. Heat half the butter in a large, heavy-bottomed saucepan and sweat the chopped shallot and the diced parsnip in it, stirring frequently, for 10 minutes. Increase the heat from low to moderate, add the wine, and cook, uncovered, until it has completely evaporated. Add the stock and a small pinch of salt and bring to a boil. Simmer, uncovered, over moderate heat for about 15 minutes or until the parsnip is very tender. Reduce the heat to low. Beat the egg yolks and cream together briefly and gradually add this mixture to the soup while beating continuously. Continue beating until the soup is very hot but do not allow it to boil or else it will curdle. Remove from the heat, beat with a hand-held electric beater until smooth; beat in the remaining, chilled butter a small piece at a time. Add a little more salt if necessary (not too much, as the bacon will be salty), freshly grated white pepper, and a little grated nutmeg. Ladle into hot individual soup bowls, sprinkle the croutons and crispy bacon over the surface, and serve at once.

———•———

PHILADELPHIA PEPPER-POT

5¼ cups chicken stock (see page 37)

3 medium-sized potatoes, diced

½ medium-sized onion, finely chopped

1 green celery stalk, finely chopped

1 large green pepper, finely chopped

2 tbsp butter

3 tbsp olive oil

¼ cup all-purpose flour

7 oz cooked honeycomb tripe, diced

½ cup heavy cream or sour cream

salt and freshly ground black pepper

Serve with:

thick slices of bread, crisped in the oven

Heat the stock. Clean, trim, and chop the vegetables. Fry the onion, celery, and pepper gently in the butter and oil for 15 minutes, stirring frequently. Add the flour and cook, stirring continuously, over a low heat for 2 minutes. Continue stirring while gradually adding the hot stock, followed by the tripe, potatoes, and plenty of coarsely ground pepper. When the liquid comes to a boil, cover and simmer gently for about 1 hour. Add salt to taste and remove from heat. Leave to stand for 5 minutes, then ladle into individual soup bowls, spoon 3 tbsp cream into the center of each serving and finish with a little more freshly ground pepper.

———— • ————

RUSSIAN BEET AND CABBAGE SOUP
Borscht

¾ lb raw beets
½ lb white cabbage
1 carrot
1 leek
1 medium-sized potato
1 green celery stalk
3 tbsp butter
¼ lb fresh or drained, canned tomatoes
¾ lb boiling beef (see method)
1 thick slice smoked bacon
1 bay leaf
1½ tbsp red wine vinegar
3 tbsp chopped fresh dill
1 cup sour cream
salt and black peppercorns

Trim, wash, and dry all the vegetables. Peel the beets, reserving ¼ lb in weight and shred all the rest, and the cabbage, carrot, leek, and celery into strips. Dice the peeled potatoes.

Heat the butter in a kettle or a very large, deep, enameled cast-iron casserole dish and fry the vegetables in it gently for about 15 minutes, stirring frequently. Add 5¼ cups water, the tomatoes cut into strips with their seeds removed, and the beef: rump pot roast or heel of round are suitable. Add the bay leaf, bacon, a few black peppercorns, and 1 tsp coarse sea salt. Bring to a boil, skimming off any scum that rises to the surface, then turn down the heat to very low and simmer gently for 3 hours, or until the beef is extremely tender.

Cut the reserved raw beets into small pieces and grate finely. Place in a piece of cheesecloth and twist tightly to force out all the juice, collecting this in a small bowl. Stir in a pinch of salt and the vinegar.

Remove the beef and bacon from the soup, spearing them with a carving fork and cut into strips. Return these to the soup. Add a little salt if necessary and freshly ground black pepper to taste. Draw aside from the heat, stir in the raw beet juice and vinegar mixture, and immediately ladle into individual soup bowls. Place 1–1½ tbsp sour cream in the center of each serving and sprinkle with the chopped dill.

———— • ————

ICED BEET SOUP

½ medium-sized onion, coarsely chopped
2 cups chicken stock (see page 37)
3 tbsp dry white wine or dry vermouth
2 medium-sized potatoes, steamed or boiled in their skins until tender
½ lb cooked beets
1 lemon
1 cup natural yoghurt

salt and pepper
Garnish with:
½ cup whipping cream, stiffly beaten
1½ tbsp chopped chives

Place the onion in a very large saucepan with 1 cup of the stock and the wine or dry vermouth. Cover and boil gently over low heat for 15 minutes. Remove the lid and increase the heat to moderately high to reduce the liquid for 2–3 minutes. Draw aside from the heat.

Peel the potatoes and beets, slice very thinly, and add to the saucepan together with another 1 cup of the stock, the lemon juice, a generous pinch of salt, and some freshly ground pepper. Beat well with a hand-held electric beater until smooth and creamy or put through a vegetable mill. Stir in the yoghurt and a little more salt to taste.

Pour into individual soup plates, cover with plastic wrap, and place in the refrigerator until just before serving. Decorate the chilled soup with a spoonful of whipped cream and some chopped chives.

———•———

CREAM OF POTATO AND LEEK SOUP
Potage Parmentier

1 lb potatoes, peeled
1 lb leeks
6 tbsp heavy cream
scant 2 tbsp unsalted butter
¼ cup finely chopped parsley
salt and white peppercorns

Cut the potatoes into small pieces and slice the leeks into thin rings (use both green and white parts but remove and discard the outer layer and the tougher, darker green leaves). Place these vegetables in a large saucepan with 6¼–6½ cups water and 1½ tsp salt, bring slowly to a boil, and simmer, partially covered to allow the steam to escape, for 50 minutes. Use a hand-held electric beater to beat until smooth and creamy or put through a blender or vegetable mill.

The soup can be prepared in advance to this point and reheated just before serving.

Stir the cream into the very hot soup off the heat; stir in the butter, adding a small piece at a time, followed by the parsley. Taste and add a little more salt if needed and some freshly ground white pepper.

———•———

CREAM OF LEEK AND ZUCCHINI SOUP

2¼ lb leeks
1½ lb zucchini
2 vegetable bouillon cubes
3 tbsp finely chopped parsley
1 cup heavy cream
salt and pepper

Prepare the leeks, removing the outer layer and ends, and use both the white and the tender green parts. Wash very thoroughly and cut into thick rings. Trim and wash the zucchini and slice into thick rings.

Place the vegetables in a large saucepan with 7 cups water and the 2 crumbled bouillon cubes. Bring to a boil, cover, and simmer over low heat for 50 minutes. Turn off the heat and beat with a hand-held electric beater until smooth, or put through a blender or vegetable mill. Return to the saucepan and reheat briefly. Add salt and freshly ground white pepper to taste and stir in the parsley.

Ladle into soup bowls. Pour the cream in a very thin stream into each bowl to form concentric circles then draw the tip of a spoon from the center outward across the circles to give an attractive spider web effect and serve. This soup is equally delicious served hot or cold.

———•———

CREAM OF POTATO SOUP WITH CRISPY LEEKS

1¼ lb potatoes, peeled

3 cups chicken stock (see page 37)

9 oz leeks, tender parts only

2 tbsp butter

1 cup light cream

1–1½ tbsp finely chopped parsley

generous pinch sugar

nutmeg

salt and pepper

Slice the potatoes very thinly and place in a large saucepan with the stock. Bring to a boil, cover, and simmer gently over low heat for 30 minutes until very tender. Put through a vegetable mill; return this purée to the saucepan.

Slice the leeks into very thin rings and sweat over very low heat in the butter until just tender and still crisp. Season with a little salt, freshly ground pepper, and a pinch each of sugar and grated nutmeg.

Stir the cream into the thin potato purée and heat to just below boiling. Add a little salt and pepper to taste. Transfer this soup to individual bowls, spoon equal amounts of the leek into the center of each serving, and sprinkle with a little parsley.

———•———

ONION SOUP

6¾ cups beef stock (see page 37)

12 thick slices French bread

1¾ lb trimmed and peeled onions

3 tbsp butter

¼ cup olive oil

generous pinch sugar

6 tbsp all-purpose flour

½ cup dry vermouth or dry white wine

¼ cup brandy

1 large clove garlic, peeled

½ cup grated Swiss cheese

salt and freshly ground pepper

Heat the stock. Preheat the oven to 320°F and bake the bread slices on a cookie tray for 30 minutes to make them dry and crisp.

Cut the onions in half from top to bottom; slice very thinly, also from top to bottom. Sweat the onions in the butter and 1½ tbsp of the olive oil over low heat for 15 minutes in a large, heavy-bottomed saucepan with the lid on. Sprinkle with 1 tsp salt and a generous pinch of sugar (to help them brown) and continue cooking over low heat, taking the lid off for frequent stirring. When the onions are golden brown, add the flour, and stir well with a wooden spoon while cooking for about 2 minutes.

Remove the saucepan from the heat and keep beating continuously and vigorously with a balloon whisk as you add the boiling hot stock gradually to prevent lumps from forming. Add the vermouth or white wine and stir.

Return the saucepan to a low heat and simmer for 40 minutes. Periodically skim off any skin that may form on the surface. Add a little more salt if wished and the brandy.

When the bread slices have baked for 20 minutes in the oven, take them out, sprinkle with the remaining olive oil, and return to the oven for another 10 minutes. Then take them out again, cut the garlic clove in half, and rub the cut surfaces firmly over the dry bread. Place 2 pieces in each of 6 heated soup bowls, ladle the soup over them, and serve, handing round the grated Swiss cheese separately. Serves 6.

ONION SOUP AU GRATIN

Ingredients as for Onion Soup on page 173 with the addition of:

¾ cup freshly grated Parmesan cheese

¼ cup grated Swiss cheese (Gruyère or Emmenthal)

2 tbsp grated raw Bermuda onion

3 tbsp olive oil

Make the onion soup (see page 173). Mix the ¾ cup of Parmesan cheese listed above with the ½ cup grated Swiss cheese listed in the recipe on page 173 and set aside. Preheat the oven to 320°F.

Pour the soup into hot, ovenproof earthenware bowls or heatproof china bowls. Sprinkle the Swiss cheese flakes and the grated raw onion evenly over the surface. Place 2 crisped bread croutons (see recipe on page 173) in each bowl of soup and sprinkle them with the grated cheese. Drizzle a little olive oil over the cheese and add some black pepper. Bake for 20 minutes. Make sure the broiler is very hot and place the bowls under it for the last 5 minutes to make the cheese topping crunchy and golden brown. Serve at once.

— • —

PROVENÇAL GARLIC SOUP
Aïgo Bouïdo

17–20 cloves garlic

2 cloves

1 fresh sage leaf

generous pinch thyme

1 bay leaf

5 sprigs parsley

3 egg yolks

6 tbsp olive oil

½–¾ cup grated Parmesan cheese

½ cup grated Swiss cheese (Gruyère or Emmenthal)

salt and pepper

Separate the garlic cloves, detaching them from the base; do not peel them. Blanch them for 30 seconds in a saucepan of fast boiling unsalted water; drain, refresh under running cold water, and peel. Put them in a large saucepan with 6¼ cups water, a large pinch of salt, and some freshly ground pepper, the cloves, sage leaf, thyme, bay leaf, and parsley sprigs. Bring to a boil then simmer for 30 minutes. Put through a vegetable mill fitted with a fine-gage disk (or strain through a fine sieve, pushing the solid ingredients through the sieve by rubbing with the back of a wooden spoon). Return the soup to the saucepan, add a little more salt if needed, heat to boiling point, and then turn off the heat. Beat the egg yolks in a bowl very briefly using a whisk or hand-held electric beater until they are smooth and creamy; continue beating as you gradually add the olive oil. Gradually beat in about ½ cup of the hot soup, then reverse the procedure and gradually beat the contents of the bowl into the saucepan containing the hot soup. Serve with the mixed grated cheeses.

— • —

BLACK SALSIFY AND FISH SOUP

1½ lb black salsify

1 lb firm white fish

juice of 1 lemon

4½ cups fish stock (see page 38)

¼ cup olive oil

1 clove garlic

1 sprig thyme

salt and pepper

Serve with:

garlic croutons (see page 170)

Black salsify discolors very quickly once peeled, so have a large bowl of cold water acidulated with half the lemon juice standing ready and drop each piece into it as soon as it is peeled. Cut the salsify into pieces 1½ in long and boil for 20 minutes in salted, acidulated water. Drain, allow to cool, then cut each piece lengthwise in half, remove the woody center, and slice the remaining, tender parts into *julienne* strips (see page 24). Cut the fish into chunks, removing all the skin. Sprinkle with a little salt. Heat the fish stock.

Fry the peeled whole garlic clove and the thyme gently for 2 minutes in the olive oil; add the fish pieces and sauté for 1 minute while stirring. Add the salsify and continue to fry gently for 2 minutes more while stirring. Add the hot fish stock, bring slowly to a very gentle boil, and allow to simmer for 30 seconds before removing from the heat. Check that the fish is cooked (simmer for a little longer if not), add a little more salt if necessary and some freshly ground pepper, and serve with the garlic croutons.

———— • ————

ICED HAWAIIAN POTATO SOUP

2 large potatoes

1½ cups vegetable stock (see page 38)

1 cup dry white wine

1 large or 2 small cucumbers

4 tender green celery stalks

1 small onion

1 cup light cream

1 lime

salt and black peppercorns

Garnish with:

1½–2 tbsp finely chopped coriander or parsley

Peel the potatoes, cut them into small pieces, and place in a large saucepan with the stock and the wine. Bring quickly to a boil over high heat, then reduce the heat, cover, and simmer for 20 minutes. While the potatoes are cooking, peel and prepare the remaining vegetables. Chop them all finely together, then add to the potatoes and continue simmering for another 10 minutes. Remove the saucepan from the heat and use a hand-held electric beater or a vegetable mill to blend the mixture until it forms a fairly smooth purée. Add salt to taste and when cool stir in the cream and season with plenty of freshly ground black pepper. Refrigerate until required. Serve very cold with the lime, cut lengthwise into quarters for each person to squeeze into the soup; decorate with chopped coriander or parsley.

———— • ————

SPAGHETTI WITH SCALLION, TOMATO, AND HERB SAUCE

1½–1¾ lb large round scallions

2 cloves garlic, finely chopped

1 bouquet garni

6 tbsp extra-virgin olive oil

½ lb ripe tomatoes, skinned and seeded

1–2 red chili peppers

12 oz spaghetti

salt

Trim the scallions, remove and discard the outer layer, cut off the leaves, and slice the bulbs into very thin rings. Place in a large enameled cast-iron fireproof casserole dish with the chopped garlic, the bouquet garni, and the olive oil. Sweat over low heat, uncovered, for 20 minutes. Bring a large pot of salted water to a boil.

Remove and discard the bouquet garni. Chop the

tomato flesh coarsely; use drained canned tomatoes if preferred. Remove the stalk and seeds from the chili peppers and chop the flesh finely. Add the tomatoes and chili to the scallions with a pinch of salt and cook gently for another 5 minutes, stirring occasionally. Remove from the heat.

Cook the spaghetti until just tender, with some "bite" left to it; reserve 1/2 cup of the cooking water when you drain it. Add the spaghetti to the scallion mixture, together with the reserved hot water, and stir briefly over low heat to coat the spaghetti.

———— • ————

PASTA AND POTATOES

1 large onion, finely chopped
1 green chili pepper, seeds removed, finely chopped (optional)
4 medium-sized potatoes
2 large ripe tomatoes
oregano
½ cup chicken stock (see page 37)
½ lb large pasta shapes (e.g. shells or large ribbed macaroni)
½ cup extra-virgin olive oil
salt and pepper
Garnish with:
sprigs of basil
6 tbsp additional diced ripe tomato flesh

Bring a large pan of salted water to a boil ready to cook the pasta. Heat ½ cup of the oil in a wide, deep fireproof casserole dish and sweat the onion and chili pepper over a low heat for 10 minutes, stirring frequently. Add the diced potatoes and continue to cook over low heat while stirring for 5 minutes. Chop the tomato flesh coarsely and stir into the saucepan with a generous pinch each of oregano,

salt, and freshly ground pepper. Stir all the ingredients while cooking for 2–3 minutes. Add the stock and stir again. Cover and simmer for 5 minutes. Add more salt to taste. Turn off the heat.

When the pasta is barely tender and still has plenty of "bite" left in it, drain, reserving about 1 cup of the cooking water. Add the pasta to the contents of the saucepan and stir over low heat for about 5 minutes. This should complete the cooking of the pasta. Add some of the reserved cooking water if the liquid in the pan reduces appreciably as the finished dish should be very moist. Turn off the heat and leave to stand, uncovered, for 10 minutes, to allow the flavors to blend, then transfer to heated bowls. Drizzle about 1 tbsp olive oil over each serving, season with a little freshly ground pepper, and garnish with sprigs of fresh basil and some diced tomato flesh.

———— • ————

PISSALADIÈRE

1 lb pizza dough (see pages 70–71)
3½ lb onions
½ cup olive oil
generous pinch sugar
20 small black olives
salt and pepper

Make the pizza dough and while it is rising, peel and slice the onions into slivers.

Heat the oil in a very wide skillet and fry the onions gently over low heat for 10 minutes, stirring continuously. Sprinkle with the pinch of sugar, season with salt and freshly ground pepper, cover, and sweat over very low heat for another 15 minutes, stirring occasionally. Remove the lid, turn up the heat a little, and fry gently, stirring all the time: the onions will turn a pale golden brown and shrink. Remove from the heat; add more salt and pepper if necessary.

Preheat the oven to 475°F. Lightly oil a shallow,

rectangular baking tray measuring approximately 14×16 in or a large quiche dish. When the pizza dough has risen sufficiently, knock it down by hand and then roll out to a rectangle large enough to line the base and shallow sides of the baking tray, or a circle to fit the quiche dish. Press the dough into the shape of the tray with your knuckles. Sprinkle the surface with a little salt and olive oil. Spread the cooked onions evenly over the surface and arrange the olives on top in a lattice design.

Bake for 15 minutes or until the pizza dough is cooked and golden brown; serve warm or at room temperature.

———•———

POTATO DUMPLINGS WITH BASIL SAUCE
Gnocchi al pesto

2¼ lb potatoes
1¾ cups all-purpose flour
1 egg
⅛ cup grated Parmesan cheese
unsalted butter
nutmeg
salt and pepper
For the Genoese basil sauce (pesto):
1 large bunch fresh basil
½ cup freshly grated Pecorino cheese
½ cup freshly grated Parmesan cheese
¼ cup pine nuts
1 small clove garlic, peeled
6 tbsp extra-virgin olive oil
sea salt

Steam or boil the potatoes. Make the basil sauce: take all the basil leaves off the stalks, rinse the leaves, and blot dry with paper towels. Place in the food processor with the cheeses, pine nuts, garlic, and a pinch of salt. Process to a paste.

Peel the potatoes and mash them while they are still boiling hot into a large mixing bowl. Use your hands to gradually mix in the sifted flour, adding a pinch of salt and pepper, a pinch of grated nutmeg, the finely grated Parmesan cheese and the egg. The resulting dough should be firm, smooth, and easy to shape.

Break off about a quarter of the dough; using the palms of your hands, roll it on a lightly floured pastry board or working surface into a long cylinder or sausage about ¾ in in diameter. Cut into 1-in lengths. Repeat this process with the remaining dough. Press these pieces one by one against the tines of a fork, forming a hollow on one side of each dumpling and a pattern on the other side. As you shape them, place on a clean cloth or board dusted with sifted flour. Sprinkle lightly with more sifted flour.

Bring plenty of salted water to an even boil in a large saucepan. Drop the gnocchi into the boiling water in batches, one by one in quick succession. If you try to cook too many at once, they may stick to one another. They will bob up to the surface when they are done; remove with a slotted spoon and place in a colander on a plate beside the saucepan. As soon as the last batch has cooked, transfer them to very hot plates, top with a little butter and the basil sauce.

———•———

GENOESE SEAFOOD AND VEGETABLE SALAD PLATTER
Cappon magro alla genovese

For the vegetable assortment:
½ lb black salsify
1 small cauliflower
4 very young, tender globe artichoke hearts
½ lb string beans

177

5 medium-sized young carrots

1 bunch celery, inner stalks only

2 medium-sized potatoes

1 medium-sized cooked beet

3–4 lemons

olive oil

salt and pepper

For the bread and seafood assortment:

1 2–3-day old large loaf white bread

1 large clove garlic

12 unpeeled, cooked jumbo shrimp

1 cooked spiny or clawed crawfish or lobster weighing approx. 2½ lb

1 cooked sea robin or mullet

1 quart fish stock (see page 38)

4 canned anchovy fillets, drained

olive oil

lemon juice

salt and pepper

For the sauce:

2 thick slices white bread

1 small bunch parsley

1 clove garlic

½ cup pine nuts

1 tbsp pickled capers

yolks from 2 hard-boiled eggs

6 large black olives, pitted

½ cup wine vinegar

1 cup extra-virgin olive oil

salt and pepper

Start the preparation for this very filling main-course salad in advance or the day before. You will need extra-virgin olive oil and wine vinegar in addition to the basic quantities listed above. Trim, prepare, and wash all the raw vegetables. See page 165 for method of cooking black salsify. Divide the cauliflower into florets and cook in salted water. Remove the outer leaves from the artichokes, cut the hearts lengthwise into quarters, then prepare and cook as described on page 185. Boil the string beans. The carrots, celery stalks, and the potatoes, meanwhile, can be boiled together until tender.

Drain the vegetables. Cut the celery and salsify into small pieces. Slice the potatoes thickly. Slice the carrots and peel and slice the hot, cooked beet. Place each vegetable in a separate serving dish and drizzle a mixture of lemon juice, olive oil, and some salt over them. Season with freshly ground pepper to taste.

Cut two thick, lengthwise slices from a 2–3-day old large white loaf and heat gently in the oven until very dry, hard, and crisp. Cut a large garlic clove in half and rub the exposed surfaces over the bread slices. Place the bread in the bottom of a very wide, deep serving dish; moisten with wine vinegar diluted with a little cold water and some salt. Allow to soften for 2–3 hours.

Peel the shrimp but leave the fan-shaped flippers attached; remove the crawfish or lobster from its shell (reserve the soft, dark meat in the head for other uses) and cut into round slices about ¼ in thick. Poach the sea robin or mullet in fish stock; leave to cool in the cooking liquid, drain, and remove the bones. Slice as neatly as possible. Drain the oil from the anchovy fillets. Keep the different types of seafood separate and dress with a mixture of olive oil, some lemon juice, a little salt, and freshly ground white pepper to taste.

Make the sauce: cut the crusts off the bread, moisten thoroughly with vinegar, and squeeze out excess moisture. Chop all the other ingredients listed for the sauce together very finely, including the moist bread and push the resulting mixture through a fine sieve into a bowl. Stir in ½ cup vinegar and then gradually beat in 1 cup olive oil. Season.

A couple of hours before assembling the salad, drizzle some olive oil over the 2 large rectangles of bread and moisten with a layer of the sauce. Leave to stand. Shortly before serving, continue layering, using one type of vegetable after another and alternating vegetables with layers of sea robin or mullet, anchovies, and crawfish or lobster slices. Each layer should be covered with a thin layer of the sauce. Decorate with the shrimp. Serves 6–8.

JERUSALEM ARTICHOKE VOL-AU-VENTS RUSSIAN STYLE

1 lb Jerusalem artichokes
2 shallots, finely chopped
2 tbsp butter
1½ tbsp olive oil
¾ cup vegetable stock (see page 38)
4 vol-au-vent puff pastry cases, 6 in in diameter
1 red chili pepper, finely chopped
½ cup tomato paste
1¼ cups heavy cream
juice of 2½ lemons
salt

Peel the Jerusalem artichokes and drop them immediately into a large bowl of cold water acidulated with the juice of 1 lemon to prevent them from discoloring. When they are all prepared, drain, dry, and cut into ¼-in dice.

Sweat the chopped shallot in 1 tbsp of the butter and 1½ tbsp oil, add the diced artichokes, sprinkle with a pinch of salt, and cook over low heat, stirring and mixing for 2–3 minutes. Add the stock, bring slowly to a boil, cover, and simmer gently for about 10 minutes or until the artichokes are tender but not at all mushy. Set aside.

Place the vol-au-vent cases in the oven, preheated to 325°F for 15 minutes. While they are heating make the sauce. Melt the remaining butter in a small saucepan and fry the chili pepper gently for 1 minute. Add the tomato paste and continue stirring over a low heat for about 2 minutes. Stir in 1 cup boiling hot water and simmer, uncovered, over slightly higher heat for 2–3 minutes. Stir in the cream and a pinch of salt; simmer, stirring continuously, for another 2–3 minutes. Add this rather thin sauce to the artichokes and place them over low heat, stirring carefully without breaking them up. When they are very hot, turn off the heat, add the remaining lemon juice, a little more salt if needed, and stir once more. Fill the pastry cases with this mixture, put their lids on top, and serve at once on very hot plates.

· · ·

SPANISH OMELET
Tortilla de papas

4 medium-sized Spanish onions
½–1 green chili pepper or ½ red or green pepper
2 tbsp butter
¼ cup olive oil
2 medium-sized cold boiled potatoes
2 slices smoked bacon, ¼ in thick
8 eggs
6 tbsp heavy cream
½ cup grated Parmesan cheese (optional)
salt and pepper
Serve with:
green salad or tomato salad

Trim and peel the onions, cut from top to bottom in quarters, and then slice very thinly in the same direction. Heat the butter and oil in a very wide, fairly deep skillet, preferably nonstick, and fry the onion gently with the chili pepper or the red or green pepper for 15 minutes over low to moderate heat, stirring frequently. When they are tender and lightly browned and have shrunk considerably, add the diced potatoes, a pinch of salt, some freshly ground pepper and stir over low heat for 2–3 minutes. Remove the rind from the bacon and cut it into small dice. This stage of the omelet preparation can be completed several hours in advance.

When you are ready to serve the omelet, sauté the diced bacon in a nonstick skillet over high heat until crisp and golden brown. Stir continuously to prevent the bacon from burning. Drain off excess fat and blot the diced bacon on paper towels.

Break the eggs into a large bowl and beat while blending the cream, some salt, freshly ground pepper, and the Parmesan cheese if used.

Add the bacon to the onion mixture over fairly gentle heat in the large skillet, distributing it evenly. When hot, pour the eggs into the skillet and tilt it from side to side a few times to let the eggs settle in an even layer. Cook for several minutes, until the omelet has set firmly on the underside. Turn and continue cooking for a few minutes. The outside should be firm, the inside still soft and creamy.

Serve hot or cold with salad.

— • —

ONION AND BACON PANCAKES

4 pancakes (recipe on page 36, use half quantity)
1¼ lb onions
2 tbsp butter
6 tbsp sunflower oil
generous pinch sugar
2 thick slices smoked Canadian bacon
1 cup heavy cream
salt and pepper

Make the batter for the pancakes the day before or at least 6 hours in advance and store, covered, in the refrigerator.

Trim and peel the onions and slice finely. Heat the butter and ¼ cup of the oil in a wide, nonstick skillet and cook the onions very gently for 20–25 minutes, stirring frequently. Sprinkle with the sugar halfway through this time, to brown them.

Cut the rind off the bacon and cut across the slices into ¼-in wide strips. Fry, stirring continuously, over high heat in a nonstick skillet until crisp and golden brown. Drain off all the fat and blot the crisp bacon on paper towels before adding it to the onions. Season sparingly with salt and generously with freshly ground pepper and set aside.

Lightly oil a very wide, nonstick skillet (approximately 10 in in diameter) with oil and heat. Cook 4 pancakes. Heat the onion mixture over low heat and correct the seasoning. Return a pancake to the very wide skillet, spread a quarter of the onion mixture over it, and heat for about 30 seconds. Fold all four "sides" of the pancake over the filling to meet in the middle, forming a square envelope. Repeat with the other pancakes and the rest of the filling. Serve on very hot plates.

— • —

PROVENÇAL PANCAKES

1 cup pancake batter (see page 36)
2 large onions
2 large yellow peppers
½–1 green chili pepper
2 medium-sized ripe tomatoes
1 small clove garlic, minced
3 tbsp finely chopped parsley
6 fresh basil leaves, torn into small pieces
¼–½ cup coarsely grated Swiss cheese
½ cup coarsely grated Parmesan cheese
½ cup olive oil
oregano
salt and pepper
Serve with:
Rice and Turnip Purée (see page 193)
Garnish with:
radish flowers (see page 28) or tomato wedges
sprigs of parsley

Make the pancake batter at least 2 hours in advance and chill in the refrigerator (see recipe on page 36). You will need only one-quarter of the volume so reduce the ingredients by three-quarters.

Make the filling: trim, peel, and thinly slice the onions. Heat 1/4 cup of the oil in a wide, fairly deep skillet and fry the onions very gently for 10 minutes, stirring frequently. Rinse and dry the peppers, remove their stalks, seeds, and white membrane and cut into very thin strips. Add to the onions with the chili pepper and fry over moderate heat for 5 minutes, stirring.

Peel the tomatoes and remove their seeds, chop the flesh coarsely, and add to the onion and pepper mixture with the garlic. Stir well, cover, and simmer gently over low heat for 15 minutes, stirring occasionally. Add a pinch of oregano and season with salt and freshly ground pepper and cook, stirring continuously over moderate heat to allow the liquid to evaporate and the mixture to thicken. Add a little more salt to taste, stir in the parsley and basil, and allow to cool to room temperature.

When cold, transfer to a large bowl. Grate both cheeses. Mix with the pepper and onion mixture and stir in the pancake batter. Lightly oil a small nonstick skillet about 7 in in diameter and heat over moderate heat for about 2 minutes or until it is very hot. Ladle about 1/2 cup of the mixture into the center of the skillet, quickly tipping it from back to front and side to side to distribute the mixture all over the surface of the skillet while still liquid. Cook for 3–4 minutes, or until set and pale golden brown on the underside, then turn carefully and cook for another minute. Keep hot in a warm oven while you cook the other pancakes. Serve without delay, with a Rice and Turnip Purée or your own choice of accompaniment. Decorate with radish flowers or tomato wedges and sprigs of parsley. If you find the pancakes tend to break up when turning them at your first attempt, increase the quantity of pancake batter used to one-third of the recipe quantity given on page 36.

———— • ————

SPICED VEGETABLES IN CREAMY SAUCE
Makhanwala

3 medium-sized onions, peeled and finely chopped
approx. ¼ lb string beans
1 cup peeled, diced potatoes
1 cup peeled, diced carrots
1 cup cauliflower florets
½ cup peas, fresh shelled, or thawed frozen
½ cup canned black beans, drained
½ cup ghee (clarified butter, see page 36)
6 tbsp coriander leaves, finely chopped
2½ tsp mild curry powder
½ tsp ground cumin
1 tsp paprika
1 cup light cream
6 tbsp tomato ketchup
salt
For the masala:
6 tbsp grated fresh coconut or creamed coconut
3 tbsp grated fresh ginger
1–2 green chili peppers, finely chopped
2 large cloves garlic
Garnish with:
3 tbsp chopped coriander leaves
Serve with:
chapatis (see page 81)
rice

Trim, prepare, and rinse the vegetables. Place all the masala ingredients in the food processor and blend well with 5–6 tbsp water.

Heat the ghee in a wide, fairly deep saucepan and fry the onions gently over low heat for 10 minutes, stirring frequently, until they are tender and pale golden brown. Add the 6 tbsp chopped coriander leaves, the curry powder, cumin, and paprika. Cook for a few seconds, stirring, then add the masala mixture, together with 1 tsp salt. Stir well and continue cooking gently for 10 minutes. Add the vegetables and stir so that they are coated and flavored with the spicy mixture. Add about 1 cup water, stir once more, cover, and simmer over low heat for 15–20 minutes. When the vegetables are almost tender but still crisp, add the cream and ketchup, and then continue cooking, uncovered, for about 10 minutes, stirring occasionally, until the vegetables are tender but still firm. Taste, add more salt if needed, then serve piping hot, sprinkled with the 3 tbsp chopped coriander leaves and accompanied by chapatis and steamed or boiled rice. Serves 6.

———•———

SCRAMBLED EGGS WITH SCALLIONS INDIAN STYLE

10 large scallions
1 green chili pepper
ghee (clarified butter, see page 36)
1½ tsp cumin seeds
3 tbsp tomato paste
8 eggs
½ cup light cream, lightly warmed
3 tbsp chopped coriander leaves
salt
Serve with:
chapatis (see page 81) or stuffed parathas (see page 49)
Spiced Lentils and Spinach (see page 238)

Cut off the root end and remove the outermost layer of the scallions; slice both the bulb and leaves into thin rings. Heat 6 tbsp ghee in a wide, nonstick skillet and gently fry the scallions and the finely chopped chili over low heat for 12 minutes, stirring frequently.

Roast the cumin seeds for about 30 seconds in a small nonstick skillet over fairly high heat, stirring continuously; when they start to jump about, they are done. Remove from the heat immediately and set aside.

Mix the tomato paste with 3 tbsp hot water and stir into the onions over moderate heat for 1 minute to allow excess moisture to evaporate.

Break the eggs into a large mixing bowl and beat briefly with a pinch of salt. With the heat on very low under the skillet containing the scallions, pour in the beaten eggs; stir until the egg mixture has thickened but is still very creamy (about 1–1½ minutes). Remove from the heat, quickly stir in the lightly warmed cream, the cumin seeds, and chopped coriander leaves. Stir for about ½–1 minute. Serve at once on very hot plates.

———•———

PERSIAN VEGETABLE AND EGG TIMBALE
Kukuye Sabzi

1½ lb spinach
2 large leeks
2 scallions
1 firm Boston lettuce
3 tbsp finely chopped parsley
28 walnut halves, shelled
8 eggs
3 tbsp all-purpose flour
olive or sunflower oil
salt and black peppercorns
For the sauce:
1 cup whipping cream
1 cup natural yoghurt
salt
Garnish with:
seeds from ½ ripe pomegranate (optional)
small mint leaves

Remove the stalks carefully from the spinach and save them for soups or for braising as you only need the leaves. Wash, dry on paper towels, chop finely, and place in a large bowl.

Trim, wash, and prepare the leeks, scallions, and lettuce. Remove the outer, tougher layers or leaves of all these, as only the tender parts are used for this recipe. Cut them all into very thin strips. Mix in the bowl with the spinach and parsley.

Preheat the oven to 325°F. Grease the inside of a nonstick ring tube or turban mold lightly with oil (approximately 10 in outer diameter, 4 in inner diameter).

Chop the walnuts very coarsely. Break the eggs into a bowl and beat them briefly with 2-1/2 tsp salt and 1/2 tsp freshly ground black pepper. Sift the flour into the bowl and beat well. Add to the prepared vegetables and mix thoroughly. Carefully ladle into the mold, tap the bottom of the mold on the work surface gently to get rid of any trapped air, and cook in the oven for 50 minutes, or until the surface is pale golden brown and the cooked mixture has shrunk away from the sides a little. Test by inserting a thin skewer or cocktail stick deep into the mixture: it should come out clean and dry. Take out of the oven and leave to stand for 15 minutes before unmolding. While the mold is cooking, make the sauce: beat the cream stiffly and fold in the yoghurt and a pinch of salt. Pour some of the sauce into a small bowl in the central well of the mold or simply spoon in enough sauce to fill this central cavity in the serving dish. Arrange the pomegranate seeds and the mint leaves decoratively on top of the creamy sauce. Serves 6.

——— • ———

LEEK AND CHEESE PIE

1¾ lb leeks
1¾ cups milk
1½ tbsp cornstarch

½ cup heavy cream

pinch grated nutmeg

1 cup grated hard cheese

¼ cup butter, softened at room temperature

salt and pepper

Bring 11 cups (5½ pints) salted water to a boil in a very large saucepan. Trim and wash the leeks, removing the tough, outer layer and the ends of the green leaves. Keeping the leeks whole, partially cut through them into quarters from the top, green end to within about 3 in of the root end to enable you to wash them thoroughly. Boil for 20 minutes then drain well. Preheat the oven to 400°F.

Bring the milk slowly to a boil in a small saucepan. Mix the cornstarch with 3 tbsp cold water and stir into the hot milk off the heat. Return to the heat and keep stirring as the milk thickens for about ½–1 minute.

Remove from the heat and gradually beat in the cream using a balloon whisk. Season with salt, pepper, and nutmeg.

Use a little of the butter to lightly grease a shallow ovenproof dish. Slice the leeks across, cutting them in half. Arrange in a single layer in the greased dish alternating a green half with a white half. Cover with the white sauce. Sprinkle with the cheese and dot with flakes of the remaining butter; bake in the oven for 25 minutes and finish by placing under a very hot broiler for 5 minutes to brown. Serve at once.

——— • ———

STEAMED VEGETABLES WITH CURRY SAUCE

4 leeks
2 potatoes
2 zucchini

2 small turnips
2 large fennel bulbs
salt
For the curry sauce (yields approx. 2¼ cups):
9 oz creamed coconut
1 large onion
½ green chili pepper
¼ cup ghee (clarified butter, see page 36)
1 piece fresh ginger, peeled and grated
1½ tbsp all-purpose flour
3 tbsp mild curry powder
3 tbsp cider vinegar or wine vinegar
½ cup chicken stock (see page 37)
1 tsp fresh lime or lemon juice
3 tbsp chopped coriander leaves
salt and pepper
Garnish with:
4–6 sprigs coriander

Make the coconut milk for the curry sauce: break up the block of creamed coconut into fairly small pieces and place in a large bowl. Add 1¾ cups boiling water and leave to stand for 30 minutes. Strain through a clean linen napkin or two layers of cheesecloth, twisting the ends of the cloth tightly to squeeze out all the moisture. Discard the solid material left inside the cloth and heat the coconut milk.

Slice the onion finely; chop the chili pepper, discarding the seeds. Cook the onion in the ghee over low heat in a wide, heavy-bottomed saucepan for 10 minutes while stirring. Add the chili pepper and the ginger and continue frying gently for another 5 minutes. Stir in the flour and cook for a few seconds. Mix the curry powder with the vinegar and stir into the onion mixture. Start steaming the vegetables at this point; they will take approximately 12–15 minutes. Remove the curry mixture from the heat and pour in the scalding hot coconut milk all at once while beating continuously with a whisk to prevent lumps from forming. Return to a low heat, beat in

the stock, and keep beating until the mixture acquires a velvety consistency.

Use a hand-held electric beater to beat the hot sauce until it is smooth. Season with salt and pepper and add a little lime juice or lemon juice to taste. Add the chopped coriander just before serving.

Transfer the steamed vegetables to heated plates, cover with the curry sauce, and garnish each serving with a sprig of coriander.

———•———

RUSSIAN CARROT MOLD

1¼ lb young, tender carrots
6 eggs
3 tbsp fresh dill, chopped
6 tbsp butter, 5 tbsp of which melted
1 cup fine fresh soft breadcrumbs
3 tbsp fine dry breadcrumbs
nutmeg
salt and pepper
Serve with:
sour cream or smetana (see page 59)
buttered peas

Preheat the oven to 350°F. Use 1 tbsp of the butter to grease a 6-cup capacity soufflé dish; sprinkle the inside with fine dry breadcrumbs. Trim and peel the carrots. (The fresher the carrots, the more flavor they will have.) Cut into small pieces and boil in a large saucepan of salted water for about 30 minutes or until very tender. Drain well and reduce to a smooth purée in the food processor or vegetable mill. Transfer to a large bowl, stir in the 6 egg yolks, the dill, salt, freshly ground white pepper to taste, and a pinch of nutmeg. Stir thoroughly, then add the melted butter and the fine fresh breadcrumbs and mix well.

Beat the egg whites with a pinch of salt until stiff but not at all dry or grainy and fold into the carrot mixture. Pour the mixture into the prepared soufflé dish, smoothing the top level with a spatula. Place in a roasting pan, pour sufficient boiling water to come about halfway to two-thirds of the way up the sides of the dish, and cook for 40 minutes.

Leave to stand and set for a few minutes before unmolding onto a hot serving dish; cut like cake and serve with the sour cream or smetana and buttered peas.

———•———

BLACK SALSIFY GREEK STYLE

1¾ cups chicken stock (see page 37)
juice of 2 lemons
2 lb black salsify
1 medium-sized onion
3 tbsp chopped parsley
6 tbsp extra-virgin olive oil
2 tbsp all-purpose flour
3 egg yolks
salt and pepper

Bring the stock to a boil. Stir the juice of 1 lemon into a large bowl of cold water. Trim, wash, and peel the black salsify one root at a time. As each root is peeled, cut it lengthwise into thick slices; cut each of these into 1½-in lengths and remove any of the hard, woody central section from each piece with a sharp knife. Drop these into the bowl of acidulated water to prevent them discoloring.

Chop the onion finely and fry gently with the chopped parsley in the oil in a large, heavy-bottomed saucepan for a few minutes, until tender.

Drain the salsify well, add to the onion, and sauté for 5 minutes. Add a pinch of salt, sprinkle with the flour, and stir for 1 minute to cook the flour and distribute it evenly. Add approximately 2 cups of the boiling hot stock and stir well. Reduce the heat to low, cover, and simmer for 10 minutes, adding a little more hot stock when needed. When the salsify is tender, make a thickening liaison by beating the

egg yolks with the juice of the second lemon and 3 tbsp of the hot stock in a small bowl. Sprinkle over the salsify and continue cooking over low heat for 1–2 minutes, stirring continuously. The egg yolk will thicken the sauce slightly and make it slightly creamy; as soon as it does, remove from the heat. This is not meant to be a thick sauce. Continue stirring off the heat for half a minute, add a little more salt and pepper if wished, and serve.

———•———

WHITE SALSIFY POLISH STYLE

vegetable stock (see page 38)
1¾ lb white salsify (also known as oyster plant)
juice of 1 lemon
4 hard-boiled eggs
¼ cup finely chopped parsley
¾ cup ghee (clarified butter, see page 36)
¼ cup fine dry white breadcrumbs
salt and pepper
Serve with:
Rösti Potatoes (see page 195)

Make the vegetable stock. Peel the salsify and drop each piece into a large bowl of cold water acidulated with the lemon juice to prevent discoloration. Prepare the vegetable as described in the preceding recipe for black salsify. Drain and boil in the special stock for about 15 minutes, or until tender. Drain again and cut the pieces of salsify crosswise into slices.

Transfer to a very hot serving dish and sprinkle with the finely chopped eggs and parsley. Keep hot while you heat the ghee until it starts to deepen in color. Add the breadcrumbs, a pinch of salt, and a little freshly ground pepper. Stir well, then sprinkle all over the salsify and serve.

———•———

185

NEAPOLITAN POTATO CAKE
Gattò di Patate

3¼–3½ lb potatoes
1½ cups milk
½ cup butter
2 eggs
¾ cup freshly grated Parmesan cheese
1 lemon
5½ oz smoked cheese, diced
¼ lb Mortadella sausage or ham, sliced
15 basil leaves, torn into small pieces
1 lb mozzarella cheese, sliced
2 oz Neapolitan-type peppery sausage, thinly sliced
fine dry breadcrumbs
salt and pepper

Wash the potatoes and steam until tender. Spear them with a carving fork and peel. Push them through a potato ricer while still hot; repeat the ricing operation, then push them through a fine sieve into a large mixing bowl. This makes them easy to prepare.

Preheat the oven to 350°F. Heat the milk together with ¼ cup of the butter and a little salt and freshly ground pepper until scalding. Stir most of the milk into the sieved potato; work in the eggs, Parmesan cheese, and the finely grated rind of the lemon. The mixture should be fairly soft; add the rest of the milk if necessary. Stir this mixture thoroughly for 2 minutes, then add the smoked cheese, the chopped Mortadella or ham, and the basil. Grease a cake pan about 10 in in diameter and at least 2 in deep. Place a quarter of the mixture in the pan and press out flat into an even layer with the palm of your hand. Spread out slices of mozzarella on the surface and cover with another quarter of the mixture. Repeat these last two layers. Spread the sausage slices on top. Sprinkle a covering of breadcrumbs over the surface and dot small pieces of the remaining butter on top. Bake in the oven for about 30 minutes,

browning under a very hot broiler for a final 5 minutes to form a crisp topping. Remove from the oven and allow to stand for 10 minutes before serving.

——— • ———

FRIED POTATOES, PEPPERS, AND BEEF

2 large potatoes
2 large yellow peppers
4 cloves garlic
1 red chili pepper
generous pinch oregano
3 tbsp finely chopped parsley
14 oz thickly sliced tenderloin of beef
6 tbsp extra-virgin olive oil
3 tbsp light olive oil
salt and pepper

This dish hails from the Isle of Capri. Peel the potatoes and cut into small cubes. Blanch in a pan of boiling salted water for 5 minutes, drain, and allow to cool. Cut the peppers in half, remove the stalks, seeds, and white membrane, and slice into wide strips. Cut these into 1½-in lengths.

Heat the extra-virgin olive oil in a wok or a large skillet and sauté the garlic cloves, crushed with the flat of a knife but still left whole, together with the peppers. Add the finely chopped chili pepper, the oregano, and half the parsley. Cook over fairly high heat for 2 minutes. Sprinkle with a pinch of salt, sauté for another minute, and add the potatoes. Stir, cover the skillet, and cook over low heat for 10 minutes, stirring every few minutes. Cut the beef tenderloin into small cubes, season these with salt and freshly ground black pepper. Heat the light olive oil over high heat in a large, nonstick skillet until very hot; add the beef and stir-fry for 1 minute, turning the beef regularly. Add to the potatoes, stir, and sprinkle with the remaining parsley.

VEGETABLE AND CHEESE BAKE

2½ lb potatoes

2 large onions

1¼ lb canned tomatoes

½–¾ cup grated hard cheese

extra-virgin olive oil

generous pinch oregano

salt and pepper

Serve with:

vegetarian nut roast

broiled meat, fish or poultry

Preheat the oven to 325°F. Peel the potatoes and cut into slices about ½ in thick. As you cut the slices, put them in a large bowl of cold water to prevent them from turning brown. Peel the onions and slice very thinly. Drain the tomatoes, reserving the juice. Cut the tomatoes open and spoon all their seeds into the sieve; drain off any more juice and keep it. Discard the seeds. Chop the flesh coarsely.

Grease a wide, shallow ovenproof dish with oil. Drain the potatoes and transfer to the oven dish together with the onions and tomatoes. Sprinkle with the reserved tomato juice. Season with salt, freshly ground pepper, and sprinkle with the oregano and the grated cheese. Finally, moisten with 1 cup boiling water and sprinkle with 3 tbsp olive oil. Bake for 1¼ hours or until the potatoes are very tender and the top is crisp and lightly browned. Allow to stand for 10 minutes before serving.

———— • ————

POTATO AND ARTICHOKE PIE

4 large potatoes

5 artichoke hearts (see method)

1 lemon

4 large cloves garlic, minced

3 tbsp finely chopped parsley

extra-virgin olive oil

salt and pepper

Serve with:

sausages

braised onions

Preheat the oven to 325°F. Peel the potatoes, cut into fairly thin slices, dropping them into a large bowl of cold water as you do so.

If you use very tender, young artichokes prepare as described on page 18, strip down to the inner leaves, cut each artichoke lengthwise in quarters, cut each quarter lengthwise in half. Drop into a large bowl of cold water acidulated with the lemon juice as you prepare them to prevent discoloration. If using larger, less tender varieties of artichoke use 10; strip off all the leaves, and remove the choke from the bottom.

Drain the vegetables and spread out evenly in a large, shallow ovenproof dish. Season with salt and pepper and sprinkle with 6 tbsp olive oil, the minced garlic, the parsley, and 1 cup boiling water. Cook for 1 hour or until tender. Alternatively, sauté the vegetables in the oil with the garlic in a large skillet for about 8 minutes. Then add the other ingredients, including the water, cover, reduce the heat, and simmer for about 20 minutes, adding a little more hot water as required.

———— • ————

POTATO SOUFFLÉ WITH ARTICHOKE SAUCE

1¾ lb potatoes

½ cup butter

½ cup finely grated Swiss cheese

4 eggs

187

small pinch cream of tartar or 2–3 drops lemon juice
nutmeg
salt and pepper
Serve with:
Artichoke Sauce (see page 77)

Grease a 6-cup capacity or slightly larger soufflé dish with 1 tbsp of the butter. Preheat the oven to 350°F. Steam the potatoes until tender. Spear with a carving fork and peel while still boiling hot. Put through a potato ricer twice or push through a fine sieve. Transfer the mashed potatoes to a large mixing bowl. Melt the remaining butter and add it to the potatoes, together with the cheese and the egg yolks, adding a pinch each of salt, pepper, and grated nutmeg.

Beat the egg whites with a pinch of salt and a small pinch of cream of tartar (or a few drops lemon juice) until they are stiff. Fold in the potato and cheese mixture. Pour into the prepared dish and bake for 35–40 minutes; the soufflé should rise up well in the dish and turn golden brown on top. Serve at once with the artichoke sauce which should be made while the soufflé is cooking.

———— • ————

SWEET POTATO PANCAKES WITH HORSERADISH AND APPLE SAUCE

1¼ lb sweet potatoes
½ cup all-purpose flour
1 small onion, finely chopped (optional)
2 eggs
1½ tbsp ghee (clarified butter, see page 36)
nutmeg
salt and pepper
For the horseradish and apple sauce:
1 2-in piece horseradish root
1 green apple
1½ tbsp fresh lime or lemon juice
1½ tsp cane sugar
1 tsp cider vinegar
½ cup low-fat natural yoghurt
½ cup heavy cream
nutmeg
salt and pepper

Heat plenty of salted water in a large saucepan. Peel the potatoes, grate them coarsely with a grater or in the food processor, and blanch them in the boiling salted water for 5 minutes. Drain, transfer to a clean cloth or linen napkin and twist the material to force out all excess moisture.

Mix the potato in a large bowl with the flour, onion, and the lightly beaten eggs, seasoned with salt, pepper, and a pinch of nutmeg. Cover the bowl and chill in the refrigerator.

Make the horseradish and apple sauce: peel the horseradish (wear sunglasses or goggles as it will make your eyes water and sting) or use vacuum packs or jars of ready-grated moist horseradish that is not as strong as fresh. Peel the apple, cut into pieces, and process with the horseradish in the food processor until smooth and creamy. Transfer to a bowl, stir in the lime or lemon juice, the sugar, 1 tsp vinegar, and a pinch each of salt and freshly ground pepper. Mix in the yoghurt and cream and add more salt to taste.

Make the pancakes: heat 1½ tbsp ghee in a small nonstick skillet (about 7 in in diameter) and when very hot, add ½ cup of the potato batter, pressing the mixture level before it sets. Fry over moderate heat for 3 minutes, then flip the pancake over and cook for another 2 minutes until browned and crisp. Transfer the pancake to paper towels to drain and keep hot while cooking the other pancakes.

These pancakes look more like large fritters or thin potato cakes. Serve as soon as the last one is cooked; pass round the horseradish and apple sauce separately.

GREEK POTATO FRICADELLES
Patatokeftédes

2¼ lb potatoes

3 eggs

1 small onion, grated

¾ cup grated hard cheese

¼ cup finely chopped parsley

fine dry breadcrumbs

light olive oil

grated nutmeg

salt and pepper

Garnish with:

parsley

lemon wedges

Serve with:

beet salad

Steam the potatoes until very tender; spear with a carving fork and peel while still boiling hot. Put them through a ricer twice (or push through a fine sieve) and place in a large mixing bowl. Work in 1 whole egg and 1 yolk (reserve the white from this second egg). Mix in the onion, cheese, parsley and a generous pinch each of salt, pepper, and grated nutmeg.

Break off clumps and shape between your palms to the size of a small egg; press gently to flatten slightly. Lightly beat 2 egg whites (you will have one yolk left over, which you can reserve for use in another recipe). Spread out about 1 cup bread-crumbs in a large plate. Dip the fricadelles in the egg whites to moisten all over and then coat thoroughly with the breadcrumbs. Arrange on a plate or chopping board, cover with Saran wrap, and chill in the refrigerator for at least 1 hour.

When it is time to cook the fricadelles, heat plenty of oil in a deep-fryer to 350°F and fry them 2 or 3 at a time until crisp and golden brown. Place in the oven in a single layer on paper towels to keep hot while you finish frying the rest. Serve without delay on a heated oval platter, garnished with sprigs of parsley and lemon wedges, accompanied by a beet salad dressed with oil, wine vinegar, salt, freshly ground pepper, and sprinkled with finely chopped parsley.

———•———

VEGETABLES WITH POTATO STUFFING

1 lb potatoes

12 large, freshly picked summer squash flowers

1 very large onion

4 zucchini

2 eggs

¾ cup freshly grated Pecorino cheese

fine dry breadcrumbs

1 tsp fresh marjoram leaves

1 clove garlic

thyme

6 tbsp olive oil

butter

salt and black pepper

Steam the potatoes until tender. Prepare the squash flowers carefully by trimming the stalk to within ½ in of the flower, removing the pistils delicately, and rinsing briefly in cold water; spread out to dry on a clean towel. Peel the onion, trim off the ends, and make deep cross-cuts in both top and bottom. Trim the ends off the zucchini and rinse. Blanch the onion and zucchini in a large saucepan of boiling salted water for 10 minutes (take the zucchini out after 5–7 minutes if they are very thin); drain and leave to cool.

Preheat the oven to 350°F. When the potatoes are very tender, spear them with a carving fork and peel while still very hot; push through a potato ricer or sieve into a large mixing bowl. Stir in 1 tbsp butter, 3 tbsp oil, 2 eggs, and two-thirds of the grated cheese. Mix the remaining grated cheese with 4 tbsp dry breadcrumbs and reserve for later use. Chop the marjoram leaves finely with the peeled garlic and stir into the potato mixture together with a generous pinch of thyme, a pinch of salt, and plenty of freshly ground black pepper. Mix thoroughly, adding a little more salt if desired.

Lightly oil a large cookie sheet. Cut the zucchini lengthwise in half, scoop out some of the flesh, leaving a layer of flesh in place next to the skin. Slice the onion lengthwise and separate out the layers to form shallow saucers. Fill the summer squash flowers carefully with 2 or 3 tsp of the mixture; do not overfill. Pinch the ends of the flowers gently to seal. Sprinkle a little salt over the partially hollowed zucchini and the onion pieces and fill with stuffing mixture; again, do not overfill. Transfer to the cookie sheet, sprinkle each stuffed item with a little olive oil and with the grated cheese and breadcrumbs. Bake for 15–20 minutes, placing the cookie sheet under a very hot broiler for the last 2–3 minutes to brown the topping. Serve at once, decorating the larger vegetables with a small sprig of fresh basil.

This mixture makes a good stuffing for most vegetables.

———•———

STUFFED VEGETABLE SELECTION

4 zucchini
2 long eggplants
1 large onion
1 large red pepper
½ cup grated Swiss cheese
½ cup lean ham, diced
1 whole egg
3 tbsp béchamel sauce (see page 36)
3 large potatoes
2 large tomatoes
1 cup fine dry breadcrumbs
olive oil or sunflower oil
¼ cup butter
salt and pepper

Start the preparation several hours in advance or the day before you plan to serve this dish by blanching the vegetables and preparing the stuffing. Bring a very large saucepan of salted water to a boil. Trim and wash the zucchini and eggplants and slice lengthwise in half. Peel the onion and cut into quarters. Add all these to the boiling water; after 5 minutes add the rinsed, whole pepper and continue blanching for another 5 minutes. Drain the vegetables.

As soon as they are cool enough to handle, scoop out most of the flesh of the zucchini and eggplants, leaving only a thin layer next to the skin and spoon it into a large mixing bowl. Spread out the hollowed zucchini and eggplants upside down on a slightly tilted chopping board to finish draining.

Process the Swiss cheese, ham, and the vegetable pulp in the food processor until smooth and evenly blended; transfer this back to the large mixing bowl and stir in the egg and the béchamel sauce. Season with salt and pepper. Refrigerate for 2 hours.

About 1¼ hours before you plan to serve this dish, preheat the oven to 320°F. Grease 1 or 2 cookie sheets with oil. Peel the potatoes and slice very thinly. Spread them out in a single layer on the cookie sheets (slightly overlapping if necessary), season with a little salt and freshly ground pepper. Fill the zucchini and the eggplants with some of the chilled stuffing mixture. Cut the pepper lengthwise into quarters, remove the stalk, seeds, and inner membrane and place some stuffing mixture in each concave quarter, smoothing it neatly. Refrigerate.

Separate the layers of onion if they have not already come apart when blanching; fill each concave slice with some stuffing mixture. Wash and dry the tomatoes; cut them horizontally in half, scoop out the seeds and some of the flesh, and stuff them with the remaining mixture.

Place all the vegetables, filling uppermost, on top of the potato slices, arranging them so that the colors alternate attractively. Sprinkle with breadcrumbs, place a piece of butter on top of each, and bake in the oven for about 50 minutes, until golden brown and crisp on top. Take out of the oven and allow to stand for 10 minutes before serving.

——— • ———

BOILED OR STEAMED VEGETABLES WITH GUASACACA DRESSING

4½ cups chicken stock (see page 37)
3 medium-sized potatoes
4 leeks
4 small tender turnips
4 carrots
For the guasacaca dressing:
2 ripe avocado pears
1 ripe tomato
1 finely chopped hard-boiled egg
1½ tbsp chopped coriander leaves
1 tbsp finely chopped parsley
½ green chili pepper (seeds removed)
1½ tbsp red wine vinegar
1½ tsp pili-pili hot relish (see page 113) or pinch cayenne pepper
¼ cup olive oil
salt and pepper
Serve with:
tortilla chips

Heat the stock. Trim, wash, and dry all the vegetables; cut into fairly large pieces and boil gently in the stock for 20–30 minutes or until they are tender. Meanwhile, prepare the sauce: cut the avocados lengthwise in half, take out the pit; peel each half and mash the flesh with a fork to a coarse purée.

Peel the tomato (blanch first if necessary) and remove all the seeds and any tough parts; dice the flesh and stir into the avocado together with the chopped egg, coriander, parsley, and the finely chopped green chili pepper. Mix in the vinegar, pili-pili relish or cayenne pepper, olive oil, and salt and pepper to taste. Serve the drained vegetables hot, handing round the sauce and the tortilla chips separately.

Avocado discolors within a short time, so do not prepare more than 30 minutes before serving.

——— • ———

BRAISED ONIONS AND RICE
Soubise

½ cup risotto rice (e.g. arborio), par-boiled
2½ lb strong onions
1 cup butter
½ cup heavy or light cream
3 tbsp grated Swiss cheese
2 tbsp finely chopped parsley
salt and pepper
Serve with:
peppered tenderloin of beef
braised globe artichokes
Belgian Endive Flemish style (see page 86)

Preheat the oven to 300°F. Bring a large saucepan of salted water to a boil, add the rice, and boil for 4 minutes, then drain. Peel and thinly slice the onions.
Melt ¼ cup of the butter in a large, fireproof casserole dish; add the onions and stir well to coat with the butter. Add the rice, ½ tsp salt, and a little freshly ground white pepper and stir well. Cover and cook in the oven for 1 hour, stirring now and then. The onions and rice should be tender and pale golden

brown when done. Add a little more salt and pepper if necessary, followed by the cream, cheese, and the parsley. Stir gently but thoroughly and serve at once on hot plates. This is a delicious accompaniment for peppered tenderloin of beef (steak *au poivre*). Braised artichokes and/or braised Belgian endive Flemish style would complete a main course.

———•———

SWEET–SOUR BABY ONIONS

14 oz peeled white baby onions

2 small pieces of orange rind

2 cups red wine vinegar

½ cup cane sugar (white or demerara or soft light brown)

salt

Serve with:

Rösti Potatoes (see page 195)

fried, baked or poached eggs or vegetable omelets

Place the onions in a single layer in two wide, fairly shallow flameproof casserole dishes. Sprinkle both with a little salt and add ½ cup water, 1 cup vinegar, and a piece of orange rind. Heat to boiling point, cover, and cook over low heat for 10 minutes.

Add half the sugar to each batch, stir, replace the lids, and simmer very gently for another 10 minutes, or until the liquid has almost completely evaporated and the surface of the onions is glazed and glossy.

———•———

TURKISH RICE WITH LEEKS
Zeytinyagli Pirasa

1 lb leeks

1 medium-sized onion, finely chopped

¼ cup olive oil

1½ tbsp all-purpose flour

1 tsp superfine sugar

½ cup long-grain rice, parboiled

salt

Garnish with:

1 lemon

Serve with:

artichoke hearts in oil

Prepare the leeks: trim off the roots and remove the tough, outer layers. Cut off most of the green part, leaving about 2 in in place. Wash thoroughly. Cut lengthwise in quarters, then slice each length into 1-in sections.

Cook the onion in the oil over low heat in a very large skillet or wide saucepan for about 10 minutes, stirring frequently until it is tender and wilted but not at all browned. Stir in the flour, scant 1 tsp salt, and 1 tsp superfine sugar and continue cooking and stirring for 2 minutes; 1–1½ tbsp tomato paste can be added at this point for extra flavor if desired. Gradually add 1¾ cups hot water, stirring continuously. Increase the heat to moderately high and keep stirring as the mixture comes to a boil; add the leeks. Mix thoroughly, turn down the heat to very low, cover tightly, and simmer gently for 10 minutes. When this time is up, sprinkle in the rice, stir well, cover tightly again, and cook for another 15 minutes, until the rice is tender but still has a little bite left in it. Serve garnished with lemon wedges for each person to squeeze over the leeks and rice. Sprinkle with chopped fresh dill or parsley if desired.

———•———

CARROTS WITH CREAM AND HERBS

1¾ lb carrots

2 finely chopped shallots

¼ cup butter

2 pinches sugar
1 cup heavy or light cream
generous pinch oregano
3 tbsp chopped fresh basil
salt and pepper
Serve with:
Spinach Soufflé (see page 78)

Wash, trim, and peel the carrots. Cut into large matchstick strips. Heat the butter gently in a very wide saucepan or skillet and cook the shallots over very low heat, stirring frequently, for 10 minutes, until wilted and tender. Add the carrots and sprinkle with a little salt, freshly ground pepper, and 2 pinches of sugar. Stir over gentle heat for 1 minute, then cover and sweat slowly for about 30 minutes or until the carrots are tender but still crisp, stirring often and adding a very little hot water when necessary.

Add a generous pinch of oregano, stir over moderately high heat for a few seconds, then add the cream. Reduce the heat and stir for 2–3 minutes. Remove from the heat and add a little more salt if desired. Serve in a heated serving dish with a sprinkling of chopped fresh basil leaves. Serve at once.

—— • ——

DEEP FRIED BLACK SALSIFY

1 egg yolk
1 cup all-purpose flour, sifted
juice of 1 lemon
1¾ lb black salsify
olive oil
salt

Make the batter: beat the egg yolk in a large bowl with ½ tsp salt and 1 cup iced water. Sift in the flour and stir just enough to get rid of any lumps. Cover and chill in the refrigerator.

Bring a large saucepan of salted water to a boil; add half the lemon juice. While it is heating, peel the black salsify and cut into pieces approximately 1½ in long. As each piece is prepared, drop it into a large bowl of cold water acidulated with the remaining lemon juice to prevent discoloration. When the water boils, drain the black salsify pieces and boil for 18 minutes or until tender but still firm. Drain well and leave to cool a little. Cut each piece in half lengthwise and remove the tough, woody central section.

Heat plenty of oil in a deep-fryer to 350°F. Dip each piece of black salsify into the batter and deep-fry a few pieces at a time, until they are crisp and golden; make sure the temperature of the oil remains constant. Remove each batch from the oil with a slotted spoon and finish draining on paper towels; keep hot while you finish the deep frying. Serve immediately.

—— • ——

TURNIP AND RICE PURÉE

3 large turnips
2¼ cups milk
1 cup long-grain rice
2 tbsp butter
2 large cloves garlic
½ tsp fresh thyme leaves
⅓ cup light cream
3 tbsp finely chopped parsley
salt and white pepper
Serve with:
Mushrooms Bordeaux style (see page 264)
Peas with Bacon and Basil (see page 234)

Peel and coarsely chop the turnips. Bring the milk slowly to a boil in a large, heavy-bottomed, fireproof casserole dish, then sprinkle in the rice and add the butter, the peeled, whole garlic cloves, the thyme

193

(use a pinch of dried thyme if fresh is not available), and 1/2 tsp of salt. Mix well. Simmer for 10 minutes, stirring every few minutes. Add the turnips and a little more milk if necessary to cover them.

Cover, reduce the heat as low as possible, and simmer for 15 minutes, stirring at intervals. When the turnips are very tender and nearly all the milk has been absorbed, remove from the heat and beat with the hand-held electric beater until the mixture forms a smooth purée (or put through the vegetable mill with a fine-gage disk). These first stages can be completed several hours in advance.

When it is time to finish cooking the purée, reheat it over low heat without a lid on, stirring occasionally until it has thickened further. Turn off the heat and stir in the cream, a little freshly ground white pepper, and the parsley. Add a little more salt if needed and serve.

———— • ————

BABY TURNIPS SAUTÉED WITH GARLIC AND PARSLEY

2¾ lb very small baby turnips
juice of 1 lemon
butter
olive oil
1 large clove garlic
¼ cup finely chopped parsley
salt and pepper
Serve with:
Peas with Bacon and Basil (see page 234)
Mushroom and Herb Omelet (see page 261)

Trim off the ends of the baby turnips and peel with the potato peeler one by one, dropping them immediately into a large bowl of cold water to which the lemon juice has been added, to prevent discoloration. Drain and dry, then cut each root from top to bottom into quarters and then into eighths. Heat 2 tbsp butter and 1½ tbsp olive oil in a wide, nonstick

skillet and as soon as the butter has stopped foaming, sauté half the root vegetables over fairly high heat for 3–4 minutes until they turn golden brown. Use a slotted spoon to transfer this first batch to a bowl. Add a little more butter and oil to the skillet and when very hot, sauté the second batch in the same way. Return the first batch to the skillet containing the second batch, add the clove of garlic (crushed but still in one piece) and a little salt and freshly ground pepper. Reduce the heat to very low, cover, and sweat gently for 8–10 minutes or until tender but still with a little bite left; stir every 2–3 minutes.

Remove from the heat, add a little more salt and pepper if needed, sprinkle with the parsley, and serve at once.

———— • ————

RADISHES WITH CREAM, CORIANDER, AND POMEGRANATE SAUCE

3 small bunches small radishes
2 large shallots, finely chopped
2 tbsp butter
¼ cup raspberry vinegar
1¼ cups crème fraîche *or heavy cream*
2½ tbsp chopped fresh coriander leaves
seeds from ½ a pomegranate
salt and pepper
Serve with:
Rösti Potatoes (see page 195), veal dishes, or vegetable omelets

Trim and wash the radishes; cut them into round slices. Heat the butter in a fairly shallow flameproof casserole dish and fry the shallots very gently for about 10 minutes until wilted and tender but not at all browned. Add the radishes, turn up the heat a little, and fry for 1 minute, stirring with a wooden spoon. Season lightly with salt and pepper. Sprinkle with the vinegar and continue cooking until the liquid has completely evaporated.

Pour the *crème fraîche* or cream over the radishes and simmer gently for 2–3 minutes, stirring occasionally. Take off the heat, add a little more salt and pepper if necessary, and sprinkle with the coriander and pomegranate seeds. Serve straight from the dish.

———— • ————

BEET, TOMATO, AND CUCUMBER SALAD

1 large beet, cooked
1 large or 2 small cucumbers
2 firm, medium-sized tomatoes
1 tsp sugar
6 tbsp grated fresh coconut flesh
3 tbsp chopped fresh coriander leaves
½ green chili, finely chopped
6 tbsp chopped unsalted peanuts
2¼ cups chilled natural yoghurt
1½ tbsp sunflower oil
1½ tbsp cumin seeds
salt
Serve with:
chapatis (see page 81)

Peel and dice the beet and the cucumber. Peel the tomatoes, remove the seeds and any tough parts, and dice. Place all the prepared vegetables in a large salad bowl, sprinkle with a pinch of salt and with the sugar, and gently stir in the coconut, coriander, chili pepper, peanuts, and yoghurt. Add a little more salt if desired.

Heat the oil in a small, nonstick skillet and when very hot, add the cumin seeds. Fry for 2–3 seconds; as soon as they start to jump about in the skillet, take off the heat and add to the salad. Stir carefully once more and serve with chapatis as a delicious, refreshing accompaniment to almost any Indian main course.

———— • ————

RÖSTI POTATOES

6 large potatoes, cooked
2 tbsp butter
salt
Serve with:
fried, poached or baked eggs with crispy bacon or cheese omelets

Grate the potatoes coarsely. Place in a large mixing bowl and season with a little salt and pepper.

Divide the butter in half and divide one half again between two wide, nonstick skillets; when it starts to sizzle and the foam has disappeared, spread out half the potato in one, half in the other, pressing down well to make the potato cakes firm and evenly spread out. Fry over moderately high heat for 4 minutes or until the edges have browned well and turned crisp. Use plates to help you turn the potato cakes: slide each one in turn onto a plate, place another plate on top, turn upside down, and then slide the potato cake back into its skillet. While you have the potato cakes on the plates, add half of the remaining butter to each skillet and heat. Cook for another 4 minutes, serving as soon as the cakes are evenly browned and crisp. To serve 4, cut each flat potato cake neatly in half. Fried eggs and crispy bacon go well with these potato cakes, as do cheese omelets.

———— • ————

RÖSTI POTATOES WITH SHALLOTS

6 large potatoes, cooked and coarsely chopped

¾ cup finely chopped shallots or Bermuda onions

2 tbsp butter

salt and pepper

The method is the same as for the preceding Rösti recipe on page 195 but, before adding the potato to the butter in the skillets, gently fry half the shallots in each skillet over low heat until soft but not at all browned. Press an even layer of the potato on top of the shallots and then proceed as directed in the previous recipe.

——— • ———

ENDIVE WITH RASPBERRY VINEGAR AND CREAM

4 heads red Treviso endive (see method)

2 shallots, finely chopped

2 tbsp butter

3 tbsp raspberry vinegar

1 cup heavy cream

salt and pepper

If you cannot obtain red Treviso endive, use Belgian endive instead.

Cut off the solid base of each, wash the leaves, dry in a salad spinner, and cut into thin strips. Sweat the shallots in the butter in a large, heavy-bottomed stainless steel or enameled saucepan. Add the prepared endive, turn up the heat slightly, and fry gently while stirring for 2 minutes.

Sprinkle with the vinegar and keep stirring over the heat until it has completely evaporated. Season lightly with salt and pepper, add the cream, and mix well. Cover and cook gently over low heat for 4 minutes. Taste, adjust the seasoning, and serve.

MACEDOINE OF VEGETABLES WITH THYME

5 large potatoes

7 large carrots

7 large zucchini

1 large shallot, finely chopped

3 tbsp clarified butter (see page 36)

1 sprig thyme

salt and pepper

Peel the potatoes and place in a large saucepan with enough cold water to cover; set them aside while you trim, peel, rinse, and dry the carrots and zucchini. Heat a large saucepan of salted water for blanching the vegetables.

Use the smaller end of a melon scoop to scoop out balls of the vegetables. There will be a fair amount of wastage as the balls should be as neat and evenly sized as possible. If preferred, cut the vegetables into small cubes or dice. Save the remaining pieces of vegetable to chop up for soups, etc. Blanch the carrot balls for 5 minutes; the potatoes for 1. Drain well. Fry the shallot gently in the clarified butter for 5 minutes while stirring, until wilted and tender but not colored. Add the sprig of thyme and the other vegetables; increase the heat to moderately high and sauté for 5 minutes until they are tender but still very firm. Season with salt and pepper and serve.

——— • ———

DEEP-FRIED POTATOES

1½ lb very large potatoes

light olive oil

salt

Peel the potatoes and cut into slices just under ¼ in thick. Rinse well, drain thoroughly, and spread out

196

on a clean cloth, cover with another cloth, and blot completely dry, using fresh cloths if necessary to ensure there are no damp patches. Heat plenty of oil in a deep-fryer to 300°F and fry the first batch of the potatoes, leaving it in the oil for 4 minutes or until the potatoes bob up to the surface and begin to puff up slightly. As each batch is fried, take it out with a slotted spoon or frying basket, and spread out in a single layer on paper towels. You can complete these first stages an hour or two in advance.

Just before you want to serve the potatoes, heat the oil in the deep-fryer again, this time to 350°F, and again fry the potatoes in batches for 1–2 minutes or until they puff up well and are crisp and golden brown. Drain well, keep hot in a very wide, uncovered heated dish while you fry the rest. Sprinkle lightly with salt and serve at once on a hot, uncovered dish or directly on heated individual plates.

— · —

POTATO MOUSSELINE

1¾ lb potatoes

2–2½ cups milk

¼ cup butter

nutmeg

salt and pepper

Serve with:

roast veal or pork, egg dishes or with Peas with Bacon and Basil (see page 234)

Peel the potatoes, cut them into fairly small pieces, and boil in salted water for 20 minutes or until very tender. Drain well and put twice through a potato ricer or push through a fine sieve. Bring the milk slowly to a boil. Melt the butter in a saucepan, add the potato and a pinch each of salt, pepper, and nut-

meg. Add the scalding hot milk to the potatoes in a thin stream, beating continuously with a balloon whisk or hand-held electric beater. Continue beating until the purée reaches boiling point, adding a little more hot milk if necessary. Taste and adjust seasoning. Take off the heat and keep beating for a minute or two longer, by which time the potato should be very light and fluffy. This is a marvelously light, digestible way of preparing potatoes.

— · —

POTATO GRATIN

1¼ lb potatoes

3 tbsp butter

2 sprigs thyme

1¾ cups milk

1 large clove garlic, peeled

½ cup grated Swiss cheese

½ cup freshly grated Parmesan cheese

nutmeg

salt and pepper

Serve with:

egg dishes or roast meat or poultry

Preheat the oven to 400°F. Peel and rinse the potatoes; slice about ¼ in thick using a mandoline cutter if you have one. Heat 2 tbsp of the butter in a wide, nonstick skillet with the thyme sprigs. When it is hot, add the potatoes and sauté for 2 minutes. Season with salt, pepper, and a pinch of freshly grated nutmeg. Pour in the milk. Heat until the milk is simmering and then continue cooking over a moderate heat for about 10 minutes, stirring frequently with a spoon to prevent the slices from catching and burning on the bottom of the skillet.

Cut the garlic clove lengthwise in half and rub the cut surfaces hard all over the inside of a wide, shallow ovenproof dish. Use some of the remaining butter to grease the inside of this dish lightly.

When the potatoes are cooked and the milk has acquired a thick, creamy consistency, season with a little more salt and pepper and transfer to the prepared dish. Mix the two cheeses and sprinkle over the potatoes; dot the surface with a few pieces of butter. Place in the oven for 15 minutes, until the cheese topping is crisp and light golden brown. Serve at once. This goes well with roasts or egg dishes and is also delicious when served with green vegetables as a vegetarian main course.

———— • ————

ALMOND CROQUETTE POTATOES

1 lb potatoes
¼ cup butter
3 eggs
½ cup flaked or slivered almonds
nutmeg
light olive oil
grated Parmesan cheese
salt and pepper
Serve with:
roast meat or poultry; foil-baked trout or Radish and Cress Salad in Cream Dressing (see page 163)

Steam the potatoes, spear and peel them while still boiling hot, and put through a potato ricer twice (or push through a fine sieve) while still hot. Transfer to a heavy-bottomed saucepan and keep stirring over moderate heat as you add the butter a small piece at a time. Continue stirring until the mixture is firm and quite dry, leaving the sides of the saucepan cleanly. Remove from the heat and gradually beat in 2 egg yolks and half a lightly beaten egg. Season with salt, freshly ground white pepper, and a pinch of nutmeg. Allow to cool completely.

Spread the almonds out in a shallow dish or plate; beat the 2 remaining egg whites with the remaining amount of beaten egg in a shallow dish with a pinch of salt.

Shape the potato mixture between your palms into balls the size of walnuts, dip in the beaten egg, drain briefly over the dish, and coat evenly with the almonds, pressing down gently so that they adhere. Place on paper towels on top of a chopping board or work surface.

Heat plenty of oil in a deep-fryer and lower the croquettes into the oil in the frying basket a few at a time. Fry for about 3 minutes or until the almond covering is golden brown. Keep hot, uncovered, on paper towels while you fry the remaining batches. Serve without delay.

———— • ————

JERUSALEM ARTICHOKES WITH CREAM SAUCE

1¼ lb Jerusalem artichokes
juice of 1 lemon
2 shallots, finely chopped
1½ tbsp butter
1½ tbsp olive oil
1 sprig thyme or small pinch dried thyme
½ cup chicken stock (see page 37)
½ cup crème fraîche or heavy cream
½ vegetable bouillon cube
3 tbsp chopped chives
salt and pepper

Rinse the Jerusalem artichokes and peel, dropping immediately into a large bowl of cold water acidulated with the lemon juice to prevent discoloration. When they are all peeled, drain them and cut into slices about ¼ in thick.

Fry the shallot very gently until wilted and tender in the butter and oil with the thyme. Add the artichokes, stir for 2–3 minutes, add the stock and simmer, uncovered for 5 minutes. Add the *crème fraîche* or cream, the crumbled 1/2 bouillon cube, and stir. Simmer uncovered, stirring now and then for another 5 minutes or until the Jerusalem artichokes are tender but still firm. Season with salt and pepper to taste, sprinkle with the chives, and serve at once.

———— • ————

CARROT CAKE

1¾ cups very finely grated carrots
5 eggs
1 cup sugar
1¾ cups ground almonds
½ cup cornstarch or potato flour
1 lemon
butter for greasing the cake pan
confectioner's sugar
1 cup whipping cream, lightly beaten
Decorate with:
1 small carrot made out of marzipan or 1 bunch redcurrants

Preheat the oven to 350°F. Beat the egg yolks with half the sugar until pale and creamy. Beat the egg whites until stiff in another bowl and gradually beat in the remaining sugar. Mix the almonds with the cornstarch or potato flour and add to the beaten egg yolks together with the carrots and the grated rind and juice of the lemon. Stir very well before folding in the sweetened beaten egg whites. Grease a 10-in diameter springform cake pan with butter; transfer the mixture to it, leveling the surface with a spatula. Bake for 40 minutes. Allow to cool to lukewarm before unmolding. Dust with sifted confectioner's sugar and decorate with a marzipan carrot or with a spray of redcurrants. Serve with the cream, beaten until it just holds its shape.

INDIAN CARROT DESSERT
Gajjar halva

2¼ lb carrots
4½ cups rich milk
8 whole cardamom pods
1 cup ghee (clarified butter, see page 36)
sugar
2 tbsp seedless white raisins
2 tbsp flaked or slivered almonds
1¼ cups whipping cream, lightly beaten
Decorate with:
few small mint leaves

Trim the carrots and peel with a potato peeler. Grate finely in the food processor or with a grater. Place in a large, heavy-bottomed saucepan or fireproof casserole dish and add the milk. Use the tip of a small knife to make a small slit in the outer covering of the cardamom pods and add to the carrots and milk; they will flavor the mixture subtly without releasing their seeds. Bring slowly to a boil over moderate heat. Reduce the heat and simmer gently for 30 minutes or until the milk has disappeared, mainly through absorption by the carrots, partly through evaporation. Stir in all but 1½ tbsp of the ghee and keep stirring while cooking over very low heat until the mixture turns a russet color. Stir in sugar to taste; mix well over the heat.
Toast the almonds with the seedless white raisins in a nonstick skillet without any fat for 1–2 minutes while stirring, then mix into the carrot mixture for 2 minutes. Remove from the heat and allow to cool slightly. Serve warm or at room temperature in individual small bowls, with cream beaten until it just holds its shape. Decorate each serving with a couple of mint leaves. This is a very filling dessert, best served in small portions. Serves 8.

———— • ————

SWEET POTATO CREAM INDIAN STYLE

4½ cups milk
3 cardamom pods
½ lb sweet potatoes
½ cup sugar
saffron threads
6 tbsp flaked or slivered almonds (toasted)
Decorate and flavor with:
3 tbsp finely chopped, unsalted pistachio nuts

Heat the milk slowly to boiling point in a large, heavy-bottomed saucepan or fireproof casserole dish. Slit open the thick, dry skin of the cardamom pods, take out all the seeds, and grind them finely, using a pestle and mortar or electric grinder. Set aside.

Peel the sweet potatoes, cut into small pieces, and process in the food processor until finely grated. Stir into the boiling hot milk and cook over low heat, stirring frequently, for about 15 minutes. Add the sugar and stir for another 2 minutes to give the sugar time to dissolve completely. Remove from the heat and allow to cool slightly.

Place a generous pinch of saffron threads in a cup, add 3 tbsp boiling water, and stir with a teaspoon until they dissolve. Stir into the tepid potato mixture, together with the ground cardamom seeds and the almonds.

Spoon into small bowls or coupes, cover with plastic wrap and chill in the refrigerator for 2 hours or more before serving. Just before serving, sprinkle with chopped pistachio nuts.

— • —

CARNIVAL FRITTERS

1 large potato weighing about 9 oz
1 cake fresh compressed yeast or ½ envelope active dry yeast
4½ cups unbleached white flour
1 egg
1 ripe juicy orange
¼ cup dark Jamaica rum
light olive oil
sugar for sprinkling
salt

Peel the potato, cut into fairly small pieces, and boil until tender; drain very thoroughly. Put through a potato ricer twice or push through a very fine sieve. Place in a large mixing bowl. Dissolve the yeast in about ¼ cup lukewarm water with ½ tsp sugar. If dried yeast is used, leave to stand for a few minutes until there is a layer of foam on the surface. Sift the flour onto the potato in the bowl with a pinch of salt and stir well, moistening with the lightly beaten egg, the juice and finely grated rind of the orange, the rum, and the gently stirred yeast mixture. Add a little more lukewarm water if necessary. The dough should be quite firm and smooth. Shape it into a ball, place in a bowl and cover with a damp cloth. Leave to rise in a warm place for about 1 hour or until it has doubled in volume.

Break off pieces of the dough and roll into long, even sausages about ½ in thick; cut these into 10-in lengths and join the ends, pressing them together firmly. Heat plenty of oil in a deep-fryer to 350°F and fry the rings, 2 or 3 at a time, until they are golden brown all over. Drain well and keep hot on paper towels while frying the rest. Sprinkle liberally with sugar and serve while still very hot.

·FRESH AND DRIED LEGUMES·

PAKORAS

p. 209

Preparation: 1 hour
Cooking time: 20 minutes
Difficulty: easy
Snack/Appetizer

FRIED STUFFED PURIS

Urd dhal puri

p. 209

*Preparation: 45 minutes + 4 hours
soaking time*
Cooking time: 25 minutes
Difficulty: fairly easy
Snack/Appetizer

VEGETABLE SAMOSAS

p. 210

Preparation: 1 hour 30 minutes
Cooking time: ½–2 minutes
Difficulty: fairly easy
Snack/Appetizer

BAKED TOMATOES WITH FAVA BEAN MOUSSE

p. 211

Preparation: 30 minutes
Cooking time: 25 minutes
Difficulty: easy
Appetizer

HUMMUS

p. 212

*Preparation: 15 minutes + 12 hours
soaking time*
Cooking time: 3 hours
Difficulty: very easy
Appetizer

BORLOTTI BEAN AND ENDIVE SALAD WITH HORSERADISH DRESSING

p. 212

Preparation: 10 minutes
Difficulty: very easy
Appetizer or accompaniment

SPLIT GREEN PEA SOUP

p. 213

*Preparation: 20 minutes + 2 hours 30
minutes soaking time*
Cooking time: 1 hour
Difficulty: easy
First course

INDIAN PEA SOUP

Hara shorba

p. 213

Preparation: 25 minutes
Cooking time: 45 minutes
Difficulty: very easy
First course

PEA AND ASPARAGUS SOUP

p. 214

Preparation: 15 minutes
Cooking time: 20 minutes
Difficulty: very easy
First course

VENETIAN RICE AND PEA SOUP

Risi e bisi

p. 214

Preparation: 25 minutes
Cooking time: 30 minutes
Difficulty: easy
First course

HARICOT BEAN SOUP

p. 215

*Preparation: 15 minutes + 12 hours
soaking time*
Cooking time: 1 hour 30 minutes
Difficulty: easy
First course

LIMA BEAN AND PUMPKIN SOUP

p. 216

Preparation: 20 minutes
Cooking time: 1 hour
Difficulty: easy
First course

*Borlotti bean and
endive salad with
horseradish dressing*

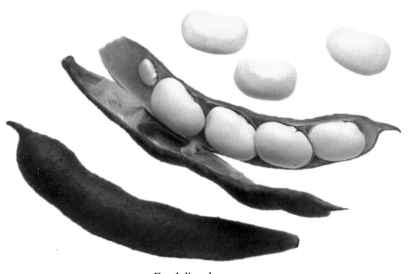

Green beans (*haricots verts*)
Phaseolus vulgaris

Fresh lima beans
Phaseolus multiflorus

PULSES

Cannellini white kidney beans
Phaseolus vulgaris

Navy or pea beans
Phaseolus vulgaris

Borlotti or cranberry beans
Phaseolus vulgaris

Black beans

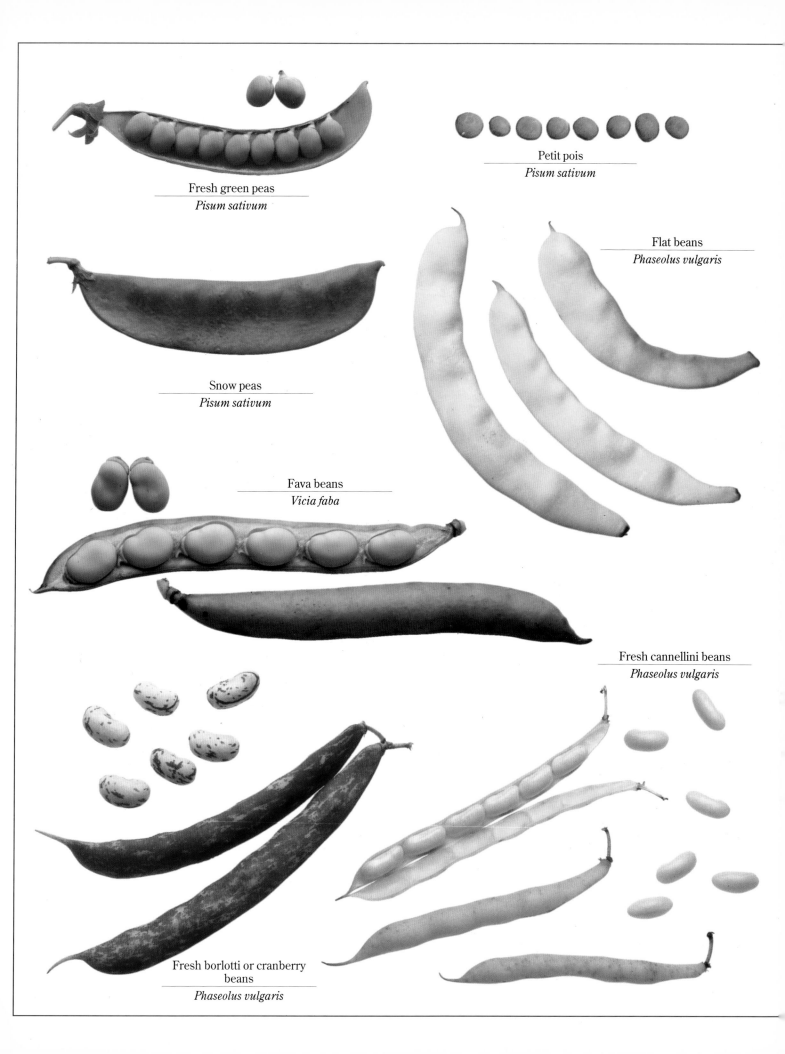

Fresh green peas
Pisum sativum

Petit pois
Pisum sativum

Flat beans
Phaseolus vulgaris

Snow peas
Pisum sativum

Fava beans
Vicia faba

Fresh cannellini beans
Phaseolus vulgaris

Fresh borlotti or cranberry
beans
Phaseolus vulgaris

MINESTRONE

p. 216

Preparation: 1 hour
Cooking time: 3 hours 30 minutes
Difficulty: easy
First course

MISO AND TOFU SOUP

Tofu no misoshiru

p. 00

Preparation: 5 minutes
Cooking time: 10 minutes
Difficulty: very easy
First course

CHICK PEA AND SPINACH SOUP

p. 214

Preparation: 15 minutes + 12 hours soaking time
Cooking time: 3 hours
Difficulty: very easy
First course

BEAN AND PUMPKIN RISOTTO

p. 218

Preparation: 40 minutes +12 hours soaking time
Cooking time: 3 hours
Difficulty: easy
First course

SPRINGTIME RISOTTO

p. 218

Preparation: 30 minutes
Cooking time: 20 minutes
Difficulty: easy
First course

PASTA WITH PEAS AND PARMA HAM

p. 219

Preparation: 10 minutes
Cooking time: 40 minutes
Difficulty: very easy
First course

PASTA AND BEANS WITH ARUGULA

p. 219

Preparation: 15 minutes
Cooking time: 3 hours 30 minutes
Difficulty: very easy
Lunch/Supper dish

BEANS CREOLE WITH RICE AND FRIED BANANAS

p. 220

Preparation: 45 minutes
Cooking time: 2 hours 30 minutes
Difficulty: easy
Lunch/Supper dish

COUNTRY LENTIL SOUP

p. 221

Preparation: 15 minutes + 1 hour soaking time
Cooking time: 1 hour 15 minutes
Difficulty: very easy
Lunch/Supper dish

TUSCAN CHICK PEA SOUP

p. 221

Preparation: 20 minutes + 12 hours soaking time
Cooking time: 2 hours 30 minutes
Difficulty: easy
Lunch/Supper dish

PEAS AND FRESH CURD CHEESE INDIAN STYLE

Panir mater

p. 222

Preparation: 25 minutes + 2 hours standing time
Cooking time: 30 minutes
Difficulty: easy
Main course

MILANESE EGGS AND PEAS

p. 223

Preparation: 10 minutes
Cooking time: 35 minutes
Difficulty: very easy
Main course

GREEN PEA MOLDS WITH EGGPLANT MOUSSELINE

p. 223

Preparation: 40 minutes
Cooking time: 60 minutes
Difficulty: easy
Main course

CASSEROLE OF PEAS AND BEANS WITH ARTICHOKES

p. 224

Preparation: 35 minutes
Cooking time: 15 minutes
Difficulty: very easy
Main course

BRAISED GREEN PEAS, ARTICHOKES, AND LETTUCE

p. 225

Preparation: 30 minutes
Cooking time: 1 hour 20 minutes
Difficulty: easy
Main course

SNOW PEAS AND SHRIMP CANTONESE

p. 225

Preparation: 5 minutes + 30 minutes marinating time
Cooking time: 7 minutes
Difficulty: easy
Main course

BEANS AND EGGS MEXICAN STYLE

Frijoles con huevos rancheros a la mexicana

p. 226

Preparation: 30 minutes
Cooking time: 3 hours
Difficulty: very easy
Main course

WARM TUSCAN BEAN SALAD

p. 226

Preparation: 10 minutes
Cooking time: 2 hours
Difficulty: very easy
Main course

STRING BEAN AND POTATO PIE

p. 227

Preparation: 30 minutes
Cooking time: 45 minutes
Difficulty: easy
Main course

CANNELLINI BEANS WITH HAM AND TOMATO SAUCE

p. 228

Preparation: 20 minutes
Cooking time: 2 hours 30 minutes
Difficulty: easy
Main course

Baked tomatoes with fava bean mousse

NORTH AFRICAN VEGETABLE CASSEROLE

p. 228

Preparation: 45 minutes + 12 hours soaking time
Cooking time: 3 hours 30 minutes
Difficulty: easy
Main course

GREEK CHICKPEA CROQUETTES

Pittarúdia

p. 229

Preparation: 20 minutes + 12 hours soaking time
Cooking time: 25 minutes
Difficulty: easy
Main course

MUNG DHAL FRITTERS

p. 229

Preparation: 30 minutes + 12 hours soaking time
Cooking time: 25 minutes
Difficulty: easy
Main course

RED LENTILS WITH RICE INDIAN STYLE

p. 230

Preparation: 30 minutes + 20 minutes soaking time
Cooking time: 30 minutes
Difficulty: easy
Main course

BORLOTTI BEAN PURÉE WITH CREAM SAUCE

p. 231

Preparation: 10 minutes
Cooking time: 2 hours 30 minutes
Difficulty: easy
Main course or accompaniment

MEXICAN BEANS WITH GARLIC AND CORIANDER

p. 231

Preparation: 15 minutes
Cooking time: 2 hours 30 minutes
Difficulty: very easy
Main course or accompaniment

FAVA BEANS WITH CREAM

p. 232

Preparation: 20 minutes
Cooking time: 20 minutes
Difficulty: very easy
Main course or accompaniment

JAPANESE STIR-FRIED RICE WITH GREEN PEAS AND MIXED VEGETABLES

Yakimeshi

p. 232

Preparation: 30 minutes
Cooking time: 6 minutes (with ready-cooked rice)
Difficulty: easy
Accompaniment or appetizer

Warm Tuscan bean salad

207

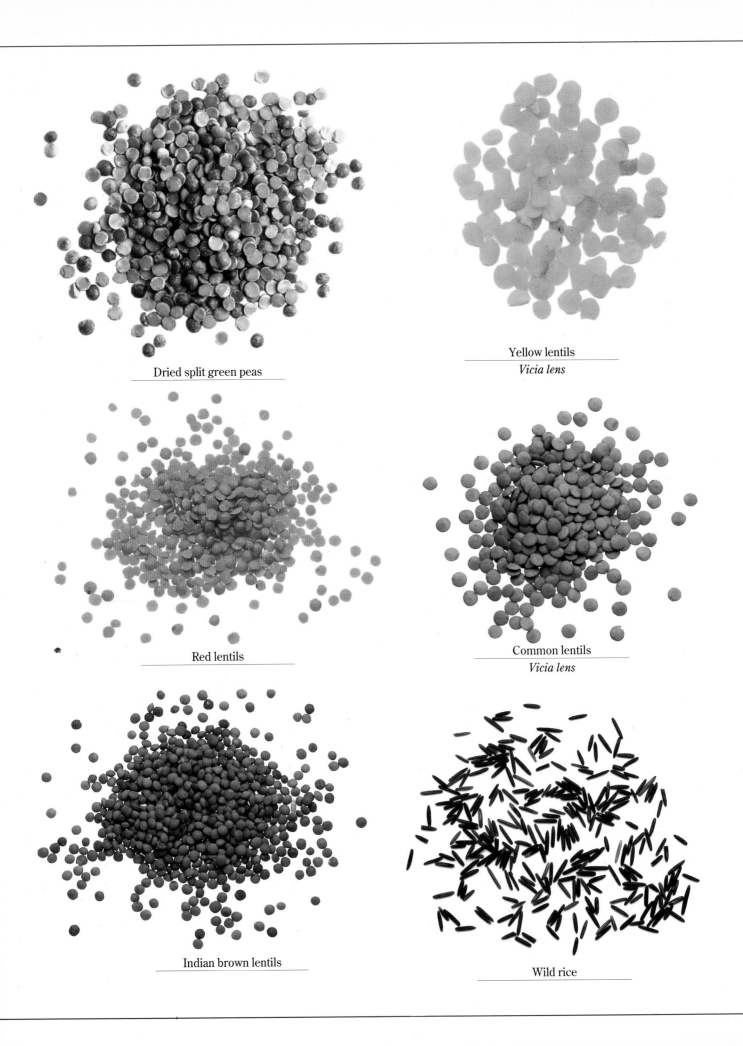

Dried split green peas

Yellow lentils
Vicia lens

Red lentils

Common lentils
Vicia lens

Indian brown lentils

Wild rice

Black-eyed beans

Dolicos melanophtalmus

Red kidney beans

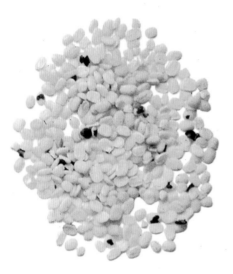

Dhal or split urd beans

Lima beans

Phaseolus multiflorus

Adzuki beans

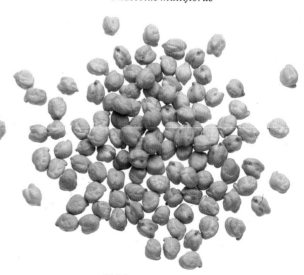

Chickpeas (garbanzos)

Cicer arietinum

Split green pea soup

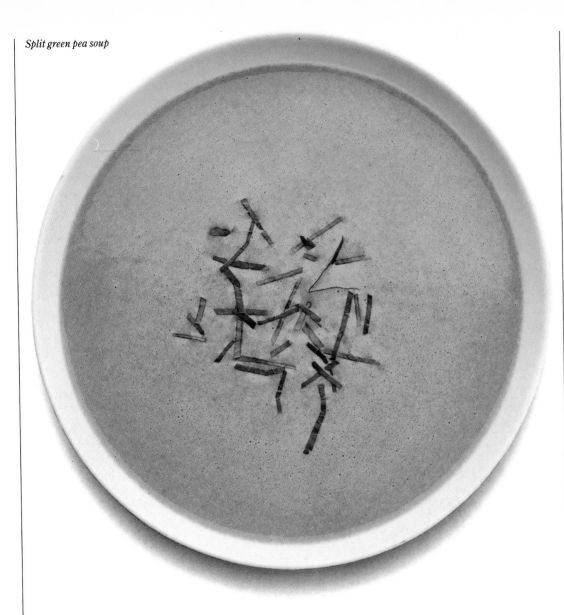

PURÉE OF PEAS

p. 233

Preparation: 20 minutes
Cooking time: 25 minutes
Difficulty: very easy
Accompaniment

CANTONESE STIR-FRIED RICE AND PEAS

p. 233

Preparation: 15 minutes
Cooking time: 6 minutes (with ready-cooked rice)
Difficulty: easy
Accompaniment

PEA AND WILD RICE TIMBALES

p. 234

Preparation: 5 minutes
Cooking time: 45 minutes
Difficulty: very easy
Accompaniment

PEAS WITH BACON AND BASIL

p. 234

Preparation: 10 minutes
Cooking time: 35 minutes
Difficulty: very easy
Accompaniment

STRING BEANS MAÎTRE D'HOTEL

p. 235

Preparation: 25 minutes
Cooking time: 15–18 minutes
Difficulty: easy
Accompaniment

BEAN AND PEANUT PILAF

p. 235

Preparation: 30 minutes + 12 hours soaking time
Cooking time: 1 hour 30 minutes
Difficulty: easy
Accompaniment

STRING BEANS WITH TOMATOES AND SAGE

p. 236

Preparation: 20 minutes
Cooking time: 35 minutes
Difficulty: very easy
Accompaniment

STRING BEANS WITH CRISPY BACON

p. 236

Preparation: 20 minutes
Cooking time: 25 minutes
Difficulty: easy
Accompaniment

STRING BEAN AND POTATO SALAD WITH MINT

p. 237

Preparation: 20 minutes + 1 hour cooling time
Cooking time: 40 minutes
Difficulty: very easy
Accompaniment

STIR-FRIED SNOW PEAS

p. 237

Preparation: 25 minutes
Cooking time: 8 minutes
Difficulty: easy
Accompaniment

SNOW PEAS IN CREAM AND BASIL SAUCE

p. 237

Preparation: 25 minutes
Cooking time: 8 minutes
Difficulty: easy
Accompaniment

SNOW PEAS WITH SHALLOTS AND BACON

p. 238

Preparation: 25 minutes
Cooking time: 8 minutes
Difficulty: easy
Accompaniment

GREEN BEANS WITH TOMATO SAUCE

p. 238

Preparation: 25 minutes
Cooking time: 25 minutes
Difficulty: very easy
Accompaniment

SPICED LENTILS AND SPINACH

p. 238

Preparation: 30 minutes
Cooking time: 1 hour 30 minutes
Difficulty: very easy
Accompaniment

BRAISED LENTILS

p. 239

Preparation: 15 minutes
Cooking time: 1 hour 15 minutes
Difficulty: very easy
Accompaniment

JAPANESE SWEET RED BEAN PASTE

Yude azuki

p. 239

Preparation: 10 minutes + 12 hours soaking time
Cooking time: 1 hour 45 minutes
Difficulty: very easy
Dessert

ADZUKI JELLY

Mizuyokan

p. 240

Preparation: 20 minutes + 12 hours soaking time + 3 hours chilling time
Cooking time: 1 hour 45 minutes
Difficulty: very easy
Dessert

PAKORAS

For the chickpea flour batter:

1¾ cups chickpea flour

½ tsp baking powder

½ tsp cayenne pepper

½ tsp ground coriander seeds

1 tsp ground cumin seeds

salt

For the coconut chutney:

1 cup grated fresh coconut flesh

½ clove garlic

3 tbsp grated fresh ginger

1 green chili pepper, finely chopped

½ cup chopped coriander leaves

¼ cup natural yoghurt

1 tbsp fresh lime juice

½ tsp black mustard seeds

1½ tbsp peanut or sunflower oil

salt

For the hot ketchup:

1 cup tomato ketchup

½–1 tsp pili-pili hot relish (see page 113)

For the fritters:

1 large onion

1 sweet potato

2 oz spinach leaves

1 medium-sized eggplant

12 okra or 3 green peppers

sunflower oil

Ask for chickpea flour at your nearest Indian grocer's shop. It is pale, creamy yellow in color. Make the batter: sift the flour, baking powder, cayenne pepper, coriander, cumin, and a pinch of salt together in a large mixing bowl. Gradually add just enough cold water to make a fairly thick coating batter, stirring continuously with a wooden spoon. Add a little more salt, cover, and leave to stand.

Make the coconut chutney: place the grated coconut flesh in the blender with ¼ cup cold water, the garlic, ginger, chili pepper, and coriander and process at high speed until smooth, adding a little more water if necessary. Spoon into a serving bowl and mix in the yoghurt, lime juice, and a pinch of salt. Heat the oil in a nonstick skillet and fry the mustard seeds for a few seconds, stirring them as they cook; as soon as they start to jump about in the skillet, remove from the heat and stir them into the coconut mixture; add more salt if needed and place in the refrigerator to chill until just before serving.

Mix the ketchup with the pili-pili relish, spoon into 4 very small bowls, and set aside.

Prepare the vegetables: peel the onion and the sweet potato; wash the spinach leaves and drain well. Cut the onion into thin rings and the eggplant, okra or peppers, and sweet potato into thin round slices. Heat plenty of sunflower oil in a deep-fryer to 350°F or use a wok. Dip the vegetable pieces one by one in the batter and then add to the very hot oil. Fry in small batches until the batter coating has puffed up and is crisp and golden brown. Remove and place on paper towels to drain while you fry the remaining batches. Serve at once, giving each person a small bowl of the coconut chutney and the hot tomato ketchup as dips.

———— • ————

FRIED STUFFED PURIS
Urd dhal puri

1 cup split black gram lentils (urd dhal or dhuli urd, see method)

1 cup ata (chapati) flour (see method) or all-purpose flour

⅓ cup ghee (clarified butter, see page 36)

1 green chili pepper, finely chopped

sunflower oil

salt

209

For the garam masala:
10 black peppercorns
2 tsp coriander seeds
2 tsp fennel seeds
2 tsp cumin seeds

The above quantities make 5 stuffed puris. You can buy black gram lentils (urd dhal) from Indian grocers and health food stores that stock a good range of beans. Check that the split gram is white, not yellow. Rinse in a sieve under running cold water; drain and then soak in a bowl of cold water for 4 hours.

Sift the ata flour (a whole wheat flour sold by Indian grocers and by many specialist food stores) into a large mixing bowl, add ½ tsp salt and 3 tbsp hot ghee, and stir in, gradually adding just enough cold water to make a firm dough. Knead well by hand on a pastry board or work surface for 10 minutes then shape into a ball, wrap in plastic wrap, and leave to rest for at least 30 minutes.

Make the garam masala: heat a nonstick skillet, add the peppercorns, coriander, fennel, and cumin seeds, and roast for 30 seconds over moderate heat, stirring continuously; they will start to release a marvelous spicy aroma. Remove from the heat; grind the seeds finely in a spice mill or use a mortar and pestle; put through a fine sieve to trap the pieces of coriander skins and discard these.

Once the lentils have finished soaking, place them in the food processor with a little cold water and process briefly until they form a coarse, thick purée (if they are too smooth, the filling will not have an interesting texture). Heat the remaining ghee in a large skillet and fry the chopped chili pepper gently over low heat for 2–3 minutes, stirring continuously; add the lentil purée and ½ tsp salt. Cook, stirring continuously, over slightly higher heat for a few minutes, until the purée darkens and thickens further. Remove from the heat and leave to cool before stirring in the garam masala.

Start making the puris about 30 minutes before you plan to serve them. Knead the dough for 2–3 minutes and divide into 10 pieces of even size; shape these into balls between your palms and put them in

a large bowl; remove 2 and cover the bowl with plastic wrap to prevent the dough from drying out. Flatten these first 2 balls with your palms and then roll out on a lightly floured pastry board with a floured rolling pin to form 2 thin, matching disks 3 in in diameter. Have a cup of cold water standing ready. Place 3 tbsp of the lentil filling in the center of one disk, dip your finger into the cold water, moisten all round the edge of the disk, then cover neatly with the other disk, pressing the edges together to achieve a good seal. Do exactly the same with the remaining dough and filling.

Heat plenty of oil in a very large skillet or in a deep-fryer and when hot but not smoking, place 3 of the puris in it and fry until they puff up and are pale golden brown; keep spooning the hot oil over the exposed surfaces of the puris to cook them if they are not immersed in oil. Take these puris out of the oil with a slotted spoon and keep hot while you fry the remaining two.

———— • ————

VEGETABLE SAMOSAS

For the pastry dough (makes 16):
2¼ cups all-purpose flour
½ cup ghee (clarified butter, see page 36)
¼ cup hot milk
sunflower oil
salt
For the filling:
1 lb boiled potatoes
½ cup ghee (clarified butter, see page 36)
1¼ cup minced onion
1½ cups thawed frozen peas
1 small piece fresh ginger, grated
1 green chili pepper, finely chopped
3½ tbsp chopped coriander leaves

1 tsp coriander seeds, ground and sifted
1 tsp ground roasted cumin seeds
1 tsp mild or medium curry powder
½ tsp pili-pili hot relish (see page 113) or pinch chili powder
¼ cup fresh lime juice
salt
For the coriander chutney:
1 large bunch coriander
1 green chili pepper, finely chopped
3 tbsp fresh lime juice
½ tsp ground cumin seeds
salt and pepper

Make the pastry: sift the flour into a large mixing bowl, add ⅓ cup of the ghee, and rub in with your fingertips until the mixture resembles breadcrumbs. Gradually work in as much of the hot milk as is needed to make a firm dough. Knead for a few minutes until the dough is smooth and elastic. Shape into a ball, brush the surface lightly with oil all over, and enclose in plastic wrap to prevent it from drying out. Leave to rest for 45 minutes.

Peel and dice the boiled and cooled potatoes. Heat 6 tbsp of the ghee in a nonstick skillet and sweat the onion for 15 minutes, stirring frequently. When it is tender but not browned, add the peas, grated ginger, chili pepper, chopped coriander leaves, and ½ cup water. Stir, adding a pinch of salt, cover, and simmer gently for about 20 minutes or until the peas are very tender. Add the diced potato, the ground coriander and cumin, the curry powder, pili-pili relish or chili powder, and fresh lime juice to taste. Stir and mix all these ingredients over low heat briefly then remove from the heat.

Make the coriander chutney: detach the leaves and place in the food processor. Remove the stalk, seeds, and inner pale parts of the chili pepper, and add to the processor with the lime juice, 1 tsp salt, ground cumin, and some freshly ground black pepper; add 1 tbsp cold water and process to a thick paste.

Assemble the samosas: divide the pastry dough into 8 equal parts and roll into balls between your palms. Take up one of them and place the others in a large bowl, covered with plastic wrap to prevent them from drying out. Flatten the ball of dough between your palms and then roll out on a lightly floured pastry board with a floured rolling pin to make a thin disk of pastry about 7 in in diameter. Cut this disk in half, moisten along the cut edge, and fold the semicircle in half, pressing the moistened edges together: you will now have a triangular pocket. Hold this pastry case in your left hand and use a teaspoon to fill it, packing the filling gently; when it is full, moisten along the inside of the top edges and pinch them together to seal. Place the samosa flat on the pastry board and press along the sealed edges with the tines of a fork to complete the sealing process and give the edges a decorative, ribbed effect. Place on a chopping board and cover with plastic wrap or place in a large, shallow plastic food container with a lid to prevent the pastry from drying out while you prepare 15 more samosas in the same way.

Heat plenty of sunflower oil in a deep-fryer to 350°F or use a wok or deep, heavy-bottomed saucepan. Fry the samosas in batches until the pastry is golden brown all over (about 1½–2 minutes). Transfer to paper towels to drain and serve warm, passing round the coriander chutney.

—— • ——

BAKED TOMATOES WITH FAVA BEAN MOUSSE

1 lb young, tender fava beans
2 lb medium-sized ripe tomatoes
1 egg
½ cup freshly grated Parmesan cheese
½ clove garlic, minced
1 canned anchovy fillet, minced
3 tbsp extra-virgin olive oil

fine breadcrumbs
salt and pepper

Bring a saucepan of lightly salted water to a boil ready to cook the fava beans. Preheat the oven to 400°F. Rinse and dry the tomatoes; slice off and discard the top of each one. Remove the seeds and the dividing sections of flesh as well as the central part of the tomatoes. Lightly oil a cookie tray and place the tomatoes in it. Sprinkle a small pinch of salt inside each tomato.

When the water in the saucepan comes to a boil, add the fava beans and cook them until very tender; drain them and remove their seed cases when cool enough to handle. This will take some time; do not worry if you break up their contents. Discard the seed cases. Place the bright green kidney-shaped beans in the food processor and add the lightly beaten egg, a little salt and pepper, the Parmesan cheese, and the garlic. Process to a very thick, smooth purée. Transfer to a large bowl. Cook the minced anchovy fillet in 3 tbsp oil over low heat, then stir well into the fava bean purée. Add sufficient fine fresh breadcrumbs to thicken the purée further: it should hold its shape firmly. Spoon it in batches into a pastry bag with a medium-sized, fluted tip. Pipe plenty of the filling into each tomato (see illustration on page 207). Bake for 7–10 minutes.

———— • ————

HUMMUS

1 lb chickpeas
¾–1 cup tahini (sesame seed paste)
4 cloves garlic, minced
½ cup fresh lemon juice
cayenne pepper
extra-virgin olive oil

salt
Serve with:
hot pita bread

Rinse the chickpeas thoroughly in a sieve under running cold water, drain, place them in a large bowl, and add sufficient cold water to cover them amply. Leave to soak for at least 12 hours. When they have finished soaking, drain them, place in a large, heavy-bottomed saucepan or flameproof casserole dish, pour in enough cold water to come about 2 in above them, and bring slowly to a boil. Turn down the heat to very low, partially cover the saucepan so as to allow the steam to escape, and simmer gently for 3 hours, by which time the chickpeas should be tender. Do not add salt until they have cooked for about 2¾ hours. Drain, reserving the cooking liquid for use in soups, etc. as it is very nutritious. Reserve about 6 whole chickpeas with which to decorate the finished hummus; purée all the rest through a vegetable mill fitted with a fine-gage disk into a large mixing bowl.

Using a whisk, beat in the tahini, garlic, and lemon juice to taste. Beat in a little of the reserved cooking liquid. Add salt to taste and transfer to a bowl.

Sprinkle the surface with a very small pinch of cayenne pepper and drizzle a little olive oil over the surface.

———— • ————

BORLOTTI BEAN AND ENDIVE SALAD WITH HORSERADISH DRESSING

1 lb cooked borlotti or cranberry beans, fresh or canned
1 large head red Treviso endive or radicchio
1½ tbsp grated horseradish mixed with 1½ tbsp wine vinegar or cider vinegar

---

½ cup olive oil

---

salt and freshly ground white or black pepper

The slightly bitter taste of the endive contrasts well with the bland, almost sweet taste of the beans and the color combination is most attractive. Separate the leaves and use as containers for the beans or cut the heads and bases into fairly small pieces with a very sharp stainless steel knife; rinse and drain well. Drain the beans and mix with the endive in a large salad bowl or arrange as shown in the illustration on page 201.

Mix the horseradish and vinegar with ½ tsp salt and plenty of freshly ground pepper, then gradually beat in the olive oil, adding a very little at a time, so that it emulsifies. Sprinkle this dressing over the salad.

— • —

SPLIT GREEN PEA SOUP

---

1 lb dried split green peas

---

1 small onion

---

1 medium-sized carrot

---

1 green celery stalk

---

1 leek

---

2 oz fairly lean bacon slices

---

bouquet garni: 1 sprig thyme, 1 bay leaf, 4 sprigs parsley

---

5 cups chicken stock (see page 37)

¼ cup fresh chervil or watercress leaves

½ cup crème fraîche or heavy cream

---

2 tbsp unsalted butter

---

3 tbsp olive oil

---

salt and pepper

Soak the peas in a large bowl of cold water for 2½ hours, drain well in a sieve, rinse well under running cold water, drain again, and place in a large saucepan or kettle with 3 cups water.

While the peas are soaking, coarsely chop the peeled onion and carrot together with the celery, leek, and bacon. Sweat all these chopped ingredients in 3 tbsp oil for 15 minutes over gentle heat, stirring frequently. Add this mixture to the presoaked split peas and water in the saucepan together with the bouquet garni. Bring to a boil, then cover with a lid, leaving only a small gap for the steam to escape, reduce the heat to low, and simmer for 40 minutes, by which time the peas should be very tender. Discard the bouquet garni; put through a vegetable mill and return to the saucepan. Add the hot stock to this purée, until the consistency you prefer is achieved. Season with salt and pepper to taste and bring slowly to a boil.

Rinse the chervil or watercress; blanch for a few seconds only in boiling salted water, drain, refresh under running cold water, squeeze out excess moisture, and reserve. Stir the cream into the soup and remove from the heat as soon as it returns to a gentle boil. Add a little more salt if wished, stir in the chervil or watercress and the butter.

— • —

INDIAN PEA SOUP
Hara shorba

1¼ cups frozen peas

---

2 potatoes, peeled and diced

---

1 onion, diced

---

2 oz spinach leaves

2½ tsp cumin seeds

½ tsp coriander seeds

---

1 piece fresh ginger, 1 in thick

5½ cups chicken stock (see page 37)

---

3 tbsp coriander leaves

---

6 mint leaves

½ fresh chili pepper, finely chopped

213

½ cup light cream
3 tbsp fresh lime juice
pinch ground cinnamon
salt and pepper
Garnish with:
coriander leaves

Clean and prepare all the vegetables. Grind the cumin seeds in a spice mill or pound to a powder with a mortar and pestle. Take out ½ tsp of the ground cumin and set aside; add the coriander seeds to the cumin and grind them too. Sift in a fine sieve to separate the coriander seed cases. Place the ½ tsp of ground cumin in a small, nonstick skillet and roast gently over low heat while stirring; as soon as the spice releases a full aroma, remove from the heat and reserve.

Place the potatoes, onion, ginger, and the mixed cumin and coriander in a large saucepan with the chicken stock. Bring to a boil, cover, and simmer over fairly low heat for 20 minutes. Add the peas, the well washed spinach, fresh coriander leaves, mint, the reserved roasted cumin, the chili pepper, and 1 tsp salt. Replace the lid and simmer for a further 25 minutes. Remove and discard the ginger. Beat with an electric beater to reduce to a purée or put through a vegetable mill and return to the saucepan. Stir in the cream, add a little more salt, lime juice to taste, a pinch of cinnamon, and pepper. Garnish with the coriander leaves.

———•———

PEA AND ASPARAGUS SOUP

5 cups chicken stock (see page 37)
1 large bunch asparagus
2 scallions
2 medium-sized new potatoes

1 small Boston lettuce
1 cup fresh or frozen peas
1 cup long-grain rice
½ cup freshly grated Parmesan cheese
3 tbsp chopped parsley
1½ tbsp olive oil
2½ tbsp unsalted butter
salt and pepper

Bring the chicken stock slowly to a boil. Pepare the vegetables: use the tender, top half of the well washed asparagus spears and cut them into pieces about ¾ in long. Remove and discard the roots and outer layer of the scallions and slice into thin rings. Peel the potatoes and dice. Use only the heart of the lettuce for this soup and shred it.

Heat 1½ tbsp of the butter with 1½ tbsp olive oil over gentle heat in a heavy-bottomed saucepan or deep, flameproof casserole dish and sweat the scallions until tender. Add the peas and lettuce and cook, stirring, for 1 minute. Pour in the boiling hot stock, stir, and allow to return to a boil before sprinkling in the rice. Add the potatoes and asparagus. Stir briefly, partially cover, reduce the heat to moderately low, and simmer for 14–15 minutes, or until the rice is tender but still fairly firm. Remove from the heat, add the Parmesan cheese, the solid remaining butter, and the chopped parsley. Stir until the butter has melted and blended with the soup. Add more salt if desired and some freshly ground pepper to taste.

———•———

VENETIAN RICE AND PEA SOUP
Risi e bisi

5 cups boiling hot vegetable stock (see page 38)
¼ cup finely chopped bacon

1 scallion, peeled and sliced into rings
4 tbsp finely chopped parsley
1¼ cups peas, fresh or frozen
1 cup long-grain rice
¼ cup butter
3 tbsp extra-virgin olive oil
½ cup freshly grated Parmesan cheese
salt and pepper

This dish is halfway between a soup and a risotto. Bring the vegetable stock slowly to a boil. While it is heating, fry the bacon, scallion, and the parsley in 2 tbsp of the butter and the olive oil in a large, heavy-bottomed saucepan or deep skillet over low heat, stirring frequently; when the onion is just tender but not at all browned, add the peas, turn up the heat slightly, and cook for 1 minute, stirring.

Pour in about 1 cup of the boiling stock, turn the heat down to low, cover, and simmer gently for 10 minutes. Add the remaining stock and when it comes to a boil, sprinkle in the rice; cook, uncovered, over fairly high heat for about 14 minutes or until the rice is tender but still has a little "bite" left to it, stirring now and then. There should be plenty of slightly thickened liquid left when the rice is done.

Remove from the heat and stir in the remaining, solid butter and the Parmesan cheese. Stir gently but thoroughly. Add salt and pepper to taste, stir once more, cover, and leave to stand for 5 minutes. Serve in heated individual soup bowls or deep plates.

— • —

HARICOT BEAN SOUP

1 lb dried white navy beans, soaked
½ tsp baking soda
extra-virgin olive oil

salt and freshly ground pepper
For the sauce:
3 tbsp extra-virgin olive oil
1 large clove garlic, minced
8 medium-sized ripe tomatoes, peeled, seeded, and chopped
½–1 dried chili pepper (optional)
pinch oregano
salt
Garnish with:
4 sprigs basil
Serve with:
garlic bread

Start preparation the day before you plan to serve this dish. Rinse the beans in a sieve and place in a bowl with enough cold water to amply cover them; add ½ tsp baking soda. Leave the beans to soak overnight or for more than 12 hours, then drain, place in a saucepan with enough water to cover them completely, and bring slowly to a boil over moderate heat. Cover, leaving a gap so that the steam can escape. Simmer very gently for about 1½ hours or until the beans are very tender. Remove the scum from the surface of the water at intervals. Add a large pinch of salt when the beans are nearly done. While the beans are cooking, make the sauce. Heat the olive oil in a saucepan over moderately low heat and add the whole minced garlic clove. After about 1 minute, when the garlic is a pale golden brown, add the chopped tomatoes. Turn up the heat, add the crumbled chili pepper and a pinch of oregano, and cook briskly, stirring, for 2–3 minutes. Set aside. When the beans are done, put them, with their cooking liquid, through a vegetable mill fitted with a fine-gage disk. Ladle the resulting purée into individual heated bowls, spoon some of the tomato mixture into the center of each bowl, drizzle a little olive oil onto each serving, and sprinkle with a generous amount of freshly ground black pepper. Garnish with small basil sprigs and serve with garlic bread.

— • —

LIMA BEAN AND PUMPKIN SOUP

½ cup dried or ¾ cup frozen lima beans

1½ lb fresh pumpkin

2 large potatoes

4½ cups milk

½ cup long-grain rice

1 vegetable bouillon cube

1 tbsp sugar

3 tbsp chopped parsley

salt and white peppercorns

Rinse the beans and soak them in a bowl of cold water overnight or for at least 12 hours. Drain them and place in a saucepan with plenty of cold water; bring slowly to a boil then simmer gently for 45 minutes, adding salt shortly before the end of this cooking time. If you prefer to use frozen beans, move onto the next step.

While the beans are cooking, peel the pumpkin, removing the seeds and filaments; cut the flesh into small pieces. Cut the peeled potatoes into pieces of the same size. Place the pumpkin and potato pieces in a saucepan and add 4 cups water and 1 tsp salt. Bring to a boil, cover, and simmer gently for 40 minutes. Remove from the heat and purée in the blender.

Return the puréed pumpkin to the saucepan, stir in the milk and beans, and place over moderate heat. When the mixture comes to a boil, add the rice and the bouillon cube, stir well, turn down the heat and simmer, uncovered, for 12–13 minutes, stirring at frequent intervals. Test the rice; it should be tender but still firm. Stir in the sugar, a little salt if needed, and plenty of freshly ground white pepper.

Ladle into soup bowls and sprinkle with the parsley.

—— • ——

MINESTRONE

2 cups peas, fresh or frozen

1 cup borlotti or cranberry beans, fresh or canned

6 ripe tomatoes, peeled, seeded, and chopped

3 large potatoes, peeled and left whole

2 tender young zucchini, sliced into rounds

2 stalks green celery, sliced thinly

2 medium-sized carrots, sliced into rounds

¼ green cabbage, cut into large pieces

½ cup chopped bacon slices

¼ cup finely chopped salt pork back fat

1 clove garlic, peeled and finely minced

¼ cup chopped parsley

1 medium-sized onion, peeled and finely minced

2 fresh sage leaves, finely chopped

10 fresh basil leaves, finely chopped

1 cup risotto rice

¾ cup freshly grated Parmesan cheese

extra-virgin olive oil

salt and pepper

Clean and trim all the vegetables. Chop the pork fat, garlic, and parsley together. Shred the bacon. Gently fry the chopped fat, garlic, and parsley in a very large, heavy-bottomed saucepan together with the onion and sage, stirring frequently. When these have cooked for 15–20 minutes and the onion is soft and very lightly browned, add all the other vegetables, including the beans and the basil, but reserving the peas and the cabbage which will be added later. Pour in 1¼ cups cold water, add 2 tsp sea salt, and bring slowly to a boil. Cover, reduce the heat to very low, and simmer gently for 1½ hours, skimming off any scum that rises to the surface. Use a slotted spoon to take the potatoes out of the soup, mash them coarsely in a bowl with a fork or potato masher and return to the saucepan. Add the peas and the cabbage and continue cooking for another 30 minutes. The soup can be prepared 24 hours in advance up to this point, then left to cool, and refrigerated until shortly before serving.

Bring the soup to a brisk boil, uncovered; sprinkle in the rice and stir. Cook, uncovered, over fairly high heat for 13–14 minutes or until the rice is just tender. Add a little more salt if needed and serve the soup warm. Pass round a bowl of grated Parmesan cheese at the table for each person to sprinkle into the soup. Serves 6.

—— • ——

MISO AND TOFU SOUP
Tofu no misoshiru

1½ tbsp dashinomoto *(instant fish stock)*

pinch ajinomoto *(Japanese taste powder) or monosodium glutamate (optional)*

6 tbsp shiro miso *(fermented white soybean paste)*

1 cake fresh tofu

½ leek, white part only

Serve with:

steamed rice (see page 126)

Make the stock in a nonmetallic saucepan (an enameled flameproof casserole dish is best) using 4½ cups water, the instant *dashi*, and a pinch of *ajinomoto*; if you do not want to use this taste enhancer, leave it out. Bring to a boil, then reduce the heat to very low. Mix the *shiro miso* in a bowl with about ½ cup of the hot stock; pour the resulting mixture back into the saucepan or casserole dish through a sieve, rubbing with the back of a wooden spoon. Stir for 2 minutes; the soup must not come to a boil once the *shiro miso* has been added.

Cut the tofu into small cubes, add to the soup, and leave to heat through for 2 minutes; remove from the heat. Ladle the soup into bowls, garnish with the shredded leek, and serve at once, with plain steamed rice. This makes a very light but nourishing supper.

CHICKPEA AND SPINACH SOUP

1 lb dried chickpeas

pinch baking soda

1 cup vegetable stock *(see page 38)*

½ lb spinach

1 small onion

1 clove garlic

1 small carrot

1 celery stalk

3 tbsp all-purpose flour

¼ cup extra-virgin olive oil

salt and black peppercorns

Start preparation a day in advance by soaking the chickpeas in plenty of cold water to which a pinch of baking soda has been added. The next day, drain them in a sieve and rinse well under running cold water; place in a saucepan with the stock. Cover, bring slowly to a boil, then simmer gently for 1½ hours.

When the beans have been cooking for about 1¼ hours, wash the spinach, drain, and shred thinly. Chop the onion, garlic, carrot, and celery finely and fry gently in ¼ cup olive oil in a large, heavy-bottomed saucepan or flameproof casserole dish for 5 minutes, stirring with a wooden spoon. Add the flour and stir continuously for 1 minute as it cooks. Add the shredded spinach leaves, stir, draw aside from the heat, and add about 1 cup of the cooking liquid from the chickpeas, beating continuously to prevent lumps forming as the flour thickens the liquid. Add the contents of this saucepan to the saucepan containing the cooked chickpeas. Stir, cover, and simmer very gently for another hour over very low heat, by which time the chickpeas should be very tender.

Remove three ladlefuls (about 1½ cups) of the beans from the saucepan and reserve. Put the remaining contents of the saucepan, liquid and beans, through a vegetable mill and return to the saucepan. Add the reserved whole beans, season to taste, and serve.

BEAN AND PUMPKIN RISOTTO

1½ cups dried lima beans

2½ lb pumpkin flesh

2 vegetable bouillon cubes

1 small onion, minced

5 tbsp butter

1 lb long-grain rice

½ cup dry vermouth or dry white wine

½ cup milk

½ cup freshly grated Parmesan cheese

2½ tbsp finely chopped parsley

salt and pepper

Soak the beans overnight or for about 12 hours in plenty of cold water. Drain in a sieve and rinse well under running cold water. Spread them out between 2 clean cloths and rub to break and loosen the skins. Discard the skins, then place the beans in a very large, heavy-bottomed saucepan with 11 cups water and bring slowly to a boil; partially cover, allowing the steam to escape, and simmer gently for 1 hour. Peel the skin from the pumpkin, removing the seeds and filaments. Dice the pumpkin flesh. When the beans have simmered for 1 hour, add the diced pumpkin and simmer for another hour, by which time the beans should be very tender. Use a hand-held electric beater to turn the contents of the saucepan into a very thin purée, or put through a vegetable mill fitted with a fine-gage disk. Add the bouillon cubes, stir well, and bring to a gentle boil over low heat, stirring with a wooden spoon. Set aside.

Fry the onion very gently in 2 tbsp of the butter for 10 minutes, stirring now and then; add the rice, cook, stirring for 1½ minutes over slightly higher heat, pour in the vermouth or wine, and continue cooking until it has completely evaporated. Add about 1 cup of the pumpkin and bean purée and continue cooking, uncovered, stirring occasionally and adding more of the purée as the rice cooks and absorbs the moisture. After 10 minutes, add the hot milk and stir. Cook for another 4 minutes, or until the rice is tender but still firm; this risotto should be very moist. Remove from the heat, add salt to taste (the beans will not have been salted), the remaining, solid butter, and the Parmesan cheese, and stir gently into the risotto. Season with freshly ground pepper, sprinkle with the parsley, and serve. Any remaining purée can be used in vegetable soups. Serves 6.

———— • ————

SPRINGTIME RISOTTO

½ cup peas, fresh or frozen

2 oz string beans

1 bunch asparagus

2 tender baby globe artichokes or 4 fresh artichoke bottoms

1 medium-sized potato

1 medium-sized carrot

3 ripe tomatoes, peeled, seeded, and chopped

1 Boston lettuce

1 onion

1 green celery stalk

2 cups risotto rice (e.g. arborio)

¼ cup dry vermouth or dry white wine

6¾ cups vegetable stock (see page 38)

½ cup freshly grated Parmesan cheese

¼ cup unsalted butter

6 tbsp olive oil

salt and pepper

Set the stock to come slowly to a boil while you trim, wash, and peel the vegetables. Cut the string beans into pieces just under ½ in long and the asparagus tips (approximately the top third of each spear) into ¾-in lengths. Chop the artichoke hearts (if very tender varieties or buds are unavailable use artichoke bottoms) into small cubes; do likewise with the potato, carrot, and tomatoes. Shred the lettuce heart into a *chiffonade* (see page 24); chop the onion and celery together finely. Heat 6 tbsp of oil and 2 tbsp of the butter in a wide, heavy-bottomed saucepan or very large skillet and gently fry the chopped onion and celery over low heat, stirring frequently. Add all the other vegetables except the peas and the asparagus. Cook gently, stirring at frequent intervals, for 10 minutes, moistening with a couple of tablespoonfuls of stock whenever the mixture starts to look a little dry. Stir in the peas and the asparagus; sprinkle in the rice.

Turn up the heat to moderately high and cook while stirring to let the rice absorb the fat and flavors in the saucepan. Add the vermouth or wine and cook until it has evaporated. Pour in 1 cup of boiling hot stock and stir. Continue cooking, uncovered, for about 14 minutes, adding more hot stock as it is absorbed by the rice. Stir at intervals. Taste the rice to see whether it is done: it should be tender but still slightly firm to the bite. Stir in the remaining, solid butter and the Parmesan cheese. Season and serve.

PASTA WITH PEAS AND PARMA HAM

1¼ cups peas, fresh or frozen
1 small onion, peeled and finely minced
1 small clove garlic, peeled and finely minced
2 thin slices Parma ham, finely chopped
5 basil leaves, finely chopped
¼ cup olive oil
1¾ cups chicken stock (see page 37)
½ lb short pasta (e.g. penne, macaroni)

½ cup freshly grated Parmesan cheese
salt and black peppercorns

Bring a large saucepan of salted water to a boil. While it is heating, chop the onion finely with the garlic, Parma ham, and basil. Fry these together very gently in the oil in a large, heavy-bottomed saucepan or flameproof casserole dish for 15 minutes, stirring now and then. Add the peas and cook for 5 minutes, still over low heat. Pour in the stock, add a pinch of salt, cover as soon as the liquid has come to a boil, and simmer for about 10 minutes, stirring at intervals, by which time the peas will be cooked and tender.

While the peas are cooking, add the pasta to the boiling salted water and cook until it is only just tender; drain and add to the peas together with some of its cooking water. The finished dish should be very moist. Add a little more salt if necessary and plenty of freshly ground pepper. Remove from the heat, stir in the Parmesan cheese, and leave to stand for 5 minutes before serving.

PASTA AND BEANS WITH ARUGULA

2¼ lb borlotti or cranberry beans, fresh or canned
1 small green celery stalk
1 small onion, quartered
2 cloves garlic
2 vegetable bouillon cubes
¼ lb thin egg noodles, broken into short lengths
extra-virgin olive oil
white wine vinegar
pili-pili hot relish (see page 113) (optional)
salt and black peppercorns
6 heads arugula

Serve with:

multi-grain bread

Place the beans in a large saucepan or casserole dish and add sufficient water to come 2 in above them. Bring to a boil over moderate heat, simmer, uncovered, for 15 minutes, then drain. Return the beans to the saucepan, add 8½ cups water, the celery, onion, and garlic, and reheat to boiling.

Turn the heat down to very low and simmer, partially covered, for about 3 hours, skimming off any scum that rises to the surface frequently. After 2½–2¾ hours of this cooking time, add the bouillon cubes; if you prefer a very plain taste, just add a little salt and some freshly ground pepper. The beans should be tender but should not have started to disintegrate.

Remove and discard the celery stalk; ladle one-third of the beans into a bowl, put the rest through a vegetable mill fitted with a fine-gage disk, and return to the saucepan; add the reserved, whole beans and their liquid. This recipe can be prepared 24 hours in advance up to this point.

Shortly before serving, bring a large saucepan of salted water to a boil and cook the pasta in it until only just tender. Reheat the beans slowly over low heat while the pasta is cooking. Add the drained pasta and then stir in 3 tbsp extra-virgin olive oil, ¼ cup vinegar, a pinch of salt, and plenty of freshly ground black pepper. Mix 1½–2 tsp of the pili-pili relish with 1½ tbsp olive oil and a pinch of salt in a small bowl and stir into the beans and pasta mixture.

Ladle into individual soup bowls, drizzle a little olive oil over the surface, finish with a generous sprinkling of freshly ground black pepper, and serve, handing round the arugula dressed with oil and vinegar and a small bowl of pili-pili sauce separately.

BEANS CREOLE WITH RICE AND FRIED BANANAS

3 cups dried black beans

1 medium-sized onion, minced

1 clove garlic, minced

1 cup chopped green peppers

1 green chili pepper, finely chopped

¼ cup olive oil

salt and pepper

For the rice:

3¼ cups chicken stock (see page 37)

½ onion, minced

½ green pepper, minced

1½ cups par-boiled Carolina long-grain rice

½ cup dry white wine

3 tbsp butter

6 tbsp chopped coriander leaves

1 lime, quartered

For the fried bananas:

5 bananas

all-purpose flour

sunflower oil

ground cinnamon

Place the beans in a sieve and rinse well under cold running water, drain, and place them in a large, heavy-bottomed flameproof casserole dish with 8¾ cups water. Bring slowly to a boil, partially cover, reduce the heat to very low, and simmer gently for 2 hours, stirring now and then. Add a large pinch of salt when the beans have been cooking for 1¾ hours. Heat ¼ cup oil in a large skillet and gently fry the minced onion, garlic, peppers, and chili pepper over low heat for 10 minutes, stirring frequently. Add a little salt and pepper, stir, and then mix into the beans. Simmer the beans for another 15–20 minutes, until tender but not mushy.

Meanwhile, cook the rice and fried bananas. First,

bring the stock to a gentle boil. Fry the minced onion and pepper in 2 tbsp butter over moderately low heat for 5 minutes, stirring frequently. When they are pale golden brown, add the rice and cook for 1 minute while stirring; pour in the wine and continue cooking until it has completely evaporated. Gradually add the boiling stock, stirring as you do so; cover and simmer gently over a low heat for 15 minutes, or until the rice is tender but still firm and has absorbed all the liquid. Remove from the heat and leave to stand, covered, for 5 minutes, then use a fork to gently stir the rice, separating the grains. Stir in the coriander and 1 tbsp solid butter.

Next, fry the bananas. Peel and cut them lengthwise in half; cut each half into 1-1/2-in pieces. Coat these all over with flour, shaking off excess. Heat plenty of oil in a skillet and when it is hot but not smoking, add the bananas and fry briefly, until they have started to crisp and brown on the outside; remove from the oil with a slotted spoon and drain briefly on paper towels; keep warm in an oven at 200°F.

Transfer the rice to a heated serving dish; arrange the bananas all round the edge and sprinkle them with a little cinnamon. Serve the beans straight from their cooking dish and garnish each helping with a fresh lime wedge for squeezing over the rice.

COUNTRY LENTIL SOUP

1 lb green or brown lentils
1 green celery stalk
1 large clove garlic
3 or 4 sage leaves
4 canned anchovy fillets
1 large can chopped tomatoes
4 thick slices bread, toasted
¾ cup freshly grated Pecorino cheese
3 tbsp extra-virgin olive oil
1 tbsp butter
salt

Place the lentils in a sieve and rinse thoroughly under cold running water; leave to soak in a bowl of cold water for 1 hour. Heat 8¾ cups of water in a large, heavy-bottomed flameproof casserole dish and as soon as it reaches the boiling point, add the lentils. Boil them gently for 1 hour.

While they are cooking, finely chop the celery, garlic, and sage and fry gently in the oil and butter for 10 minutes over very low heat, stirring frequently. Remove from the heat, add the anchovy fillets, and mince them into the oil and butter, using the back of a wooden spoon. Return the skillet to a low heat, add the tomatoes, and cook, uncovered, for 5 minutes while stirring to reduce and thicken.

When the lentils are tender, take 1 cup of the cooking liquid out of the casserole dish and discard. Stir the tomato mixture into the lentils and allow to return to a very gentle boil. Add a little salt to taste. Place the crisp, dry bread slices in the bottom of the soup bowls and sprinkle the cheese over them; ladle in the soup and serve immediately.

TUSCAN CHICKPEA SOUP

2 cups chickpeas
8¾ cups chicken stock (see page 37)
1 sprig rosemary, bound with kitchen thread (see method)
2 cloves garlic, unpeeled
1½ tbsp tomato paste
baking soda
For the relish:
3 cloves garlic, peeled and minced
¼ cup extra-virgin olive oil
1 red chili pepper, finely chopped

221

3 canned anchovy fillets
1 cup canned chopped tomatoes
salt and black peppercorns
Serve with:
mixed salad
whole wheat bread

Rinse the chickpeas and soak overnight or for at least 12 hours in a large bowl of cold water with 1 tsp baking soda. The next day, drain them and place in a large, heavy-bottomed flameproof casserole dish with 8¾ cups stock and the rosemary (unless you wind kitchen thread round it tightly or tie it up in a cheesecloth bouquet garni bag, the tough leaves will come off during cooking and spread through the soup). Add the unpeeled garlic cloves and the tomato paste.

Bring slowly to a boil, skimming off any scum that rises to the surface. Partially cover and simmer very gently for 2¼ hours, or until the chickpeas are very tender. Take 2¼ cups of the chickpeas out of the casserole dish and reserve. Remove the rosemary and the garlic cloves and discard. Put all the rest of the soup through a vegetable mill or blender to reduce to a purée and return to the dish. Add salt to taste. The soup can be prepared 24 hours in advance up to this point.

Shortly before serving, reheat the soup gently over low heat. Make the relish: gently fry the 3 minced garlic cloves and the chili pepper in ¼ cup oil in a nonstick skillet for 1 minute while stirring; remove from the heat and add the anchovy fillets. Mince these into the oil with the back of a wooden spoon. When they have broken up, stir in the tomatoes and cook over moderate heat for 3–4 minutes, stirring to reduce.

Mix the contents of the skillet with the chickpeas and leave to stand for 10 minutes before serving in individual bowls with a generous sprinkling of freshly ground pepper.

— • —

PEAS AND FRESH CURD CHEESE INDIAN STYLE
Panir mater

1 large onion, minced
2 cloves garlic, minced
1 ¾-in thick piece fresh ginger, peeled and grated
1 green chili pepper, finely chopped
2 large ripe tomatoes, peeled, seeded, and chopped
1½ cups frozen peas, thawed
1 tsp cumin seeds
pinch saffron
1½ tbsp chopped coriander leaves
¾ cup sunflower oil
salt
For the fresh curd cheese or *panir* (makes approx. 6 oz):
8¾ cups milk
1 tsp salt
2–3 tbsp wine vinegar or cider vinegar

Make the curd cheese: bring the milk slowly to a boil in a large saucepan. Stir in the salt, reduce the heat to very low, and gradually add the vinegar to the gently simmering milk while stirring continuously. Stop adding vinegar when the milk curdles and thickens. Line a sieve with a piece of cheesecloth and pour the curdled milk into it. Gather up the edges of the cheesecloth, twist them round and force out all the liquid from the curd. Place this fresh curd cheese, still in the cheesecloth, in a large bowl and flatten slightly by hand. Place a small plate on the curd with a weight on top and leave to stand for 2 hours. Drain off the liquid (whey), unwrap the cheese, and place on a pastry board. Cut into dice with a large knife.

Heat ½ cup of the oil in a nonstick skillet and fry the diced fresh curd cheese over moderate heat, turning the pieces carefully; when they are pale golden brown all over, take them out of the skillet and drain on paper towels.

Heat ¼ cup oil in a large, heavy-bottomed saucepan

or flameproof casserole dish and fry the onion gently for 15 minutes, stirring frequently. When the onion is pale golden brown, add the garlic, ginger, and chili pepper and cook gently, stirring, for about 30 seconds. Add the chopped tomatoes, the peas, cumin, saffron, and a pinch of salt, cover, and cook for 10 minutes over very low heat, until the peas are tender, adding a little hot water when necessary to moisten. Add the *panir* cheese cubes, cook gently just long enough to heat through, then draw aside from the heat. Add a little more salt if needed. Transfer to a serving dish, sprinkle with chopped coriander leaves, and serve very hot, with hot chapatis.

———— • ————

MILANESE EGGS AND PEAS

2 large scallions, outer layer removed

2 tbsp butter

1 slice cooked York ham, approx. ¼ in thick

1¼ cups fresh or frozen peas

½ cup chicken stock (see page 37)

4 eggs

salt and pepper

Serve with:

Potato Mousseline (see page 197)

Slice the trimmed scallions into thin rings and sweat in the butter for 10 minutes in a very large, nonstick skillet. Chop the ham, add to the skillet, and cook gently, stirring, for 1 minute. Add the peas, stock, and a little salt and pepper. Bring to a gentle boil, cover, and simmer gently for 15 minutes, or until the peas are tender. Add a little more salt and pepper if needed and continue cooking over a slightly higher heat until the liquid has evaporated.

Reduce the heat to very low and with the back of a wooden spoon, make 4 evenly spaced, fairly large,

shallow depressions in the contents of the skillet and break an egg into each one. Season the eggs with salt and pepper. Cover and cook for 2–3 minutes or until the whites have set. If you like your egg yolks partially set, continue cooking for another 1–2 minutes. Serve at once.

———— • ————

GREEN PEA MOLDS WITH EGGPLANT MOUSSELINE

1 small shallot, minced

butter

½ cup frozen peas

2 basil leaves

½ cup vegetable stock (see page 37)

2 egg yolks

2 whole eggs

½ cup heavy cream

½ cup milk

½ cup freshly grated Parmesan cheese

nutmeg

salt and pepper

For the eggplant mousseline:

2 medium-sized eggplants

2 large tomatoes, peeled, seeded, and chopped

6 basil leaves

3 tbsp lemon juice

½ cup olive oil

½ clove garlic

½ dried chili pepper, seeds and stalk removed

salt and pepper

Garnish with:

sprigs of basil

223

Preheat the oven to 400°F; place the eggplants in a baking dish and bake for 25–30 minutes, turning them halfway through this time.

Make the pea molds: sweat the chopped shallot in 1 tbsp butter for 10 minutes in a heavy-bottomed saucepan, then add the peas, basil, and the stock. Cover and simmer gently for about 15 minutes or until the peas are very tender, stirring and adding a little more stock now and then. Add a little salt and pepper and continue cooking over slightly higher heat while stirring until the liquid has evaporated. Put the contents of the saucepan through a vegetable mill fitted with a fine-gage disk, collecting the purée in a large bowl.

When the eggplants have finished cooking, take them out of the oven and leave to cool completely. Turn the oven temperature down to 350°F.

Grease four 1-cup capacity timbale molds or ramekin dishes with butter (or use six of 1/2-cup capacity) and bring plenty of water to a boil in a tea kettle. Beat the 2 egg yolks briefly with the whole eggs and a pinch each of salt, pepper, and nutmeg. Lightly beat in the cream, milk, puréed peas, and cheese. Add a little more salt and pepper to taste and pour into the molds; the mixture should come close to their rims. Place the molds carefully in a roasting pan and pour sufficient boiling water into the pan to come two-thirds of the way up the sides of the molds. Cook in the oven for 30 minutes.

While they are cooking, make the sauce. Peel the eggplants, cut lengthwise into quarters, scoop out the central seed-bearing section if there are any seeds discernible, and cut the remaining flesh into small pieces. Place these in the food processor or blender with the coarsely chopped tomatoes, 6 basil leaves, the lemon juice, the olive oil, garlic, crumbled chili pepper, and a pinch of salt and plenty of freshly ground pepper. Process at high speed until very smooth and creamy. Pour this thick sauce into a bowl. (This can also be served well chilled with sesame wheat crackers as a dip.)

When the molds have had 30 minutes in the oven, take them out and leave to stand for 5 minutes. Turn out onto hot plates, spoon some of the eggplant mousseline to one side, and garnish with a sprig of basil. Serves 4–6.

CASSEROLE OF PEAS AND BEANS WITH ARTICHOKES

½ cup peas, fresh or frozen
¾ cup fava beans, shelled and seed cases removed
6 very young, tender artichoke hearts or 12 fresh artichoke bottoms
juice of 1 lemon
1 small head Romaine lettuce
1 medium-sized onion
1 clove garlic
1 sprig thyme
6 tbsp extra-virgin olive oil
salt and pepper

If you are using frozen peas, partially defrost them. It is best to remove the casing from each fava bean unless they are very young and tiny as they tend to be leathery and tough once cooked. If you are using tender artichokes, see page 18 for preparation; cut the hearts lengthwise in quarters. Older, tougher artichokes should be stripped right down to their dish-shaped bottoms; these can then be cut horizontally into 2 or more slices. In both cases drop each artichoke as soon as it is prepared into a bowl of cold water acidulated with the lemon juice. Separate the lettuce leaves, removing the hard base, wash and dry them, and shred into a *chiffonade* (see page 24). Cut the peeled onion lengthwise into quarters and then slice very thinly.

Drain the artichokes and place in a wide, fairly shallow flameproof casserole dish or saucepan with the other vegetables, the whole but lightly smashed garlic clove, and the sprig of thyme. Season with a little salt and pepper. Add ½ cup water and the extra-virgin olive oil.

Bring to a boil over moderate heat, cover tightly, turn down the heat, and simmer very gently for about 15 minutes or until the vegetables are tender but not at all mushy or overcooked. Correct the seasoning and serve.

BRAISED GREEN PEAS, ARTICHOKES, AND LETTUCE

2 lettuces

juice of 1 lemon

6 young, tender artichoke hearts (see previous recipe)

1 small onion, peeled and finely minced

1 sprig thyme

1½ tbsp butter

3 tbsp olive oil

1 cup young peas or petits pois, *fresh or frozen*

1 cup chicken stock (see page 37)

3 tbsp chopped parsley

salt and pepper

Serve with:

mozzarella cheese, dressed with olive oil, chopped fresh basil, salt and pepper; seafood dishes; roast meat

Preheat the oven to 350°F. Choose a well flavored lettuce variety for this recipe; the crisphead type is not suitable. Trim the lettuces, remove and discard the outer leaves, remove the base, separate and wash the inner leaves, drain well, and shred into a *chiffonade* (see page 24). Have a bowl full of cold water acidulated with the lemon juice standing ready; as you prepare the artichokes, drop them into it to prevent discoloration. If you are using artichoke bottoms, prepare 12 for this recipe.

Sweat the onion in the butter and oil in a large, shallow flameproof casserole dish with a tight-fitting lid with the sprig of thyme for about 10 minutes, stirring frequently. Add the well drained artichokes, turn up the heat a little, and fry them for 15 minutes, stirring from time to time. Add the peas, season lightly with salt and freshly ground pepper, and cook with the lid on over low heat for 5 minutes. If using frozen peas, cook for several minutes longer, until they have thawed completely. Add the shredded lettuce and the stock, cover tightly again, and place in the oven to cook for ½ hour. When this time is up, taste to see whether a little more salt and

pepper is needed. Sprinkle with the parsley and serve directly from the casserole dish or transfer to a heated serving plate.

———•———

SNOW PEAS AND SHRIMP CANTONESE

1 lb jumbo shrimp, net weight with heads removed

1¾-in piece fresh ginger, peeled and minced

6 tbsp all-purpose flour or cornstarch

1½ cups snow peas, washed

4 scallions, outer layers removed, sliced

1 clove garlic

¼ cup Chinese rice wine, dry vermouth or dry sherry

½ cup crustacean stock (see page 38)

½ cup sunflower oil

salt

Serve with:

Cantonese Rice (see page 233)

Bamboo Shoot and Asparagus Soup (see page 118)

Peel the shrimp; make a neat incision near the end of the tail to get to the black "vein" or intestinal tract running up the back of each shrimp; lift this up so that you can pull it out cleanly.

Place the shrimp in a bowl and add 1 tsp salt and the finely minced ginger; using your fingertips, mix thoroughly. Leave to marinate for 30 minutes, then mix in the flour or cornstarch, using your fingertips again, to coat all the shrimp.

While the shrimp are marinating, heat a large saucepan of salted water and blanch the snow peas for 2 minutes. Drain in a sieve or colander and refresh, holding the sieve under cold running water to rinse thoroughly. Drain once more. Cut the inner part of the scallion (both bulb and leaves) into 1½-in lengths and then shred these lengthwise, into *julienne* strips (see page 24).

225

Heat 1/4 cup of the oil in a wok or large skillet and when it is very hot, add the garlic clove (left whole, but flattened with the flat of a large knife blade) and the snow peas, and stir-fry for 3–4 minutes. Remove from the wok with a slotted spoon and keep warm on a hot plate while you cook the shrimp.

Add the remaining oil to the wok and stir-fry the scallions for 1 minute; add the coated shrimp and stir-fry for 2 minutes. Sprinkle with the rice wine and 1/2 cup of the crustacean stock and stir; add the snow peas, reduce the heat, and continue stirring for 1 minute over lower heat as the sauce thickens. Simmer for a few seconds longer, then remove from the heat and serve immediately, with Cantonese rice, followed, or preceded, by the bamboo shoot and asparagus soup.

———•———

BEANS AND EGGS MEXICAN STYLE

2 cups dried red kidney or pinto beans
1 small onion, minced
1¼ cups canned tomatoes, drained and chopped
1–2 fresh or dried red chili peppers, finely chopped
1 small garlic clove, minced
1½ tbsp butter
3 tbsp olive oil
salt and pepper
For the eggs:
1 medium-sized onion, minced
1 green chili pepper, finely chopped
1¼ cups canned tomatoes, drained and chopped
4 eggs
3 tbsp olive oil
salt and pepper

Rinse the beans thoroughly in a sieve under running cold water and place in a large, heavy-bottomed flameproof casserole dish with 7 cups water, half the minced onion, 6 tbsp of the chopped tomatoes, and the chili pepper; bring to a boil. Skim off any scum that rises to the surface; reduce the heat to low, cover, leaving a gap for the steam to escape, and simmer for 2½ hours or until the beans are very tender, stirring now and then. Season with salt and pepper.

Sweat the remaining minced onion with the garlic in the butter and olive oil for 10 minutes over very low heat, then add the remaining chopped tomatoes and a pinch of salt. Increase the heat to high and cook, uncovered, for 3 minutes. Add about 1 cup of the cooked beans and crush them coarsely with a fork or the back of a wooden spoon. Stir well, add another cup of cooked beans, and crush these. Continue doing this until all the beans have been added. Stir while cooking for a few minutes, to thicken the mixture. Cover and keep warm.

Cook the eggs: sweat the onion until tender with the chili pepper in a wide nonstick skillet with 3 tbsp oil. Add the 1¼ cups chopped tomatoes and a little salt and pepper to taste and boil, uncovered, over fairly high heat for 2–3 minutes while stirring, to reduce and thicken. Make 4 hollows or spaces with the back of a wooden spoon in the tomato mixture and break the eggs into them; season the eggs with salt and pepper, turn down the heat, cover, and cook gently for 2–3 minutes. Serve immediately with the beans.

———•———

WARM TUSCAN BEAN SALAD

2 cups fresh or dried white beans, Boston beans or pea beans
1 small onion
1 medium-sized carrot, peeled

2 bay leaves
1 sprig rosemary, bound with kitchen string (see method)
2 thick slices bacon
1½ tbsp chopped parsley
olive oil
red wine vinegar
salt and black peppercorns

If using dried beans, soak them in plenty of cold water with 1 tsp baking soda overnight or for over 12 hours. Drain then rinse well. If using fresh beans, simply rinse and drain. Place them in a large, heavy-bottomed flameproof casserole dish with the onion, carrot, the bay leaves, and the sprig of rosemary with kitchen thread wound tightly round it to prevent the tough leaves from dropping off during cooking. Add a few whole black peppercorns and the bacon. Tie the rosemary and peppercorns in a cheesecloth bag if preferred. Add sufficient water to cover all the ingredients easily and bring to a boil, skimming off the scum that rises to the surface. Partially cover and simmer gently over low heat for about 2 hours, stirring now and then, until the beans are tender. Add salt to taste just before they are done.

Transfer portions of the beans to individual plates and spoon a little of the liquid over them. Sprinkle sparingly with wine vinegar, plenty of freshly ground black pepper, chopped parsley, and a little oil.

—— • ——

STRING BEAN AND POTATO PIE

1 lb potatoes
1½ cups string beans, fresh or frozen
1 large onion, peeled and finely minced
2 eggs
½ cup freshly grated Parmesan cheese
1 tbsp chopped fresh marjoram leaves

1 clove garlic, minced
6 slices Italian Mortadella sausage
¼ cup butter
3 tbsp olive oil
¾ cup fine dry breadcrumbs
nutmeg
salt and pepper
Serve with:
green or tomato salad
tomato salad

Wash the potatoes under running cold water but do not peel them; steam for 30–40 minutes until very tender. Boil the trimmed beans in salted water for 20–25 minutes, until they are very tender. Drain.

Preheat the oven to 400°F. Sweat the onion in 2 tbsp of the butter and 1½ tbsp of the oil for 20–25 minutes over extremely low heat, stirring frequently, until it has almost disintegrated and is pale golden brown. As soon as the potatoes are done, spear them with a carving fork and peel them while they are still boiling hot; put through a potato ricer twice or push through a fine sieve into a large bowl. Make sure you have drained every drop of water from the beans, place them in the food processor, and process to a smooth purée. Mix this purée with the potatoes, adding the eggs and onion. Place the cheese, marjoram, garlic, and Mortadella in the food processor and process until completely blended; stir into the bean and potato mixture, seasoning with a little salt, plenty of freshly ground pepper, and a pinch of nutmeg.

Oil the inside of a 9-in shallow cake pan or quiche dish (about 1½ in deep) and fill it with the prepared mixture, leveling the surface with a knife. Sprinkle the top with a layer of breadcrumbs, dot small pieces of the remaining butter over the surface, and bake in the oven for 10 minutes; place under a very hot broiler for about 3 minutes to crisp and brown the surface.

—— • ——

CANNELLINI BEANS WITH HAM AND TOMATO SAUCE

1¼ cups dried cannellini beans

pinch powdered rosemary

1 garlic clove, unpeeled

3 tbsp olive oil

2 scallions, minced

1 small clove garlic, minced

1 large slice ham

3 tbsp chopped basil

3 tbsp chopped parsley

1 lb firm ripe tomatoes, peeled, seeded, and chopped

salt and pepper

Serve with:

a selection of mild goat's cheeses

Rinse the beans under running cold water in a sieve, place in a large, flameproof casserole dish, add sufficient cold water to cover them by about 2 in, bring slowly to a boil, and then simmer over moderate heat for 3 minutes. Drain in a sieve and rinse well under cold running water; drain again and replace in the casserole dish, covering amply with water once more. Add the rosemary and the first garlic clove, whole and still in its skin. Bring to a boil, turn down the heat to very low, partially cover, and simmer gently for 2 hours or until the beans are tender; add a little salt after about 1¾ hours.

Make the ham and tomato sauce: heat 3 tbsp olive oil in a very large heavy-bottomed saucepan or flameproof casserole dish and gently fry the scallions, the second, minced clove of garlic, and the finely chopped lean and fat of the ham for 10 minutes, stirring frequently. Add the basil, parsley, and chopped tomatoes and a little salt and pepper. Turn up the heat and cook, uncovered, for about 15 minutes, stirring frequently to allow the sauce to reduce and thicken.

Drain the cooked beans, reserving about ½ cup of their cooking water, and add to the tomato sauce, together with some of the reserved cooking water. Stir over very low heat for a few minutes, adding a very little more cooking water if necessary. The mixture should be very moist. Add a little more salt to taste and some freshly ground pepper.

---•---

NORTH AFRICAN VEGETABLE CASSEROLE

2 cups dried chickpeas

2 large eggplants

3 onions

2¼–2½ lb ripe tomatoes

baking soda

olive oil

salt and pepper

Soak the chickpeas overnight in a large bowl of cold water. The next day, drain them and place in a large, flameproof casserole dish; add sufficient cold water to cover them and a pinch of baking soda. Bring to a boil quickly over high heat, then reduce the heat to very low; partially cover and simmer gently for 2 hours or until they are tender, topping up the water level with some more boiling water if necessary. Add salt when the chickpeas have nearly finished cooking. When they are done, drain and set aside.

Preheat the oven to 350°F. Peel off lengthwise strips of skin from the eggplants about ½ in wide, leaving strips of the same width between them; cut the eggplants lengthwise into quarters and then cut these sections across into pieces about 1½ in long. Heat sufficient olive oil to amply cover the bottom of a medium-sized nonstick skillet and when hot, fry the pieces of eggplant in batches over moderately high heat, stirring, until they are browned. Remove from the skillet and finish draining on paper towels; sprinkle with a little salt and pepper. Eggplants absorb a great deal of oil so you may need to add more to fry the later batches.

Slice the onion thinly. Fry gently for 10 minutes, stirring, in the oil left over from frying the eggplants; when the onion is soft and a very pale golden brown, remove from the heat.

Blanch the tomatoes in boiling water for 10 seconds, peel them, cut into quarters, and remove all the seeds and any tough parts. Chop coarsely, transfer to a bowl, and season with a pinch of salt and pepper. Spread out the eggplant pieces in a single layer in a shallow, flameproof casserole dish. Cover with the onions, sprinkle with a little salt and pepper, cover the onion layer with the chickpeas, and top with a layer of tomatoes. Add 1-1/4 cups water, pouring slowly into a corner of the dish, bring to a gentle boil over moderate heat then place in the oven and bake for about 45 minutes, or until the vegetables are all very tender. Serves 6.

—— • ——

GREEK CHICKPEA CROQUETTES
Pittarúdia

1¼ cups dried chickpeas
2 cups canned tomatoes, drained and chopped
1 small onion, minced
2 tbsp ground cumin
2 tbsp dried mint
¾ tsp freshly ground black pepper
1 cup all-purpose flour
1 tsp baking powder
6 tbsp fresh soft breadcrumbs
2 tbsp chopped parsley
light olive oil
salt
Serve with:

Tsatsiki (see page 149)
hot pita bread

Place the chickpeas in plenty of cold water to soak overnight or for at least 12 hours. Drain them very thoroughly, transfer to the food processor and process until coarsely chopped. Transfer to a large mixing bowl and stir in the chopped tomato, the grated onion, cumin, mint, pepper, and 1 tsp salt. Sift in the flour and baking powder together and stir in the breadcrumbs and parsley. The mixture should be soft, but capable of holding its shape when drawn into a peak. Leave to rest for 30 minutes.

Pour enough olive oil into a very wide nonstick skillet to form a layer about ¾ in deep or heat plenty of oil in a deep-fryer to 350°F. When the oil is very hot but not smoking, drop heaped tablespoonfuls of the chickpea mixture into it, making sure they do not touch one another; if shallow-frying, turn once when the underside is crisp and golden brown; the croquettes should puff up as they cook. Remove the first batch from the oil with a slotted spoon and finish draining on paper towels on a hot plate while you fry the other batches. Serve as soon as they are all cooked, with Tsatsiki and hot pita bread.

—— • ——

MUNG DHAL FRITTERS

1 cup mung dhal (dried split mung beans)
5 small green peppers, finely chopped
generous pinch baking powder
¼ cup natural yoghurt
½ cup cooked green peas
1 thin piece fresh ginger, peeled and minced
1 green chili pepper, finely chopped
2 tbsp ghee (clarified butter, see page 36)
1 tsp cumin seeds

3 tbsp chopped coriander leaves

½ cup sunflower oil

salt and pepper

Soak the mung dhal (dried split mung beans) in cold water overnight. Drain. Remove the stalks, seeds, and inner membrane from 3 of the peppers, chop, and process with the mung dhal in a food processor with 3 tbsp cold water to a purée. Stir in the baking powder, yoghurt, and a pinch of salt.

Make the green pea filling. Mash the peas coarsely. Chop the ginger with the 2 remaining peppers and the chili pepper. Heat the ghee in a small skillet and fry the cumin seeds over gentle heat for a few seconds while stirring. Add the peppers, chili, and ginger, fry gently for 1 minute, stirring, then add the mashed peas and a small pinch of salt. Stir over low heat for 1 minute. Add the chopped coriander, mix well, and remove from the heat. Heat ¼ cup oil in a nonstick skillet about 6 in in diameter and when it is very hot, spoon ¼ cup of the lentil mixture into the skillet; spread the mixture out to form a thick round cake about 3 in in diameter. Make a small hole in the center before the cake has time to set. Place 1 heaping tsp of the filling in the center of the cake and press gently to flatten (see illustration on page 275). Fry for 2 minutes then turn and fry for another 1–1½ minutes. Take out of the skillet and keep hot on paper towels while you repeat the procedure until all the lentil mixture and filling is used up.

———•———

Red Lentils with Rice Indian Style

1 cup red lentils

3 thin pieces fresh ginger, peeled

½ tsp ground turmeric

1 tsp coriander seeds

¼ cup ghee (clarified butter, see page 36)

1 tsp cumin seeds

3 tbsp chopped coriander leaves

generous pinch cayenne pepper or chili powder

salt

For the rice:

1½ cups Basmati rice

1½ tbsp butter

salt

For the saffron butter:

generous pinch saffron threads

6 tbsp ghee (clarified butter, see page 36)

Cook the rice by the absorption method: measure out just under 1¾ cups cold water. Rinse the rice in a sieve under cold running water until the water runs out clear. Drain, place in a bowl, add sufficient cold water to cover, and leave to soak for 20 minutes, then drain again. Spread the rice out in the bottom of a heavy-bottomed saucepan or flameproof casserole dish and add the measured quantity of cold water, ½ tsp salt, and the butter. Heat until the water starts to boil, cover tightly, and cook gently over very low heat for up to 20 minutes (when all the water has been absorbed and the rice is tender but still firm, it is done).

Meanwhile, prepare the lentils. Rinse them well under cold running water; drain and place in a saucepan or flameproof casserole dish with 3 cups cold water and bring to a boil, skimming off any scum that rises to the surface. Add the pieces of ginger and stir in the turmeric. Reduce the heat to low, partially cover, and simmer gently for 20 minutes until the lentils are tender; stir from time to time to prevent them from burning on the bottom of the saucepan or casserole dish. Remove and discard the ginger, add salt to taste, stir, and remove from the heat.

Pound the coriander seeds with a mortar and pestle (or grind in the food processor) and sift to get rid of the tough seed cases.

Heat ¼ cup ghee in a small saucepan and fry the

cumin seeds for a few seconds. Add 2 tbsp chopped fresh coriander leaves and the cayenne pepper or chili powder, stir well, and then mix with the lentils. Add a little more salt if necessary, spoon into a serving dish and sprinkle with 1 tbsp chopped coriander. Make the saffron butter: place the saffron threads in a small bowl, add 1½ tbsp boiling water and stir well, mincing the threads with the back of a wooden spoon. Heat 6 tbsp ghee in a small saucepan, stir in the saffron and water, and mix briefly. Transfer the rice to a heated serving dish and pour the saffron butter all over it.

—— • ——

BORLOTTI BEAN PURÉE WITH CREAM SAUCE

| 3 cups dried borlotti or cranberry beans |
| 1 sprig or pinch powdered rosemary |
| 1 clove garlic, whole and unpeeled |
| ¾ cup crème fraîche or heavy cream |
| 2 tbsp unsalted butter |
| salt and pepper |

Cook the beans by the quick method described in the recipe for Cannellini Beans with Ham and Tomato Sauce on page 228; drain when cooked, reserving about 1 cup of the cooking water. Alternatively, use 2 medium-sized cans borlotti beans and reserve the liquid when you drain them. Put the beans through a vegetable mill fitted with a fine-gage disk; reheat the purée, adding some or all of the reserved liquid; it should be very thick but not

stiff. Stir continuously as you heat it over very low heat and when it comes to a boil, keep stirring as you gradually pour in the *crème fraîche* or cream. Continue cooking and stirring for 1 minute after all the cream has been added to thicken, then take off the heat. Season to taste and stir in the solid unsalted butter until it has melted.

—— • ——

MEXICAN BEANS WITH GARLIC AND CORIANDER

| 2½ cups dried red kidney or pinto beans |
| 1 ¾-in piece fresh ginger, unpeeled |
| salt |
| For the sauce: |
| 4 garlic cloves, sliced |
| ⅓ cup olive oil |
| 2 medium-sized ripe tomatoes, peeled, seeded, and chopped or canned tomatoes, drained |
| 1 tsp pili-pili hot relish (see page 113) |
| 2–2½ tbsp chopped coriander leaves |
| salt and pepper |

Rinse the beans and place in a flameproof casserole dish with sufficient water to completely cover them and the unpeeled slice of ginger. Bring to a boil and then simmer, partially covered, for 2½ hours. Add salt when they are nearly done. Drain, reserving about 1 cup of the cooking water. Discard the ginger. Put the beans through a vegetable mill fitted with a fine-gage disk and return the purée to the casserole dish; stir in some or all of the reserved liquid so that the purée is thick but not stiff. Set aside, ready for reheating.

Make the sauce: blanch the sliced garlic cloves for 3 seconds in a small saucepan containing some boiling salted water; drain at once and rinse in a sieve

231

under running cold water to refresh. Blot dry with a piece of paper towel. Heat the oil in a heavy-bottomed saucepan and fry the garlic gently until pale golden brown. Increase the heat slightly and add the tomatoes; cook for 1 minute while stirring gently. Add a pinch of salt and remove from the heat; stir in the pili-pili relish and keep warm.

Heat the bean purée over low heat, stirring continuously; when it comes to a boil, add more salt if needed and pepper to taste. Spoon into bowls, then add the tomato mixture to the middle of each bowl of bean purée and distribute an even number of garlic slices on each serving. Sprinkle with the chopped coriander and serve.

— • —

FAVA BEANS WITH CREAM

4 cups shelled fava beans

1 clove garlic, partially minced

2 tbsp butter

1 cup crème fraîche *or heavy cream*

salt and pepper

Serve with:

broiled chicken

egg dishes

Heat plenty of salted water in a large saucepan. When it comes to a boil, add the fava beans, turn down the heat, partially cover, and simmer for 10 minutes. Drain the beans in a colander, refresh under cold running water, then remove the seed cases, breaking up the bright green cotyledons inside as little as possible.

Heat the butter and garlic clove in a wide, heavy-bottomed skillet, add the beans, and cook over gentle heat for 2 minutes, stirring gently. Add the *crème fraîche* or cream and cook, uncovered, over moderate heat for 3–4 minutes to reduce and thicken the cream. Add salt and freshly ground white pepper to taste and serve at once.

JAPANESE STIR-FRIED RICE WITH GREEN PEAS AND MIXED VEGETABLES
Yakimeshi

1 cup steamed rice (see page 126)

½ cucumber

¼ cup cooked green peas

2 scallions, sliced

½–1 fresh green chili pepper, sliced into rings

1 young tender carrot, diced

½ red pepper, diced

1 egg

generous pinch dashinomoto *(instant fish stock)*

pinch granulated sugar

6 tbsp sunflower oil

salt

Serve with:

Miso and Tofu Soup (see page 217)

Cook the rice in advance; allow to cool quickly, then refrigerate in a covered container until 30 minutes before you plan to serve this dish.

Prepare all the vegetables: cut off the skin of the cucumber together with a ¼-in layer of the flesh immediately beneath it and dice these thick strips. Reserve the remains of the inner part of the cucumber for other dishes. Beat the egg briefly in a small bowl with the *dashinomoto* (if you do not have any of this instant Japanese fish stock, use a pinch of crumbled fish or vegetable bouillon cube), 3 tbsp cold water, and a pinch each of sugar and salt.

Heat 1 tsp oil over low heat in a wide nonstick skillet and pour in the beaten egg, tipping it this way and that so that it spreads out into a very thin omelet; cook gently until it has completely set but do not allow to brown at all. Remove the omelet from the skillet, spread out on a chopping board, and leave to cool. Chop finely and transfer to a bowl.

Heat the remaining oil in the wok and stir-fry all the vegetables over high heat for 2 minutes. Add a pinch of salt, followed by the rice. Reduce the heat and stir-fry with the vegetables for about 3 minutes, by which time it should be hot. The vegetables should still be very crisp. Sprinkle with another teaspoon of salt, mix in the chopped omelet, and serve.

———•———

PURÉE OF PEAS

2 medium-sized potatoes

1½ cups frozen peas

1 shallot, minced

3 basil leaves

2½ cups chicken stock (see page 37)

1 egg yolk

½ clove garlic, minced

½ cup freshly grated Parmesan cheese

3 tbsp olive oil

nutmeg

salt and pepper

Serve with:

hot vegetable molds or timbales, fried eggs or fresh cheese

Peel and dice the potatoes; place in a large, heavy-bottomed saucepan with the peas, shallot, basil leaves, and stock. Bring to a boil, cover, and simmer over low heat for about 30 minutes or until the potatoes are very tender. Drain off and reserve the liquid remaining in the saucepan; purée the cooked mixture by putting it through a vegetable mill fitted with a fine-gage disk, or rub through a sieve. Return the purée to the saucepan; stir in ¼ cup of the reserved liquid and heat to just below boiling point over low heat.

Beat the egg yolk into the purée for a few seconds

only, just enough to blend well, using a whisk. Remove from the heat, season to taste with a little salt, some freshly ground pepper, and a pinch of nutmeg; stir in the garlic and then the grated cheese. Keep beating as you add the olive oil a very little at a time. Serve at once.

———•———

CANTONESE STIR-FRIED RICE AND PEAS

1½ cups long-grain rice for steaming or boiling (see page 126)

3 eggs

pinch sugar

pinch monosodium glutamate (optional)

4 large scallions, thinly sliced

2 celery stalks, trimmed and diced

½ cup cooked green peas

1 thick slice ham, diced

⅓ cup sunflower oil

salt

Steam or boil the rice in advance, place in a sealed container when cool, and refrigerate until 30 minutes before preparing this dish.

Beat the eggs with a pinch each of salt, sugar, and monosodium glutamate (if used). Heat 1½ tbsp of the oil in a large, nonstick skillet over low heat and make a very large, thin omelet; it should be completely set but not at all browned. Remove from the skillet, spread out on a chopping board, and when completely cold, chop coarsely.

Remove the roots, leaf tips, and outer layer of each scallion; slice the leaves and bulbs into thin rings; heat the remaining oil in a large wok and stir-fry the scallions with the diced celery for 1 minute. Add the peas and the diced ham and stir-fry for 1½ minutes. Add a small pinch of salt and the cold rice; reduce

233

the heat and stir-fry for a few minutes, separating any rice grains that stick together as you do so. Sprinkle with a generous pinch of salt and continue to stir-fry for another 3 minutes, or until the rice has heated through. Add the chopped omelet, mix evenly with the rice and vegetables, and remove from the heat.

This dish makes a good foil for almost any Chinese dish. Serve with soy sauce.

———•———

PEA AND WILD RICE TIMBALES

½ cup wild rice

½ cup brown rice

1 shallot, minced

1 sprig thyme

½ cup peas, fresh or frozen

1 cup vegetable stock (see page 37)

3 tbsp butter

salt and pepper

Serve with:

Lobster Catalan (see page 140), Egg and Artichoke Fricassée (see page 129) or Pisto Manchego (see page 139)

Place the wild rice in a small, heavy-bottomed saucepan and add 2½ cups water. Bring to a boil, cover tightly, reduce the heat to very low, and cook for 30 minutes. Remove from the heat and stir in the brown rice and ½ tsp salt. Bring to a boil again quickly, cover tightly, reduce the heat to very low again, and continue cooking for about 14 minutes more, or until both the wild rice and the brown rice are tender but still firm and all the moisture has been absorbed.

While the rice is cooking, sweat the shallot in 1 tbsp of the butter with the sprig of thyme for 5 minutes, stirring. Add the peas and the stock, bring to a boil, cover, and then simmer gently for about 25 minutes,

by which time the peas should be very tender. Remove and discard the sprig of thyme and stir the peas over moderate heat until the liquid has been considerably reduced and the peas are just moist. Add salt and pepper to taste. Add the peas to the rice with ½ tbsp butter and stir until the butter has melted.

Grease 4 small timbale molds or ramekin dishes with the remaining butter and fill them with the rice and pea mixture, pressing down on each layer gently with the back of a wooden spoon to pack in evenly and firmly. Keep hot, in a *bain-marie* of very hot water for up to 30 minutes if they are not being served immediately. Turn out onto heated plates. These molds go very well with seafood, vegetarian dishes, veal or poultry. A delicious and unusual light meal can be made by mixing 4 egg yolks with a pinch of salt and pepper and a little fresh lime juice and stirring this with the rice mixture before packing into the molds; cover each mold with foil or waxed paper and bake in a preheated oven at 320°F in a *bain-marie* for about 15 minutes, then serve with bananas lightly fried in butter, sprinkled with a little cinnamon, and garnished with lime wedges and sprigs of coriander.

———•———

PEAS WITH BACON AND BASIL

2 shallots, minced

2 tbsp butter

4 slices bacon, diced

1¼ cups frozen peas or petit pois

1 cup chicken stock (see page 37)

2 basil leaves

½ cup crème fraîche *or heavy cream*

2 tbsp chopped basil

salt and pepper

Serve with:

hot vegetable molds or timbales; potato pies; roast meat or poultry

Sweat the shallot in the butter in a large, heavy-bottomed flameproof casserole dish for 5 minutes, stirring frequently. Add the bacon to the shallot and fry gently over low heat, stirring for 2–3 minutes. Add the frozen peas, the stock, basil leaves, a small pinch of salt, and some freshly ground pepper. When the liquid comes to a boil, cover and simmer gently for 20–25 minutes, stirring occasionally. When the peas are tender, stir in the cream and cook while stirring to allow it to reduce and thicken a little. Add a little more salt and pepper if needed, stir in the chopped basil, and serve immediately.

———•———

STRING BEANS MAÎTRE D'HÔTEL

1¾ lb string beans

2¼ cups béchamel sauce (see page 36)

¼ cup grated Swiss cheese

1 tsp chopped mint leaves

3 tbsp chopped parsley

salt

Serve with:

hot vegetable molds; omelets; broiled fish or meat

Heat plenty of lightly salted water in a large saucepan. Trim and rinse the beans and add to the boiling salted water to cook, uncovered, until they are tender but still crisp.

While the beans are cooking, make a béchamel sauce of pouring consistency; take off the heat and stir in the cheese. Drain the beans and spread out in a wide, shallow flameproof dish. Pour the cheese sauce all over them, heat to a gentle boil, and serve, sprinkled with the mint and parsley.

———•———

BEAN AND PEANUT PILAF

¼ cup dried black-eyed beans

¼ cup dried red kidney or pinto beans

¼ cup dried chickpeas

¼ cup unsalted whole raw peanuts

2 cups cooked long-grain rice

½ cup ghee (clarified butter, see page 36)

2 tbsp mustard seeds

1 large onion, minced

1½ green chili peppers, finely chopped

2 large tomatoes, peeled, seeded, and chopped

1½ tsp mild curry powder or garam masala

3 oz bean sprouts

¼ cup fresh lime juice

salt

Garnish with:

Cucumber Fans (see page 29)

The day before you plan to serve this dish, rinse all the dried beans and the chickpeas, place them in a large bowl, and add enough cold water to come well above them; soak overnight or for at least 12 hours. Drain, transfer to a large, heavy-bottomed saucepan, add sufficient water to cover them, and bring to a boil. Skim off any scum that rises to the surface; reduce the heat to low, partially cover, and simmer for 45 minutes. Add the peanuts and a generous pinch of salt and continue simmering until all the dried beans and chickpeas are tender. Drain and set aside.

While the beans are cooking, steam the rice and leave to cool at room temperature, then refrigerate in a sealed container until shortly before you start the final cooking stages of the pilaf. The beans can also be refrigerated if you prefer to complete preparation up to this point several hours or a day in advance.

Shortly before serving the pilaf, heat 6 tbsp of the ghee in a large deep skillet and fry the mustard seeds gently for a few seconds; as soon as they start to jump about in the skillet, add the onion and chili pepper and fry gently over low heat for 10 minutes, stirring frequently. Turn up the heat, add the tomatoes (use canned tomatoes if you prefer) and a pinch of salt. Cook, stirring for 2–3 minutes. Stir in the curry powder or garam masala, the bean sprouts, the cooked beans, chickpeas, peanuts, and the lime juice. Cover and cook gently over low heat for 2–3 minutes.

Heat the remaining ghee in another large skillet and stir-fry the rice for a few minutes, until it is heated through. Sprinkle with a small pinch of salt and combine gently with the vegetables in the larger skillet. Serve very hot, garnished with cucumber fans.

———•———

STRING BEANS WITH TOMATOES AND SAGE

1¼ lb string beans
1 small onion, minced
2 cloves garlic, minced
3–4 sage leaves
2 cups chopped tomatoes (fresh or canned)
¼ cup olive oil
salt and pepper

Trim the beans; cut them into 1¼-in lengths, rinse, and drain. Heat the oil in a large skillet or fairly shallow saucepan and fry the onion, garlic, and sage gently over low heat for about 10 minutes, stirring frequently. When the onion is just tender, add the beans and chopped tomato, a pinch of salt, and some freshly ground pepper. Stir well and add just enough cold water to cover the beans.

Bring to a very gentle boil, cover, and simmer very gently for about 25 minutes or until the beans are tender but still fairly crisp. Tiny, thin string beans will only take about 15 minutes. Check whether there is a lot of liquid left when the beans are nearly done, and if there is an appreciable amount, simmer, uncovered, for the last 5 minutes cooking time to reduce. Taste and add a little more salt and pepper if necessary; serve very hot. This is a very versatile vegetable dish; serve on its own, as an accompaniment to meat dishes or with crusty bread and goat cheese for a nourishing light meal.

———•———

STRING BEANS WITH CRISPY BACON

½ lb string beans
16 slices smoked bacon
2 tbsp butter
salt and pepper
Serve with:
omelets, hot vegetable molds or vegetable pancakes

Heat plenty of lightly salted water in a large saucepan. Preheat the oven to 350°F. Top and tail the beans, rinse well, and boil, uncovered, for 12–15 minutes or until tender but still crisp. Drain and leave to cool a little.

Grease a wide, shallow ovenproof dish with a little butter. Divide the beans into 8 equal bundles and wrap 2 bacon slices round each one; do not have the slices overlapping one another or they will not crisp well. Place these bundles neatly in the dish, dot small pieces of butter here and there over the beans where they are not covered by the bacon and bake in the oven for 10 minutes or until the bacon is crisp. Serve piping hot, allowing 2 per person.

———•———

STRING BEANS AND POTATO SALAD WITH MINT

1¼ lb string beans

1¼ lb new potatoes

juice of 1 lemon

1 sprig mint

olive oil

salt and pepper

Serve with:

stuffed tomatoes

lamb kabobs

Trim the beans, rinse, drain, and boil, without the lid on, for about 15 minutes or until they are tender but still crisp. Drain in a colander, refresh under cold running water, and leave to cool.

Scrub the potatoes but do not peel. Steam or boil until tender, and leave until completely cold.

Peel the potatoes if desired and cut into pieces, then chop the beans and mix both together in a large salad bowl or serving dish, adding salt, pepper, olive oil, and lemon juice to taste. Tear the mint leaves into small pieces and sprinkle over the salad. Serve as part of a summer cold lunch.

———— • ————

STIR-FRIED SNOW PEAS

1¾ lb snow peas

¼ cup butter

salt and pepper

Serve with:

hot vegetable molds or egg dishes

Set up a steamer ready for use. Prepare the snow peas as described on page 22, taking care to pull off any strings along their edges. Rinse, drain, and steam for 3 minutes.

Heat the butter in a large saucepan or skillet and once the foam has subsided, stir-fry the snow peas for 3–4 minutes; they must remain crisp.

Season with a little salt and pepper and serve at once on hot plates as a vegetable accompaniment to almost any dish. Snow peas go particularly well with rich, rather fatty meats such as roast duck or goose.

———— • ————

SNOW PEAS IN CREAM AND BASIL SAUCE

1¾ lb snow peas

2 tbsp butter

1 cup crème fraîche *or light cream*

2–3 tbsp finely grated Swiss cheese (Gruyère or Emmenthal)

10 basil leaves, torn into small pieces

salt and pepper

Serve with:

chicken, fish or egg dishes

Set up a steamer ready for use. Trim the snow peas as described on page 22, rinse and drain, and steam for 3 minutes. Heat the butter in a large saucepan or skillet and stir-fry over high heat for 2 minutes. Reduce the heat, add a pinch of salt and some freshly ground pepper, and the cream. Cook for 5 minutes over low heat, stirring gently to coat the vegetables with the cream. Stir in the Swiss cheese and basil and serve on hot plates. This accompaniment is particularly good with egg dishes, such as omelets, and provides a crunchy contrast in texture to soufflés.

———— • ————

Snow Peas with Shallots and Bacon

1¾ lb snow peas

4 thick slices bacon

2 large shallots, minced

1 sprig thyme

¼ cup butter

salt and pepper

Serve with:

Rösti Potatoes (see page 195) and omelets, or with Cauliflower Timbale (see page 83)

Set up a steamer ready to cook the snow peas. Prepare these as described on page 22. Rinse, drain, and steam for 3 minutes. Choose bacon that has a good proportion of lean to fat and dice the slices. Heat a nonstick skillet and when it is very hot, add the diced bacon and fry until crisp and lightly browned. Take out of the skillet, leaving as much fat behind as possible, and drain on paper towels.

Heat the butter in a large saucepan or skillet and fry the shallot very gently with the sprig of thyme for 10 minutes, stirring frequently. When the shallot is tender but not browned, turn up the heat, add the snow peas, and stir-fry for 4 minutes or until they are tender but still crisp. Season with a very little salt and some freshly ground pepper, stir in the crisp bacon, and then take off the heat. Serve at once.

—— • ——

Green Beans with Tomato Sauce

1 lb flat green beans

1 small onion, minced

1 clove garlic, crushed

1 lb canned tomatoes, drained and chopped

pinch sugar

pinch oregano

1½ tbsp chopped parsley

1½ tbsp chopped basil leaves

¼ cup olive oil

salt and pepper

Serve with:

roast chicken

broiled fish

Fill a large saucepan two-thirds full of water, add 1 tsp salt, and bring to a boil. Prepare the beans as described on page 22, rinse, drain, and add to the boiling water. Boil fast, uncovered, for 10 minutes or until just tender but still very crisp. Drain in a colander and refresh under cold running water.

Heat the olive oil in a large, flameproof casserole dish and fry the onion very gently with the garlic for about 10 minutes, stirring. Add the tomatoes with a pinch each of sugar and salt and some freshly ground pepper; increase the heat and bring to a boil, stirring. Simmer, uncovered, for about 10 minutes so that the liquid reduces and the sauce thickens. Add the oregano, parsley, and the beans and stir. Reduce the heat to low, cover, and simmer gently for 4–5 minutes, or until the beans are as tender as you like them. Add a little more salt and pepper if needed, sprinkle with the basil, and serve hot.

—— • ——

Spiced Lentils and Spinach

1¼ cups brown lentils

2¾ lb spinach

½ cup ghee (clarified butter, see page 36)

1 ¾-in piece fresh ginger, peeled and finely minced

1½ green chili peppers, finely chopped

¼ cup chopped coriander leaves
¼ cup fresh lime juice
salt and pepper
Serve with:
chapatis (see page 81)
Eggplants with Yoghurt Dressing (see page 147)

Rinse the lentils well in a sieve under running cold water. Place in a heavy-bottomed saucepan with 4½ cups cold water and bring slowly to a boil, skimming off any scum that rises to the surface. Reduce the heat to very low, partially cover, and simmer gently for 1 hour.

Trim the spinach, removing any large tough stalks (these can be used in soups and other dishes); only the leaves and tender stalks are used for this recipe. Wash very thoroughly in several changes of cold water; drain.

Heat the ghee in a large, heavy-bottomed flameproof casserole dish and fry the ginger and chili pepper briefly over moderately low heat; add the coriander, stir briefly, then add the spinach. Continue stirring until the spinach has wilted. Add the lentils, together with their cooking water, 1 tsp salt, and some freshly ground black pepper. Stir, cover tightly, and reduce the heat to very low; simmer for about 30 minutes, stirring occasionally, until the lentils are very tender. Taste and add a little more salt and pepper if needed, mix in the lime juice, and continue cooking for 2–3 minutes before removing from the heat. Serve as part of an Indian meal or as the main course for a light lunch, with a refreshing salad and chapatis.

———•———

Braised Lentils

2 cups lentils
1 small onion
1 medium-sized carrot

1 leek, green part only
4 slices bacon, chopped
½ cup dry white wine
1 cup canned tomatoes, drained
5–5½ cups beef stock (see page 37)
6 tbsp olive oil
salt and pepper
Serve with:
good-quality, highly flavored sausages

Rinse the lentils thoroughly in a sieve under cold running water. Peel the onion and the carrot; remove the tough, outermost green layers of the leek and wash the inner part thoroughly; chop these three vegetables together very finely. Heat the olive oil in a large, heavy-bottomed flameproof casserole dish and sweat the chopped vegetable mixture with the bacon for 15 minutes, stirring frequently.

Turn up the heat, add the lentils, and cook for 2–3 minutes while stirring. Pour in the wine and when it has almost completely evaporated, add the chopped tomatoes. Pour in sufficient stock to cover the lentils. Simmer very gently with a tightly fitting lid for about 1 hour, or until the lentils are tender.

Add a little salt and some freshly ground pepper. If there is a lot of liquid remaining in the casserole dish, cook, uncovered, over moderately high heat for a few minutes to reduce and thicken. Serve with full-flavored or spicy broiled or boiled sausages or poultry dishes.

———•———

Japanese Sweet Red Bean Paste
Yude adzuki

1¼ cups adzuki beans or red soy beans
baking soda
cane sugar
peanut oil

3 tbsp vanilla sugar or few drops vanilla extract

salt

Decorate with:

small mint leaves

Serve with:

vanilla ice cream, sliced kiwi fruit or fruit salad

Soak the beans overnight or for at least 12 hours in plenty of cold water; when they have finished soaking, rinse in a sieve under cold running water. Place in a large, heavy-bottomed saucepan or flameproof casserole dish with 6¾ cups cold water. Bring quickly to a boil over high heat, turn down the heat to low, add a pinch of baking soda, stir, and partially cover. Simmer gently for 1½ hours until the beans are very tender. Add a small pinch of salt when they are nearly done. Remove from the heat and put the beans and any remaining liquid through a vegetable mill fitted with a fine-gage disk.

Transfer the purée back to the saucepan or casserole dish, measuring the quantity in a cup. You will need a quarter of this amount in sugar and an eighth of the amount in oil. Stir in these quantities of sugar and oil, then add the vanilla sugar or scant 1 tsp pure vanilla extract. Cook the purée over low heat, stirring continuously until it thickens to a paste. Leave to cool at room temperature before packing into sterilized glass canning or jam jars with airtight seals or lids. This red bean paste goes very well with vanilla ice cream served with kiwi fruit or a mixed fruit salad; use a small scoop to add 1 or 2 portions of the paste to each ice cream serving and decorate with mint leaves.

—— • ——

ADZUKI JELLY
Mizuyokan

1¼ cups adzuki beans or red soy beans

baking soda

cane sugar

3 tbsp vanilla sugar or few drops vanilla extract

sugar

3 tbsp (scant 3 envelopes) gelatin powder

salt

Garnish with:

small mint leaves

Serve with:

fruit salad or vanilla ice cream

Follow the directions given in the previous recipe for Japanese Sweet Red Bean Paste for soaking and cooking the adzuki beans. Sweeten the purée but do not add any oil. Simmer the purée until it has reduced to a paste. In the early stages it will only need stirring occasionally but as it thickens you will need to stir almost continuously, otherwise it may catch and burn because no oil has been added.

Remove the red bean paste from the heat. Pour ½ cup water into a small, heavy-bottomed saucepan, add 1½ tbsp sugar and the gelatin powder, and simmer gently for a few minutes while stirring until the sugar and gelatin have completely dissolved. Stir into the hot bean paste, mixing very thoroughly. Rinse the inside of a small, rectangular dish with cold water (you will need a dish of about 3-cup capacity) and fill with the mixture. Tap the bottom of the dish gently on the work top to make it settle and smooth the surface level with a spatula. Leave to set, then chill in the refrigerator unless you wish to serve it soon after it has jelled.

To serve the jelly, run the blade of a knife around the inside of the dish or container to loosen, and turn out onto a pastry board; cut it into fairly thick slices and serve, garnishing with mint leaves.

·MUSHROOMS AND TRUFFLES·

MUSHROOM AND PARMESAN SALAD

p. 249

Preparation: 20 minutes
Difficulty: very easy
Appetizer

MUSHROOM SALAD WITH SWISS CHEESE AND CELERY

p. 249

Preparation: 30 minutes
Difficulty: very easy
Appetizer

MUSHROOM AND TRUFFLE SALAD

p. 249

Preparation: 15 minutes
Difficulty: very easy
Appetizer

HOT MUSHROOM TERRINE WITH CELERY ROOT AND TOMATO SAUCE

p. 249

Preparation: 40 minutes
Cooking time: 50 minutes
Difficulty: easy
Appetizer

MUSHROOM AND JERUSALEM ARTICHOKE SALAD

p. 250

Preparation: 35 minutes
Difficulty: easy
Appetizer

MUSHROOM, RICE AND TRUFFLE SALAD WITH SHERRY MAYONNAISE

p. 250

Preparation: 30 minutes
Cooking time: 20 minutes
Difficulty: easy
Appetizer

MUSHROOM AND NOODLE SALAD WITH SESAME SAUCE

Rien pan san su

p. 251

Preparation: 30 minutes
Cooking time: 15 minutes
Difficulty: easy
Appetizer

HONEY MUSHROOMS WITH ASPARAGUS TIPS AND BASIL DRESSING

p. 252

Preparation: 35 minutes
Cooking time: 15 minutes
Difficulty: easy
Appetizer

*Truffled eggs
with polenta*

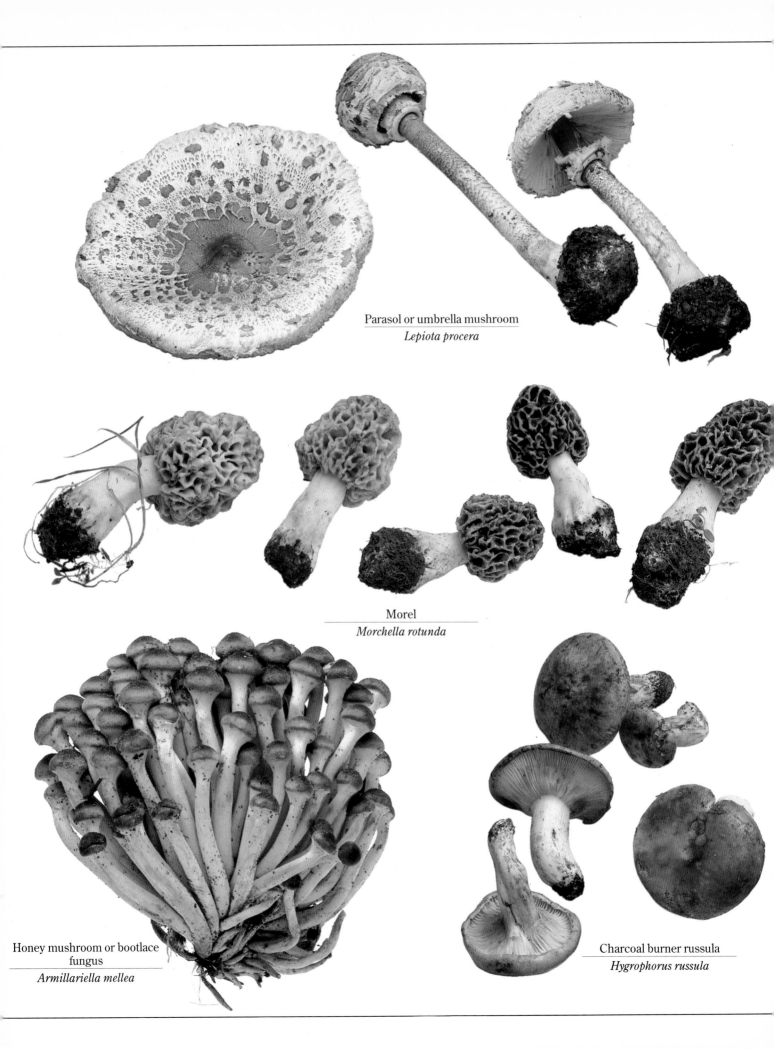

Parasol or umbrella mushroom
Lepiota procera

Morel
Morchella rotunda

Honey mushroom or bootlace
fungus
Armillariella mellea

Charcoal burner russula
Hygrophorus russula

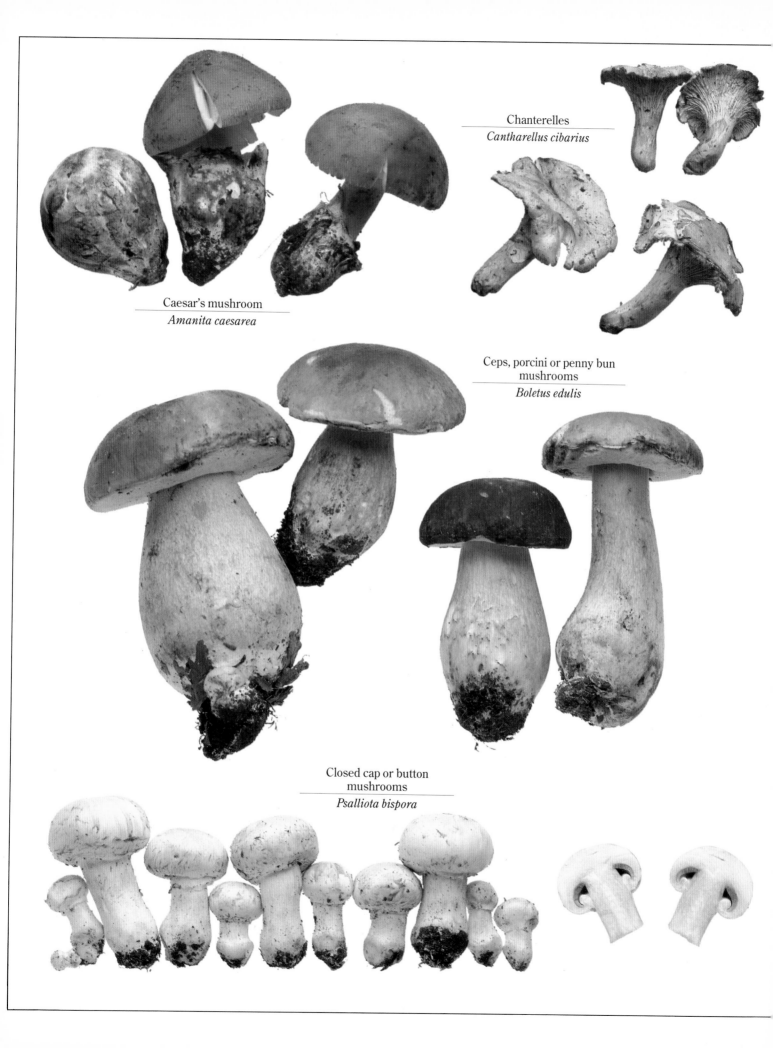

Caesar's mushroom
Amanita caesarea

Chanterelles
Cantharellus cibarius

Ceps, porcini or penny bun
mushrooms
Boletus edulis

Closed cap or button
mushrooms
Psalliota bispora

CHICKEN AND TRUFFLE SALAD WITH ANCHOVY DRESSING

p. 252

Preparation: 15 minutes
Cooking time: 10 minutes
Difficulty: easy
Appetizer

MUSHROOM SOUP

p. 253

Preparation: 20 minutes
Cooking time: 15 minutes
Difficulty: easy
First course

CHINESE CHICKEN AND MUSHROOM SOUP

p. 253

Preparation: 25 minutes + 30 minutes
soaking time
Cooking time: 5 minutes
Difficulty: easy
First course

MUSHROOM RISOTTO

p. 254

Preparation: 20 minutes
Cooking time: 25 minutes
Difficulty: easy
First course

FRIED RICE WITH MUSHROOMS

p. 254

Preparation: 20 minutes
Cooking time: 4 minutes (with ready-cooked rice)
Difficulty: easy
First course

TRUFFLE RISOTTO

p. 255

Preparation: 15 minutes
Cooking time: 25 minutes
Difficulty: easy
First course

PASTA WITH MUSHROOMS

p. 255

Preparation: 20 minutes
Cooking time: 10 minutes
Difficulty: easy
First course

TAGLIATELLE WITH TRUFFLE

p. 255

Preparation: 5 minutes
Cooking time: 3 minutes
Difficulty: very easy
First course

PASTA WITH MUSHROOMS AND PINE NUTS

p. 256

Preparation: 15 minutes + 30 minutes soaking time
Cooking time: 30 minutes
Difficulty: very very easy
First course

MUSHROOM VOL-AU-VENTS POLISH STYLE

p. 256

Preparation: 25 minutes
Difficulty: easy
First course

POTATO GNOCCHI WITH MUSHROOM SAUCE

p. 257

Preparation: 1 hour
Cooking time: 25 minutes
Difficulty: fairly easy
First course

MUSHROOM AND POTATO BAKE

p. 257

Preparation: 30 minutes
Cooking time: 40 minutes
Difficulty: easy
Main course

MUSHROOM AND ASPARAGUS PIE WITH FRESH TOMATO SAUCE

p. 258

Preparation: 45 minutes
Cooking time: 20 minutes
Difficulty: easy
Main course

MUSHROOM AND SPINACH STRUDEL WITH RED WINE SAUCE

p. 258

Preparation: 40 minutes
Cooking time: 30 minutes
Difficulty: easy
Main course

MUSHROOM AND CHEESE PUFF PASTRY PIE

p. 259

Preparation: 35 minutes
Cooking time: 25 minutes
Difficulty: easy
Main course

HOT MUSHROOM MOLDS WITH FRESH TOMATO SAUCE

p. 260

Preparation: 30 minutes
Cooking time: 45 minutes
Difficulty: easy
Main course

Chinese chicken and mushroom soup

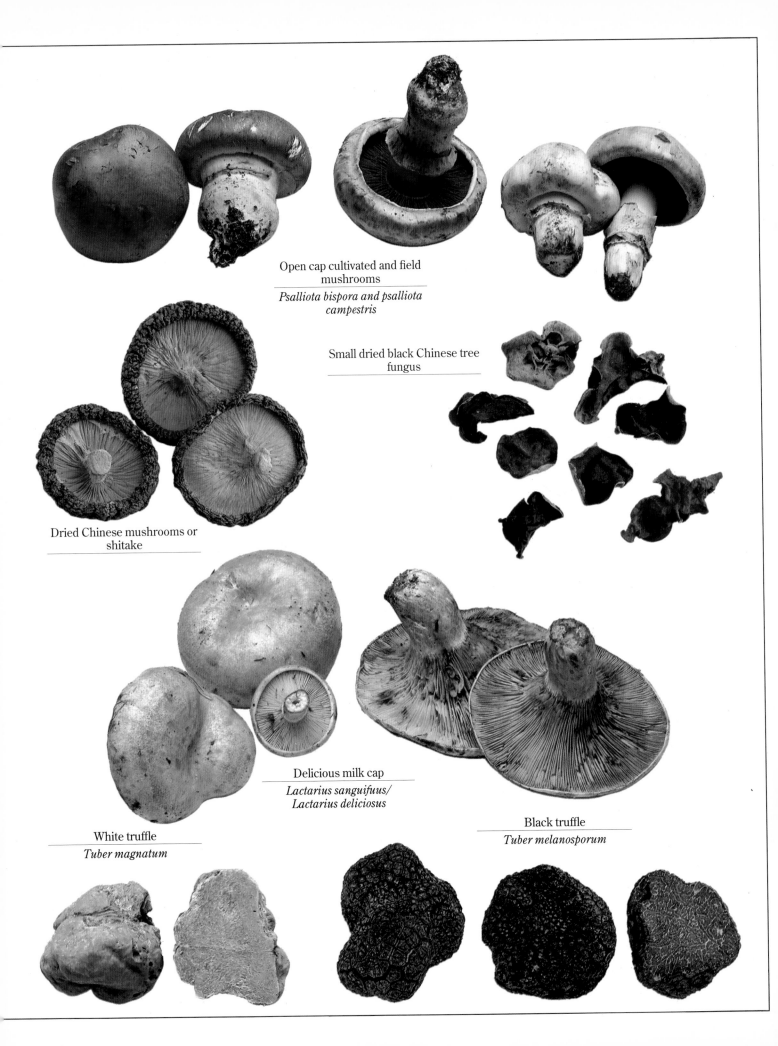

Open cap cultivated and field mushrooms

Psalliota bispora and psalliota campestris

Small dried black Chinese tree fungus

Dried Chinese mushrooms or shitake

Delicious milk cap

Lactarius sanguifuus/ Lactarius deliciosus

Black truffle

Tuber melanosporum

White truffle

Tuber magnatum

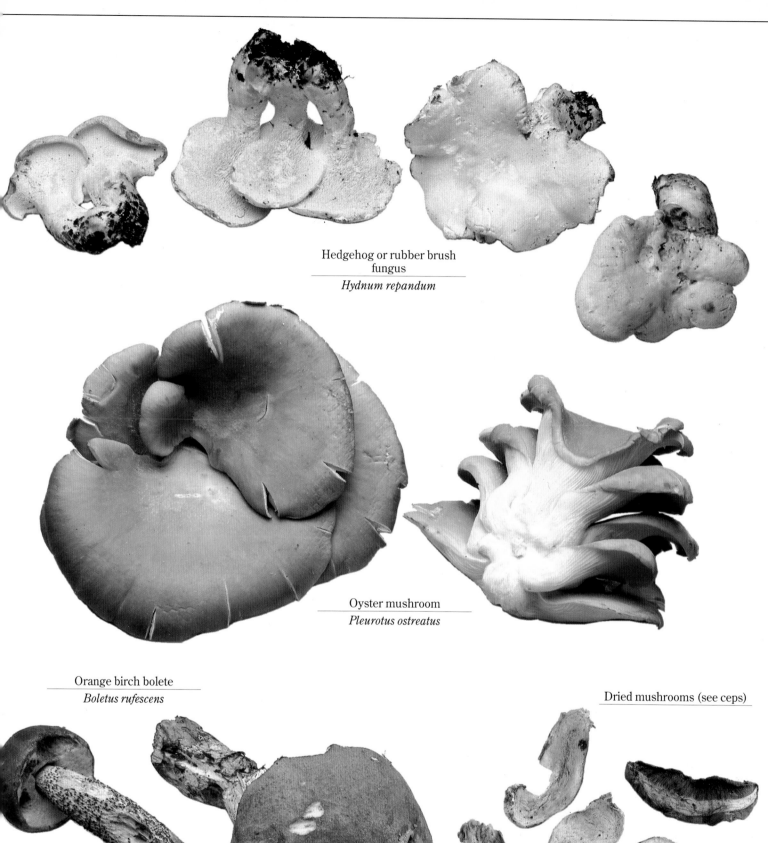

Hedgehog or rubber brush
fungus

Hydnum repandum

Oyster mushroom

Pleurotus ostreatus

Orange birch bolete

Boletus rufescens

Dried mushrooms (see ceps)

*Gnocchi with
mushroom sauce*

DEEP-FRIED MUSHROOMS WITH TARTARE SAUCE

p. 261

*Preparation: 25 minutes
Cooking time: 15 minutes
Difficulty: easy
Main course*

MUSHROOM AND HERB OMELET

p. 261

*Preparation: 20 minutes
Cooking time: 15 minutes
Difficulty: easy
Main course*

POLENTA WITH MUSHROOM TOPPING

p. 262

*Preparation: 30 minutes
Cooking time: 1 hour
Difficulty: easy
Main course*

FRIED BREADED MUSHROOMS

p. 262

*Preparation: 15 minutes
Cooking time: 25 minutes
Difficulty: easy
Main course*

MUSHROOM RISSOLES

p. 263

*Preparation: 25 minutes
Cooking time: 30 minutes
Difficulty: easy
Main course*

BAKED TRUFFLED EGGS WITH POLENTA

p. 263

*Preparation: 5 minutes
Cooking time: 40 minutes
Difficulty: easy
Main course*

MUSHROOMS BORDEAUX STYLE

p. 264

*Preparation: 25 minutes
Cooking time: 10 minutes
Difficulty: easy
Accompaniment*

MOREL MUSHROOMS WITH ASPARAGUS AND MOUSSELINE SAUCE

p. 264

*Preparation: 30 minutes
Cooking time: 12 minutes
Difficulty: easy
Accompaniment or appetizer*

MUSHROOM AND PARMESAN SALAD

4 cups mushrooms (see method)

1 piece Parmesan cheese with no rind

olive oil

salt

Use freshly gathered young field mushrooms or the freshest small brown mushrooms or white, closed cap cultivated mushrooms. Wipe the cap with a damp cloth and trim off the ends of the stalks. Just before serving, cut the mushrooms into thin slices, arrange on individual plates, and sprinkle with a small pinch of salt and a little oil. Scatter pieces of Parmesan cheese over the mushrooms and serve.

———•———

MUSHROOM SALAD WITH SWISS CHEESE AND CELERY

4 cups mushrooms (see method)

3 thin slices Swiss cheese

2 green celery stalks

olive oil

salt

See preceding recipe for choice and preparation of mushrooms. Prepare them but do not slice until just before serving. Cut the cheese into thin strips. Wash, dry, and trim the celery stalks, and run the potato peeler down the curved, outer sides to get rid of any strings. Chop finely.

Mix the cheese and celery in a bowl with a little oil and salt. Transfer to individual plates. Slice the mushrooms thinly and lay over the cheese and celery. Sprinkle with a little more oil and salt and serve at once.

MUSHROOM AND TRUFFLE SALAD

4 cups button mushrooms

1–2 canned truffles

extra-virgin olive oil

1 tsp fresh lemon juice

salt

Prepare the mushrooms as described for Mushroom and Parmesan Salad. Save the liquid from the can of truffles. Canned pieces of truffle are cheaper than whole ones. Slice the mushrooms very thinly into a bowl just before serving. Sprinkle with a little oil, 1 tsp lemon juice, a pinch of salt, and the reserved truffle liquid. Slice the truffles wafer thin and scatter all over the salad.

———•———

HOT MUSHROOM TERRINE WITH CELERY ROOT AND TOMATO SAUCE

1¼ cups fresh, trimmed cleaned ceps (boletus edulis)

3 tbsp olive oil

½ clove garlic, finely chopped

3 tbsp finely chopped parsley

1 cup heavy cream

1 cup milk

2 eggs

4 extra egg yolks

nutmeg

1½ tbsp freshly grated Parmesan cheese

3 tbsp all-purpose flour

salt and pepper

For the sauce:

1 celery root

6 tbsp olive oil

249

2 sprigs thyme
2 large ripe tomatoes or 4 canned tomatoes
4–6 large sprigs basil
1 tsp lemon juice
1½ tbsp unsalted butter
salt and pepper

Preheat the oven to 350°F. Wipe the mushrooms and trim off the ends of their stalks and slice them thinly. Cook for a few minutes in the oil with the garlic and half the chopped parsley. Season; process one-third of the mixture in the blender with the cream, milk, 2 whole eggs and the extra egg yolks, the Parmesan, salt, pepper, and nutmeg. Transfer to a bowl and mix in the reserved mushrooms and the rest of the parsley.

Grease a 2½-pint (5-cup) capacity lidded terrine dish with butter and line the bottom with greased waxed paper. Fill with the mixture, cover with the lid, sealing the join with a luting paste made with 3 tbsp flour mixed and a little cold water; leave the small hole in the lid free for the steam to escape. Place in a roasting pan, add sufficient boiling water to come halfway up the sides of the terrine, and place in the oven. Cook for 50 minutes.

Make the sauce. Peel the celery root, slice very thinly using a mandoline cutter or food processor, and cut these slices into *julienne* strips (see page 24). Heat the oil in a wide, fairly deep skillet with the thyme and stir-fry the celery root over moderately high heat until tender but still fairly crisp. Add the peeled, seeded, and coarsely chopped tomatoes and the basil leaves torn into thin strips. Season with salt and pepper and simmer for 2–3 minutes, stirring. Set aside until the mushroom terrine is cooked, then reheat the sauce just before serving, adding 1 tsp lemon juice. Stir in the butter.

Turn the hot, cooked terrine out onto a flat plate or chopping board; slice fairly thickly and serve the slices at once on heated plates with the sauce.

———•———

MUSHROOM AND JERUSALEM ARTICHOKE SALAD

1¼ cups fresh ceps (boletus edulis*) or substitute (see method)*
1 lb Jerusalem artichokes
1 lemon
olive oil
salt and pepper

Substitute large, closed cap cultivated mushrooms or freshly gathered field mushrooms for the ceps if necessary. Wipe with a damp cloth and trim off the ends of the stalks. If preparing several hours in advance, wrap in a damp cloth and chill in the salad compartment of the refrigerator. Peel the artichokes, dropping each one into a large bowl of cold water mixed with the freshly pressed lemon juice, to prevent discoloration. This stage can be completed several hours in advance.

Shortly before serving the salad, slice the mushroom caps and stalks thinly. Cut the well drained and dried Jerusalem artichokes into very thin slices, using a mandolin cutter or the slicer of your food processor. Mix the 2 vegetables gently in a large salad bowl, adding 1–2 tbsp oil and a little salt and freshly ground white pepper. Serve without delay or the artichoke slices will start to discolor.

———•———

MUSHROOM, RICE, AND TRUFFLE SALAD WITH SHERRY MAYONNAISE

2 cups risotto rice (e.g. arborio)
1 cup dry white wine
½ cup asparagus tips, steamed
1 cup chopped white celery

1 cup ceps (boletus edulis)
1 fresh or canned white or black truffle
For the sherry mayonnaise:
1 yolk from a large, hard-boiled egg
3 tbsp lemon juice
3 tbsp dry sherry
1 tsp dark, mild French mustard
1 tsp paprika
½ cup light olive oil
¼ cup light or heavy cream
salt and freshly ground white pepper

Preheat the oven to 350°F. Sprinkle the rice into a saucepan of fast boiling salted water; boil fast for 8 minutes. Drain, transfer to a casserole dish greased with butter, and sprinkle the wine all over it, together with the liquid from canned truffles if used. Cover tightly and cook in the oven until it has absorbed all the wine. Uncover and allow to cool. Rinse.

Make the sherry mayonnaise: mince the hard-boiled egg yolk in a bowl with a fork and gradually work in the lemon juice, sherry, mustard, paprika, and a little salt and pepper. Gradually add the oil, and the cream. Combine the rice with the celery, asparagus, and sliced mushrooms, together with the dressing. Slice the truffle into wafer-thin pieces and scatter all over the surface.

— • —

MUSHROOM AND NOODLE SALAD WITH SESAME SAUCE
Rien pan san su

1 packet mung bean noodles or cellophane noodles
4 dried mushroom caps (shitake, or Chinese dried mushrooms)

2 eggs
pinch sugar
pinch ajinomoto (Japanese taste powder) (optional)
2 thick slices lean ham, shredded
1 cucumber
sunflower or peanut oil
salt
For the sesame sauce:
½ cup sesame oil
¼ cup sweet Japanese vinegar
1 tsp Japanese wasabi powder
¼ cup Japanese light soy sauce
¼ cup mirin or sweet saké

small pinch ajinomoto (Japanese taste powder) (optional)

Bring a large saucepan of lightly salted water to a boil ready to cook the Japanese mung bean noodles or Chinese cellophane noodles. Soak these in a bowl of hot water for 30 minutes. Also soak the dried mushrooms in hot water for 30 minutes, drain, reserving ½ cup of water, and squeeze excess moisture out of the mushrooms. Shred the caps into thin strips, discarding the stalks. Beat the eggs lightly with a pinch each of salt and sugar and a small pinch of *ajinomoto* taste powder (optional). Heat 1½ tbsp sunflower oil in a wide nonstick skillet and add the beaten egg mixture, tipping the skillet this way and that to spread it out thinly over the bottom. Cook over low heat for a few minutes until completely set but not at all browned. Slide the very thin omelet out onto a large plate or chopping board and shred into strips. Peel, rinse and dry the cucumber; cut off both ends. Shred into thin strips.

Make the sauce by mixing all the ingredients thoroughly in a bowl, diluting with the strained reserved water from the soaked mushrooms.

Add the drained noodles to the boiling water, cook for 3 minutes only, then drain in a sieve and hold the sieve under cold running water to rinse and cool quickly. Cut into short lengths, transfer to a serving bowl, and arrange little multi-colored bundles of the

251

ham, cucumber, mushroom, and omelet strips over the surface. Cover the bowl with plastic wrap and chill in the refrigerator for 30 minutes. Serve very cold with the sesame sauce.

———•———

Mushrooms with Asparagus Tips and Basil Dressing

16 large asparagus spears
1 lb honey mushrooms or substitute (see method)
1 large shallot or ½ Bermuda onion, finely chopped
1 clove garlic, minced
½ cup vegetable stock (see page 38)
3 basil leaves
¼ cup light olive oil
salt and pepper
For the basil dressing:
6 tbsp pesto sauce (see page 177)
¼ cup light cream
salt
Garnish with:
diced tomato

Prepare the asparagus spears as directed on page 17; double the number of spears if you are using thin, green asparagus. Cut off about 2 in from the larger, tougher ends of the stalks. Make up 4 equal bundles and tie each securely but not too tightly with two pieces of string; steam or boil in salted water for about 12 minutes. If they are ready before you have finished preparing and cooking the mushrooms, drain and keep hot.

If you cannot gather wild honey mushrooms (also known as bootlace fungus), use very fresh baby but-ton mushrooms. If using honey mushrooms, separate the caps from the stalks, and cut off the tough ends of the stalks; rinse and drain the caps and stalks well. Cut the larger stalks lengthwise.

Heat the oil in a very wide nonstick skillet and stir-fry the mushrooms with the shallot or Bermuda onion and the garlic over high heat for 2–3 minutes. Season with a little salt and pepper, add the stock and the coarsely chopped basil leaves. Cover and simmer gently for 6 minutes or until the mushrooms are cooked.

Stir the cream into the pesto sauce. Untie the asparagus bundles and arrange the spears on each individual plate. Spoon a little of the basil dressing over the tips, place the mushrooms to one side of them, and decorate with diced tomato.

———•———

Chicken and Truffle Salad with Anchovy Dressing

2 boneless skinned chicken breasts
1½ tbsp lemon juice
½ lb mixed young green leaf salad (see method)
½ cup chicken stock (see page 37)
½ cup dry white wine
2 finely chopped shallots or ½ mild Bermuda onion
2 sprigs thyme
1 black truffle (canned in juice or preserved in oil)
salt and pepper
For the anchovy dressing:
2½ tsp liquid or oil from canned or bottled truffle
3 tbsp fresh lemon juice
3 tbsp white wine vinegar
½ cup olive oil
2 small canned anchovy fillets
1½ tbsp chopped chives
pinch pepper

Season the chicken breasts lightly with salt and pepper and sprinkle with lemon juice on both sides. Leave to marinate while you trim and wash the salad greens (use very small-leaved, young greens such as arugula, curly endive, lamb's tongues, etc.). Dry these well and mix; arrange a bed of them on each individual plate. Place the chicken breasts and any unabsorbed juice in a fairly small nonstick skillet with the stock, wine, shallot, and thyme. Heat to just below boiling point, cover, reduce the heat further, and barely simmer for 6 minutes, turning after 3 minutes. Test by inserting a knife into the thickest part; if the juice runs out at all pink it is not cooked, so continue simmering until done.

Make the anchovy sauce by placing all the ingredients in the blender and processing at high speed until smoothly blended. Slice the chicken breasts. Fan some of these slices out, slightly overlapping, on each bed of salad greens, cover them with a thin coating of anchovy sauce, and sprinkle with wafer-thin slivers of truffle.

———•———

MUSHROOM SOUP

½ lb ox-tongue or beefsteak fungus or field mushrooms (see method)
1 quart beef stock (see page 37)
2 shallots, finely chopped
1 clove garlic, minced
1 tomato, peeled, seeded, and diced
1½ tbsp chopped parsley
4 thick slices coarse white bread
salt and freshly ground pepper

Heat the stock while you prepare the mushrooms, brushing and wiping them clean with a damp cloth and trimming off the ends of their stalks. You may prefer to use brown or white cultivated mushrooms for this recipe. Cut the mushrooms into fairly small dice. Heat ¼ cup olive oil in a wide, shallow saucepan and fry the shallots and garlic gently for 5 minutes, stirring frequently. Turn up the heat, add the diced mushrooms and a pinch of salt, and stir-fry for about 2 minutes. Add the boiling stock and the diced tomato. Simmer, uncovered, for 8 minutes or until the mushrooms are done. Add a little more salt if needed and sprinkle in the parsley.

Place the slices of bread in a slow oven until crisp and dry. Place 1 slice in each soup bowl and ladle in the soup. Season to taste.

———•———

CHINESE CHICKEN AND MUSHROOM SOUP

¼ cup dried Chinese mushroom caps (shitake)
2 black frilly Chinese fungus (Jew's ears or tree ears)
½ cup raw chicken breast, shredded
¼ cup fillets of very fresh white fish (e.g. Dover sole, flounder)
¼ cup peeled shrimp
1 quart chicken, crustacean or vegetable stock (see pages 37–38)
3–4 canned bamboo shoots, drained
4 trimmed spinach leaves, cut lengthwise in half
salt
For the marinade:
pinch monosodium glutamate (optional)
pinch freshly ground white pepper
1½ tbsp Chinese rice wine or dry sherry
1½ tbsp cornstarch
1 tsp salt

Soak both sorts of Chinese mushrooms separately in warm water for at least 30 minutes. Cut off and discard the stalks from the shitake and trim off any tough pieces from the frilly black fungus. Rinse and squeeze out excess moisture. Cut into strips and set aside. While the mushrooms are soaking, mix all the marinade ingredients in a large bowl. Add the chicken, fish, and shrimp and mix thoroughly by hand or with a wooden spoon. Leave to stand for 15 minutes.

Bring the stock to a boil, and stir in all the ingredients except the spinach. Simmer gently for 3 minutes and then add the spinach. Simmer for another minute, then remove from the heat. Serve piping hot at the beginning of a Western-style meal or as the last course in a traditional Chinese meal.

———•———

MUSHROOM RISOTTO

1¼ cups fresh ceps (boletus edulis) or 1 cup dried ceps
¼ cup clarified butter (see page 36)
For the basic risotto:
1½ quarts chicken stock (see page 37)
¼ cup butter
2 shallots or 1 Bermuda onion, finely chopped
1½ cups risotto rice (e.g. arborio)
¼ cup dry white wine or dry vermouth
6 tbsp grated Parmesan cheese
salt and freshly ground white pepper

If you are using dried mushrooms, soak them in a bowl of warm water 30 minutes before starting preparation. Boil the stock over moderate heat. While it is heating, melt 2 tbsp of the butter in a fairly deep, wide heavy-bottomed skillet or a large, shallow saucepan and cook the shallot or onion over low heat for 5 minutes, stirring frequently. Add the rice and cook, stirring, over slightly higher heat for 1½ minutes. Sprinkle in the wine or vermouth, cook briefly, until it has evaporated, and then add 1 cup of boiling hot stock. Cook until the rice has absorbed almost all the liquid, stirring occasionally. Add more stock

and continue cooking, adding more boiling liquid whenever necessary and stirring occasionally; risotto rice absorbs a great deal of moisture. Test the rice after 14 minutes; the grains should be tender but still have a little resistance left. If necessary, cook for another minute. This risotto should be very moist.

While the rice is cooking, prepare the mushrooms. Wipe carefully with a slightly dampened cloth and cut off the stalks. Rinse briefly if desired, dry immediately with paper towels, and slice. If you are using dried mushrooms, drain them, squeeze out excess moisture, and cut into small pieces. Heat the clarified butter in a nonstick skillet and stir-fry the mushrooms for 2–3 minutes over fairly high heat. Season lightly with salt and freshly ground pepper. As soon as the risotto is done add the remaining solid butter, the Parmesan cheese, and the mushrooms.

———•———

FRIED RICE WITH MUSHROOMS

4 cups cold cooked rice
2 cups button mushrooms
3 large cloves garlic, minced
2–3 sage leaves
1½ tbsp butter
¼ cup light olive oil
1½ tbsp finely chopped parsley
salt and pepper

Steam the rice until tender the day before you plan to serve this dish and chill in the refrigerator in an airtight container; take it out of the refrigerator 30 minutes before cooking to bring to room temperature. Trim, rinse, and dry the mushrooms with paper towels. Heat the oil and butter for 30 seconds in a wok or large skillet with the garlic and sage leaves. Turn up the heat to high, add the rice and 1 tsp salt, and stir-fry for 2 minutes. Reduce the heat and break up any rice that has stuck together; continue doing this for 4–5 minutes while the rice heats through. Season to taste with a little more salt and plenty of freshly ground pepper. Stir well and spoon into small Chinese bowls if you have them.

Place the sliced mushrooms on top of the rice just as they are, uncooked, and sprinkle with parsley. The raw mushrooms make a delicious contrast in taste and texture with the hot, crunchy rice.

— • —

TRUFFLE RISOTTO

1 small white or black fresh or canned truffle
For the basic risotto:
1 quantity Mushroom Risotto (see page 254); use an extra ¼ cup freshly grated Parmesan and omit the mushrooms and garlic

Make the basic risotto. Brush the fresh truffle carefully with a soft brush and slice some wafer-thin over the risotto. Stir the truffle (and the liquid if canned) into the risotto at the same time as you add the grated Parmesan and remaining solid butter.

— • —

PASTA WITH MUSHROOMS

½ lb fresh ceps (boletus edulis*)*
8 oz fresh thin green ribbon noodles

2 cloves garlic, minced
¼ cup olive oil
3 tbsp chopped parsley
1 tbsp butter
salt and pepper

Boil a large saucepan of salted water to cook the pasta. Trim and wipe the fresh mushrooms, rinse briefly, and dry well with paper towels; slice about ⅛ in thick. Heat the oil in a wide, fairly shallow saucepan or deep skillet and fry the mushrooms with the garlic over moderate heat for 2 minutes. Lower the heat; add the parsley, a little salt and pepper, and 3 tbsp of the boiling salted water. Cover and simmer for 3–4 minutes. Remove from the heat, add a little salt to taste, and stir in the solid butter, until it has just melted.

Add the pasta to the fast boiling water; the noodles will only take a few minutes to cook. Drain briefly. Add the mushrooms and mix well.

— • —

TAGLIATELLE WITH TRUFFLE

½ lb fresh tagliatelle
¼ cup butter
¼ cup freshly grated Parmesan cheese
1 small white or black truffle
salt

Bring a large saucepan of salted water to a full boil, add the pasta, and cook for 2–3 minutes. Packaged, dry pasta will take a little longer and you will need only about ½–⅔ of this weight for an appetizer. Drain when tender but still firm, leaving 2–3 tbsp of water behind in the saucepan with the pasta. Stir in the solid butter and cheese over very low heat just long enough to melt the butter. Mix well. Serve at once, with wafer-thin slices of fresh truffle.

———— • ————

PASTA WITH MUSHROOMS AND PINE NUTS

½ cup dried ceps (boletus edulis)

½ lb fresh pasta shapes

1 baby carrot

½ small onion

½ small celery stalk

1 clove garlic

3 tbsp chopped parsley

1 bay leaf

¼ cup olive oil

3 tbsp finely chopped tomato

1 cup chicken stock (see page 37)

2 tbsp pine nuts

⅓ cup freshly grated Parmesan cheese

salt and pepper

Soak the dried mushrooms for at least 30 minutes in warm water. Heat plenty of salted water in a large saucepan for the pasta. Drain the mushrooms, discarding the water. Squeeze out excess moisture and chop them coarsely. Heat the olive oil in a small saucepan and sweat the finely chopped carrot, onion, celery, garlic, parsley, and bay leaf in it over very low heat for 10 minutes with the lid on, stirring occasionally. Turn up the heat, add the mushrooms and a little salt and pepper, and stir for 2–3 minutes. Add the chopped fresh tomato and the stock and simmer for 20 minutes, partially covered, so the sauce can reduce and thicken gradually. After this time has passed, remove the lid and cook over slightly higher heat to thicken further. Cook the

pine nuts with a very little olive oil in a nonstick skillet for 1–2 minutes, until golden. Set aside.

Add the pasta to the fast boiling water and cook until just tender; drain briefly, leaving ¼ cup of the cooking liquid with them in the saucepan. Place on a very hot serving dish, stir the pine nuts quickly into the sauce, add this to the pasta, and stir briefly. Sprinkle with Parmesan cheese and serve at once.

———— • ————

MUSHROOM VOL-AU-VENTS POLISH STYLE

8 vol-au-vent cases

1¼ cups fresh ceps (boletus edulis) or ox-tongue or beefsteak fungus

½ cup finely chopped onion

1 shallot, finely chopped

3 tbsp butter

¼ cup dry sherry or dry white wine

¼ cup vegetable stock (see page 37)

2 tbsp paprika

pinch cayenne pepper or chili powder

1½ tbsp all-purpose flour

½ cup milk

½ cup sour cream

2 tbsp finely chopped parsley

salt and pepper

Preheat the oven to 350°F; heat the vol-au-vent cases in it for 10 minutes before you fill them.

Make the filling: trim and wipe the mushrooms. Chop coarsely. If you use dried *boletus edulis* mushrooms, allow about 4 oz and soak for at least 30 minutes in lukewarm water before draining and squeezing out excess moisture. Chop fresh or dried pre-soaked mushrooms coarsely.

Cook the onion and shallot in the butter for 10 minutes. When wilted and tender, add the mushrooms and sauté over higher heat for 2–3 minutes. Season with a little salt and pepper and add the sherry and stock. Stir while cooking for 2–3 minutes to allow the liquid to reduce. Lower the heat, sprinkle in the paprika, cayenne or chili powder, and the flour, stir well, then gradually stir in the milk and cream. Cover and cook gently over low heat for 4 minutes, then remove the lid and reduce the sauce over moderate heat for a few minutes. Add the parsley. Fill the hot vol-au-vent cases with this mixture and serve.

———•———

POTATO GNOCCHI WITH MUSHROOM SAUCE

1¼ lb mixed wild and cultivated mushrooms

1 package plain potato gnocchi (see method)

1 medium-sized onion, finely chopped

2 sage leaves

1 small can tomatoes

3 tbsp oil

½ small clove garlic, minced

2 tbsp finely chopped parsley

⅛ cup freshly grated Parmesan cheese

1 tbsp butter

salt and pepper

If you wish to make your own gnocchi, follow the recipe on page 177. Trim and wipe or rinse the mushrooms, cutting the medium-sized and large caps into slices. Leave any tiny mushroom caps whole. Cut the larger stalks lengthwise in half.

Cook the onion and sage leaves in the oil over low heat for 10 minutes, in a fairly deep, very wide skillet or large, shallow saucepan. Stir frequently. Turn up the heat and add the mushrooms. Sprinkle with a pinch of salt and a generous amount of freshly ground pepper and sauté for 2–3 minutes. Dilute the gravy with 3 tbsp boiling water and add to the mushrooms, together with the canned tomatoes. Stir, reduce the heat, cover, and simmer gently for about 10 minutes or until tender. Keep hot.

A few minutes before it is time to serve this dish, add the potato gnocchi to a very large saucepan of fast boiling salted water. Remove and drain as soon as they bob up to the surface. While they are cooking, take the hot sauce off the heat, stir in the garlic, parsley, Parmesan and the solid butter. Pour the sauce over the gnocchi and serve at once.

———•———

MUSHROOM AND POTATO BAKE

4 medium-sized potatoes

8 large open cap mushrooms

¼ cup finely chopped parsley

1–2 cloves garlic, minced

olive oil

salt and freshly ground pepper

Preheat the oven to 350°F. Grease a roasting pan with olive oil. Peel the potatoes and cut into slices about ½ in thick. Spread these out in a single, slightly overlapping layer and season with salt, pepper, 3 tbsp olive oil, and 1 cup water. Cut the stalks off the mushrooms. Chop the stalks with the parsley and garlic and then mix in a bowl with 2 tbsp olive oil, salt, and pepper.

Place the mushroom caps upside down and spread an equal quantity of the mixture neatly over each one; spread out in a single layer on top of the potatoes and cover with greased foil. Bake for 40–45 minutes.

MUSHROOM AND ASPARAGUS PIE WITH FRESH TOMATO SAUCE

¼ lb fresh ceps (boletus edulis)*, wild field mushrooms or large closed or open cap cultivated mushrooms*

18 large asparagus spears

⅛ lb celery root

1 large shallot

1 sprig thyme

1½ tbsp chopped basil

¼ lb Swiss cheese

¼ lb smoked cheese

1 8-oz package frozen puff pastry, thawed

1 egg

1½ tbsp poppy seeds

butter

3 tbsp light olive oil

salt and pepper

For the tomato sauce:

generous 1 lb ripe tomatoes

1–1½ tbsp capers

1 small anchovy fillet

½ clove garlic

generous pinch oregano

3 tbsp extra-virgin olive oil

4–5 basil leaves, finely chopped

1 lemon

salt and freshly ground pepper

Preheat the oven to 400°F. Trim and wipe or rinse and dry the mushrooms. If dried mushrooms are used, you will need 1½–2 oz; soak in warm water for at least 30 minutes then drain, squeeze out excess moisture, and blot dry. Prepare the asparagus (see page 17) and steam or boil until they are just tender. Peel the celery root, grate it fairly coarsely, and immediately sauté it in a large nonstick skillet or saucepan over a moderate heat in 2 tbsp butter, until it is wilted and tender but still has a hint of crispness when tested. Season and set aside.

Cook the finely chopped shallot in 3 tbsp light olive oil in the nonstick skillet with the mushrooms and thyme until just tender. Add the chopped basil and season lightly with salt and pepper. Stir and remove from the heat. Grate the two types of cheese, keeping them separate. Roll out the pastry into a thin, square sheet on a lightly floured board or work surface. Cut the square in half and place one half on a greased cookie sheet or in a jelly roll pan. Beat the egg lightly in a small bowl or cup and brush half of it over the surface of this first pastry sheet. Spread out the mushroom mixture evenly over the glazed surface, stopping ½ in short of the edges all the way round. Arrange half the asparagus spears head to tail, spacing them out evenly in a single layer on top of the mushroom mixture. Cover them evenly with the celery root. Top with the Swiss cheese. Arrange another layer of the remaining asparagus and top with the grated smoked cheese. Cover with the other half of the pastry and seal the edges well by pressing them together all the way round with the tines of a fork.

Brush the remaining beaten egg all over the surface of the pie and sprinkle evenly with the poppy seeds. Bake for 20 minutes.

Make the sauce. Blanch, peel, and seed the tomatoes and place them in the blender. Add the capers, anchovy, garlic, oregano, 3 tsp olive oil, salt, and freshly ground pepper and process until smooth. Stir in the chopped basil and a little lemon juice to taste.

MUSHROOM AND SPINACH STRUDEL WITH RED WINE SAUCE

2 cups ceps (boletus edulis) *or cultivated mushrooms*

3 tbsp oil

1 clove garlic

2 tbsp finely chopped parsley

1–1¼ lb spinach

1½ tbsp butter

2 shallots, finely chopped

1 cup ricotta cheese

2 eggs

⅓ cup freshly grated hard cheese

1 8-oz package frozen puff pastry, thawed

1½ tbsp white sesame seeds

nutmeg

salt and pepper

For the sauce:

4 oz ham

½ cup dry red wine

2 tbsp marsala

1 shallot, finely chopped

1 sprig fresh thyme

1 bay leaf

1 cup beef stock (see page 37)

1 tsp cornstarch or potato flour

black peppercorns

Heat plenty of water in a large saucepan, adding a generous pinch of salt. Preheat the oven to 400°F. Trim the mushrooms, wipe with a damp cloth or rinse briefly and dry. Substitute closed cap field mushrooms or large, open cap horse mushrooms or cultivated mushrooms if necessary. Slice the mushrooms fairly thinly and sauté in the oil with a whole, unpeeled clove of garlic. Sprinkle with a small pinch of salt and a generous amount of freshly ground pepper, stir in the parsley, and set aside. Trim and wash the spinach very thoroughly; use only the leaves for this recipe. Add to the large saucepan of boiling salted water and blanch for 3 minutes; drain well. Heat the butter in a skillet, fry the shallot gently until tender, add the spinach, and sauté over higher heat for 2–3 minutes. Add a pinch of freshly grated nutmeg and set aside. Beat the ricotta smooth with a wooden spoon, then beat in 1 whole egg and the

hard cheese, a pinch of nutmeg, salt, and pepper and a very little more garlic if wished, finely minced.

Roll out the pastry into a large square, cut in half, and place one half on a cookie sheet. Spread all the spinach over the pastry, keeping well clear of the edges; cover with the mushroom mixture and then top with the ricotta, using a pastry bag with a large tip to make this easier. Cover with the other half of the pastry, sealing the edges very securely by pressing them together with the tines of a fork. Beat the remaining egg lightly and brush over the surface of the pastry; sprinkle the surface with the sesame seeds. Bake for about 20 minutes or until the pastry is crisp, well risen, and golden brown.

Use this baking time to make the sauce. Shred half the ham into strips. Place the rest of the ham in the blender or food processor and reduce to a smooth purée. Pour the wine into a small saucepan with the marsala, the finely chopped shallot, sprig of thyme, bay leaf, and 3 or 4 black peppercorns. Boil gently, uncovered, to reduce by one-third. Mix the cornstarch or potato flour well with the cold stock and stir into the wine mixture followed by the puréed ham. Stir continuously as the mixture heats. Once it has reached boiling point, simmer gently, uncovered, to reduce for about 10 minutes. Strain this rather thin sauce into another saucepan and keep warm. Take the cheese and spinach pie out of the oven when it is done. Reheat the sauce if necessary and stir in the reserved ham strips.

———•———

MUSHROOM AND CHEESE PUFF PASTRY PIE

1½ lb ceps (boletus edulis) or your own choice of mushrooms

1 clove garlic, finely chopped

3 tbsp finely chopped parsley

¼ cup olive oil

1 pint thick béchamel sauce (see page 36)

½ lb mild semisoft cheese, very thinly sliced
1 8-oz package frozen puff pastry, thawed
butter
all-purpose flour
1 egg, lightly beaten
2 tbsp sesame seeds
salt and pepper

Preheat the oven to 400°F. Trim and wipe the mushrooms. Cut into thin slices. Heat the oil in a wok or large skillet and stir-fry the mushrooms with the garlic and parsley over moderately high heat for a maximum of 5 minutes. Add a pinch each of salt and pepper. Simmer briefly then set the wok aside.

Make a béchamel sauce and when it is ready, remove from the heat and stir in the cheese. Return to a low heat, stirring until the cheese melts.

Cut the thawed pastry in half; roll both halves out on a lightly floured pastry board into thin disks or circles about 10 inches in diameter. Use one to line a 9-in shallow tart pan or quiche dish greased with butter and lightly dusted with flour. Lightly beat the egg and using a pastry brush, brush just over half to glaze this pie shell with a thin coating. Cover evenly with two-thirds of the cheese sauce, spread the mushrooms out on top, and then cover them with the remaining sauce. Place the other pastry disk on top as a lid, pinching the edges. Brush the remaining beaten egg over the lid and make a small, neat hole in the center. Sprinkle with the sesame seeds and bake for 20–25 minutes or until golden brown. Serves 6.

———— • ————

HOT MUSHROOM MOLDS WITH FRESH TOMATO SAUCE

¼ lb ceps (boletus edulis) or mushrooms of choice
1 shallot, finely chopped
1 sprig thyme
½ cup vegetable stock (see page 38)
1½ tbsp butter
½ cup heavy cream
½ cup milk
4 eggs
¼ cup grated Parmesan cheese
½ cup Fresh Tomato Sauce (see page 258)
salt and pepper

Preheat the oven to 350°F. Trim, wipe, and thinly slice the mushrooms. If you cannot buy or gather fresh ceps (*boletus edulis*) or closed cap field mushrooms, use very fresh brown or white small cultivated mushrooms. Sweat the shallot in the butter with the sprig of thyme for 5 minutes, increase the heat, add the mushrooms with a pinch of salt, and sauté for 2–3 minutes. Add the stock, cover, and simmer gently for 15 minutes (simmer for about 5–8 minutes if using field or cultivated button mushrooms). When the mushrooms are very tender, add more salt if desired and plenty of freshly ground white or black pepper. Remove the thyme and push the mushrooms through a fine sieve, collecting the purée in a bowl (or use a food processor). Blend in the cream and milk. Separate 2 of the eggs and beat their yolks lightly with the remaining 2 whole eggs, a pinch of salt and pepper, and the Parmesan cheese. Blend into the mushroom mixture.

Grease four 1-cup capacity timbale molds or ramekin dishes with butter. Fill them with the mushroom mixture, place well spaced out in a roasting pan or similar receptacle, and pour in enough boiling water to come about half to two-thirds of the way up the sides of the molds. Cook in the oven for 30 minutes; the mushroom mixture will gradually thicken and set to a firm custard. Make the sauce while the molds are cooking. When they are ready, turn off the oven and leave them inside for 10 minutes, still in their *bain-marie* with the door ajar. Take the molds out of the hot water, unmold, and serve with the sauce.

DEEP-FRIED MUSHROOMS WITH TARTARE SAUCE

1–1¼ lb field mushrooms or small, open cap cultivated mushrooms

3 eggs

2 cups fine fresh breadcrumbs

sunflower oil for frying

salt and pepper

For the tartare sauce:

4 hard-boiled eggs

1½ tbsp mild mustard

1 cup light olive oil

2 tbsp lemon juice

3 gherkins, finely chopped

3–4 tbsp capers, finely chopped

3 tbsp chopped chives

3 tbsp chopped parsley

generous pinch salt

Garnish with:

lemon wedges

small sprigs of parsley

Make the sauce: this is best made by hand using a whisk or hand-held beater rather than in a blender or processor. These quantities will yield about 1½ cups. Use gherkins and capers that have been pickled in a mild wine or cider vinegar; if they have been preserved in acetic acid the sauce will be unpleasantly sharp. Take the yolks out of the hard-boiled eggs and mince very thoroughly with a fork in a bowl; work in the mustard and a generous pinch of salt, ensuring that the mixture is absolutely smooth, with no lumps. Add the olive oil very gradually, as for a classic mayonnaise, starting off by beating in a few drops at a time with a whisk, gradually increasing this to a trickle. When this mixture is thick and light, stir in lemon juice to taste. Place the chopped gherkins and capers in a piece of clean cloth, gather up the edges, and twist round hard to squeeze out all excess moisture, then stir these into the sauce, with the herbs. Beat in the hard-boiled egg whites now if wished. Add a little more salt to taste and season with freshly ground white pepper. Cut the stalks off the mushrooms flush with the underside of the caps. Wipe them with a slightly dampened cloth. Leave whole. If using very large horse mushrooms, cut into thick slices. Beat the eggs lightly in a bowl with a pinch of salt and some freshly ground pepper. Heat plenty of oil in a deep-fryer to 350°F. Dip the mushrooms into the egg, drain briefly, and then roll in the breadcrumbs, coating them all over. Deep-fry a few at a time for 2–3 minutes, until the coating is crisp. Keep hot while you fry the rest.

———— • ————

MUSHROOM AND HERB OMELET

1 lb chanterelles (see method)

2 shallots or ½ Bermuda onion, finely chopped

2 tbsp butter

1½ tbsp oil

1½ tbsp chopped summer savory or tarragon

1½ tbsp chopped parsley

6 very fresh eggs

¼ cup grated Parmesan cheese

salt and pepper

If you cannot obtain chanterelles, cultivated mushrooms will be a perfectly acceptable substitute for this omelet. Heat the butter and oil in a very wide, preferably nonstick skillet and cook the shallot or Bermuda onion over very low heat until soft, stirring frequently. Turn up the heat, add the mushrooms, and sauté for about 6 minutes or until tender and most of the moisture they release when cooked has evaporated. Just before they are done, season with a pinch of salt, some freshly ground pepper, and stir in the herbs.

Beat the eggs lightly in a bowl with a pinch of salt and pepper; beat in the Parmesan cheese, for which you can substitute another finely grated hard cheese if desired, provided it is not too strong. Pour this mixture into the skillet, tipping this way and that to ensure that the mixture spreads out evenly over the bottom. Reduce the heat a little and cook for 2–3 minutes until the underside has set. Run a spatula round the edges and tip the uncooked, runny egg toward the sides; fold over before sliding out of the skillet. Serve at once.

———— • ————

POLENTA WITH MUSHROOM TOPPING

1–1¼ lb mixed edible wild and cultivated mushrooms

2 large shallots or ½ Bermuda onion, finely chopped

6 tbsp light olive oil

3 tbsp finely chopped parsley

1½ tbsp all-purpose flour

1 cup chicken stock (see page 37)

3 tbsp grated hard cheese

2 tbsp butter

salt and pepper

For the polenta:

3 cups yellow polenta or cornmeal

2 quarts or 8 cups water

salt

Make the polenta: heat the water to boiling with 1½ tbsp salt in the top of a double boiler over moderate heat. Then place this over the bottom of the double boiler containing boiling water before adding the cornmeal. Sprinkle in the cornmeal or

pour in a small, steady stream while stirring continuously with a wooden spoon. Keep stirring as the mixture thickens. If you use a double boiler, you can leave it to cook, stirring at intervals; it can take as long as 40 minutes to cook thoroughly if very coarse. Alternatively, buy a quick-cook packet of polenta and stir continuously until it is cooked (5–10 minutes). Cover and keep hot over the hot water.

Trim and wipe the mushrooms. Separate the stalks from the caps and cut the latter into fairly thick slices. Cut the larger stalks lengthwise in half. Heat the oil in a very large, fairly deep skillet and cook the shallot gently while stirring for 5 minutes. Stir in the parsley and cook for a few seconds. Turn up the heat and add the mushrooms.

Sauté the mushrooms over high heat for 2–3 minutes, sprinkle with the flour, stir well, and continue stirring as you add the stock and allow to come to a boil. Reduce the heat to very low, cover, and simmer gently for 5–10 minutes, stirring occasionally. Remove from the heat and stir in the solid butter and the hard cheese until the butter has melted. Serve the piping hot polenta on heated plates, spooning a topping of the mushrooms onto each serving.

———— • ————

FRIED BREADED MUSHROOMS

4 very large horse mushrooms (wild or cultivated) or wild parasol mushrooms

2 eggs

all-purpose flour

breadcrumbs

butter

olive oil

salt and pepper

Garnish with:

lemon wedges

sprigs of parsley

Trim and clean the mushrooms. Handle the more delicate varieties of mushrooms gently as you prepare them, to avoid damaging or breaking them. Beat the eggs lightly in a bowl with a pinch of salt and pepper. Dredge the mushrooms with flour, shaking off excess, then dip in the beaten eggs, drain briefly, and coat all over with the breadcrumbs (use fine dried or stale, fresh breadcrumbs).

Heat 1-1/2 tbsp butter and 3 tbsp oil in a large, nonstick skillet; The fat and oil will be hot enough when the butter has stopped foaming. Fry the mushrooms one at a time for about 3 minutes on each side, until the coating is crisp and golden brown. Drain on paper towels and keep hot, uncovered, in the oven while you fry the rest. Add the same quantity of butter and oil to the saucepan, waiting for it to reach the correct heat before frying the next mushroom.

—— • ——

MUSHROOM RISSOLES

1–1¼ lb chanterelles, field mushrooms or cultivated mushrooms
1 sprig thyme
1 clove garlic
2 whole eggs
¼ cup freshly grated Parmesan or other hard cheese
6 tbsp fine fresh breadcrumbs
3 tbsp finely chopped parsley
light olive oil
salt and freshly ground pepper
Garnish with:
sprigs of parsley
tomato rosebuds (see page 27)
Serve with:
Fresh Tomato Sauce (see page 258)
steamed asparagus or broccoli

Make the tomato sauce; and set aside while you prepare the rissoles. Trim and wipe or rinse and dry the mushrooms; cut them in fairly large pieces and place in a large, nonstick skillet with 3 tbsp oil, the thyme, and the whole peeled garlic clove, lightly crushed with the flat of a heavy knife. Sprinkle with a little salt and pepper, cover tightly, and cook over very low heat for about 20 minutes or until the mushrooms are very tender. Uncover and cook over high heat so that all the moisture evaporates. Remove and discard the thyme and garlic. Chop the mushrooms finely with a knife or in the food processor (do not overprocess).

Drain off any remaining liquid from the mushrooms and mix them in a bowl with the lightly beaten eggs, a little more salt and pepper, the grated cheese, breadcrumbs, a very little minced garlic if desired, and the parsley. Blend into a dense, even consistency that will hold its shape, adding more breadcrumbs if necessary. Divide the mixture into even portions and shape these into flattened rounds about 2 inches in diameter. Heat sufficient oil in a deep-fryer until it reaches a temperature of 350°F and deep-fry the rissoles 2 or 3 at a time until they are crisp and golden brown on the outside. Alternatively, shallow fry them, turning once. Drain well and keep hot on a paper towel while you finish frying the rest. Serve without delay, garnished with parsley sprigs and tomato rosebuds or garnish of your choice, accompanied by the fresh tomato sauce and steamed asparagus or broccoli.

—— • ——

BAKED TRUFFLED EGGS WITH POLENTA

1 quantity polenta (see page 262)
salt and pepper
8 very fresh eggs
2 tbsp butter

1 small fresh or canned truffle or canned truffle pieces or peelings

Make the polenta and keep hot. Divide the butter in half and melt one piece in each of two small nonstick skillets over very low heat. Break 4 eggs carefully into each pan or dish so that they sit neatly beside one another. Cook gently for about 2–3 minutes. Season the whites with a little salt and pepper when they have just set.

Spoon the piping hot polenta onto heated plates; carefully slide 2 eggs onto each polenta portion. Sprinkle with the truffle, sliced wafer-thin.

——— • ———

MUSHROOMS BORDEAUX STYLE

¾ lb cultivated mushrooms

3 tbsp light olive oil

2 tbsp butter

3 shallots, finely chopped

¼ cup fine fresh breadcrumbs

3 tbsp chopped parsley

1½ tbsp chopped chives

salt and freshly ground pepper

Trim and wipe the mushrooms and cut each mushroom in quarters. Heat the oil and butter in a large skillet or shallow saucepan until the butter has stopped foaming. Add the mushrooms and sauté over moderate heat for 3–4 minutes. Season with salt and pepper. Add the shallots and the breadcrumbs, stir, and continue frying for a further 2–3 minutes. Stir in the parsley and chives, and serve.

——— • ———

MOREL MUSHROOMS WITH ASPARAGUS AND MOUSSELINE SAUCE

*24 fresh or dried soaked morel mushrooms (*morchella elata *or* morchella esculenta*)*

16 large, asparagus spears

1½ tbsp butter

¼ cup dry white wine

1 cup mousseline sauce (see page 107)

salt

Serve with:

Cauliflower Mold (see page 83) or poached fish

If you feel uncertain about gathering your own morels (the edible type is readily distinguishable from the false morel which should not be eaten), buy dried ones and soak them in warm water for 20–30 minutes before using. Stir them around in the soaking water and rinse briefly to make sure that every grain of grit has been extracted from their heads. Fresh morels have more flavor and should be brushed meticulously with a soft, dry brush, not washed. This recipe can also be prepared with ordinary field mushrooms or fresh cultivated mushrooms.

Prepare the asparagus (see page 17) and boil in salted water for 12 minutes, or steam, until tender. Drain and keep hot. Trim and clean the mushrooms. Separate the stalks from the heads or caps, cut the stalks into thick sections, and cut the caps into quarters. Cook gently in the butter for about 8 minutes, sprinkling with a little salt and pepper. Sprinkle the wine over them and continue cooking for a minute or two, or until they are tender. Keep hot while making the mousseline sauce. Work quickly once this is ready: arrange the asparagus spears on heated plates so that 2 pairs of spears bisect one another, distribute the morels in the 4 empty sections left on each plate by this arrangement, and spoon in a little sauce. Serve as a vegetable accompaniment or on its own as an appetizer.

·MENUS FOR ENTERTAINING·

WINTER SUPPER

ENDIVE, FENNEL, AND PEAR
SALAD

—•—

ONION SOUP AU GRATIN

—•—

STUFFED LETTUCE

—•—

RÖSTI POTATOES

—•—

FLOATING ISLANDS

—•—

Brandy Coffee

With a little advance planning, entertaining a few friends with a convivial winter supper need not entail a lot of work. You can prepare a large proportion of this menu the day before and avoid spending much of the evening in the kitchen. These are easy dishes, made with everyday ingredients you are likely to have in your refrigerator or cupboard.

The meal starts with Endive, Fennel and Pear Salad, which is simplicity itself and takes only a few moments to put together. It then becomes more sustaining, with steaming hot onion soup served with large, cheese-covered bread slices crisped and browned under the broiler. The rösti potatoes will satisfy the heartiest appetites. They can be prepared in advance and cooked shortly before you plan to eat them and do not need too much skill or concentration. Stuffed lettuces and tomato sauce go very well with the rösti, providing a contrast in taste and texture.

The vanilla custard for the floating islands dessert can be made the day before, and the meringue cooked several hours in advance. Finally, a deliciously strong hot brandy coffee will bring the meal to an agreeable conclusion.

1	2	3	4

STUFFED LETTUCE	p.73
ONION SOUP AU GRATIN	p.174
ENDIVE, FENNEL, AND PEAR SALAD	p.54
RÖSTI POTATOES	p.195

FLOATING ISLANDS

Preparation: 35 minutes
Cooking: 30 minutes
Dessert
12 egg whites
1½ cups superfine sugar
½ lb white sugared almonds or praline, processed to a coarse powder
For the vanilla sauce:
4½ cups milk
12 egg yolks
1 cup granulated sugar
1 tbsp vanilla extract
unsalted butter

Preheat the oven to 350°F. Beat the egg whites with 1/2 cup of the sugar until they are very stiff, then continue beating while gradually adding the remaining sugar a little at a time; the meringue will look shiny and glossy. Fold the processed sugared almonds or praline into the meringue, mixing gently but thoroughly. Lightly grease the inside of 2 very large charlotte molds or 4 smaller molds; sprinkle the inside with extra granulated sugar, tipping out excess, and fill with the meringue. Cover loosely with foil, place the charlotte molds in a roasting pan and add boiling water to come about a third of the way up the sides of the molds. Cook in the oven for 30 minutes, then take the molds out of the water and leave to cool. If preferred, poach batches of 4 or 5 small mounds (2–3 heaping tablespoonfuls) of the meringue in barely simmering milk; remove with a slotted spoon and drain on paper towels.

While the meringues are cooking, make the vanilla custard sauce on which they will float. Add the vanilla extract to the milk, stir well, and heat very slowly until almost boiling. Beat the egg yolks with the sugar until very pale and creamy; beat in 1 tsp cornstarch or potato flour if desired to stabilize the mixture and help prevent the eggs from scrambling. Gradually add the hot milk in a thin stream, beating continuously. Pour this custard mixture into the top of a double boiler or into a heavy-bottomed saucepan and cook over very gentle heat, stirring continuously until it thickens; do not let the custard come close to boiling or it will spoil. When you draw your finger down the coated back of the spoon, the trace should remain completely clear and distinct. This custard sauce, known as *crème anglaise*, is of a fairly thick, pouring consistency. Remove from heat.

Place the top of the double boiler, or the saucepan, in a large bowl of cold water to which some ice cubes have been added. Take care that the water only comes about 3/4 of the way up the sides of the receptacle. Continue stirring until the sauce has completely cooled; this prevents a skin from forming. Turn the meringue soufflés out into a wide shallow dish and surround with the custard sauce. If you have poached small "islands" of meringue, pour the sauce into the dish first, then place them on the surface. Serves 8–10.

BRANDY COFFEE

Preparation: 5 minutes

After-dinner drink

2 cups boiling hot freshly made coffee

8 jiggers Italian grappa *or* marc

2 jiggers Cognac

1 dash anisette

½ cup light cream

¼ cup superfine sugar

1 small piece orange or lemon rind

Place the freshly made coffee in a saucepan, add all the other ingredients, stir, and heat to just below boiling point. Serve in warmed punch glasses with handles.

Floating Islands

269

RUSSIAN NEW YEAR'S SUPPER

RUSSIAN CARROT AND GREEN
APPLE SALAD

⎯ • ⎯

CARROT BREAD (BABKA)

⎯ • ⎯

BAKED POTATOES WITH
CAVIAR AND SOUR CREAM

⎯ • ⎯

RUSSIAN BEET AND CABBAGE
SOUP (BORSCHT)

⎯ • ⎯

JERUSALEM ARTICHOKE VOL-
AU-VENTS RUSSIAN-STYLE

⎯ • ⎯

DAME BLANCHE

⎯ • ⎯

Chilled vodka or dry white wine

Champagne

Black Russian

A New Year's celebration at home with family and a few old friends it is an opportunity to include some indulgent, high-calorie dishes as a treat before the following day's new year resolutions. Serve this leisurely meal by soft candlelight to create a relaxed atmosphere. Begin with a healthy fresh salad of carrots and green apples accompanied by a sweet carrot bread. Follow this restrained beginning with sumptuous baked potatoes served with sour cream and the best Beluga caviar, if you are feeling extravagant, or with the more humble but equally delicious black lumpfish roe. Chilled vodka, served in tiny glasses, is the obvious accompaniment to this course but many may prefer a full-bodied dry white wine. Borscht is a robust and sustaining soup that provides a light and refreshing interlude before the novel flavor of meltingly light Jerusalem artichokes vol-au-vents.

The last course in this Russian feast is a delicate almond dessert with a subtle and interesting flavor and texture. By the time everyone has enjoyed this light and snowy mousse, it will probably be time to open the best champagne and begin toasting the New Year.

1   2   3   4   5

CARROT BREAD (BABKA)   p.161

BAKED POTATOES WITH CAVIAR AND SOUR CREAM   p.166

RUSSIAN BEET AND CABBAGE SOUP (BORSCHT)   p.171

JERUSALEM ARTICHOKE VOL-AU-VENTS RUSSIAN STYLE   p.179

RUSSIAN CARROT AND GREEN APPLE SALAD   p.167

DAME BLANCHE

Preparation: 15 minutes + 2 hours to chill and set

Cooking time: 5 minutes

Easy

Dessert

1½ cups blanched, peeled whole almonds

1 cup milk

½ cup granulated sugar

2½ envelopes gelatin powder

2¼ cups whipping cream

Decorate with:

redcurrant jelly

toasted slivered almonds

fresh or crystallized red soft fruit

Chop the almonds coarsely in the food processor, transfer to a bowl, and add 1¾ cups warm water. Leave to stand for 15 minutes. Line a sieve with a clean cloth or large piece of cheesecloth and strain the almonds and liquid; gather up the edges of the cloth and twist round to squeeze out all the moisture from the almonds.

Heat the milk slowly with the sugar to the boiling point in a small saucepan, stirring continuously to dissolve the sugar completely. Remove from the heat and while still very hot but not boiling, sprinkle in the gelatin powder; when it has completely dissolved, stir in the almond milk. Strain the mixture into a large mixing bowl through a fine sieve and leave to cool completely. Beat the cream until stiff and fold into the almond mixture.

Ladle the mixture into crystal or frosted glass coupes or small bowls and chill in the refrigerator for at least 2 hours, until set.

Using a pastry bag fitted with a fine-gage plain tip, pipe redcurrant jelly over the surface of each serving in concentric circles or in a design of your choice. Sprinkle with a few toasted slivered almonds just before serving and place some fresh or crystallized red fruit in the center. Serves 8.

BLACK RUSSIAN

After-dinner cocktail

2 parts vodka

1 part Kahlúa or coffee-flavored brandy

Allow 3 tbsp vodka and 2 tbsp Kahlúa per person. Place some ice cubes in a cocktail glass. Mix the vodka and Kahlúa in a cocktail shaker and pour into the glass.

Dame Blanche

273

NEW YEAR'S BRUNCH

CELERY AND GREEN APPLE
JUICE

———•———

POPOVERS

———•———

SWEET POTATO PANCAKES
WITH HORSERADISH AND
APPLE SAUCE

———•———

MUNG DHAL FRITTERS

———•———

PUMPKIN AND AMARETTI
CAKE

———•———

English breakfast tea

Colombian Coffee

Pomegranate Fizz

An informal New Year's Day brunch is a very enjoyable way to entertain friends and relations. Those who stayed up until the early hours will be delighted to get a late start and have both their breakfast and lunch prepared for them. Those who find an elaborate New Year's Eve dinner rather daunting can return hospitality simply and successfully with this most flexible and versatile of meals. The menu is as simple or grand as you choose; this one begins by replacing the everyday orange juice with an alcohol-free cocktail of celery, carrot, and apple juice, followed by hot, feather-light popovers with butter and a choice of honey, maple syrup, preserves or marmalade. Prepare a strong, full-bodied English breakfast tea or Colombian coffee for those who like; others may prefer champagne. Mythology and folklore associate the pomegranate with good fortune, so a mixture of champagne and fresh pomegranate seeds will look beautiful and be the perfect drink with which to toast the new year and wish everyone luck. Sweet Potato Pancakes are served with horseradish and green apple relish to lend a sharp, invigorating contrast to their bland, sweet flavor. In some countries it is traditional to serve lentils on New Year's Day to guarantee prosperity, so they are included here in the guise of spicy fritters, accompanied by cooling and refreshing coriander chutney. After this, everyone will be ready for something sweet and the Pumpkin and Amaretti Cake will make a delicious, light, yet sustaining end to the meal.

1   2   3   4

SWEET POTATO PANCAKES WITH HORSERADISH AND APPLE SAUCE	p.188
CELERY AND GREEN APPLE JUICE	p.49
MUNG DHAL FRITTERS	p.229
POPOVERS	p.227
PUMPKIN AND AMARETTI CAKE	p.152

POPOVERS

| Preparation: 15 minutes |
| Cooking: 35 minutes |
| Easy |
| 1 cup all-purpose flour |
| 1 cup milk |
| 1½ tbsp melted butter |
| 2 eggs |
| generous pinch salt |

Serve with:

butter

honey or maple syrup

your choice of jams or preserves

The above quantities will make 10–12 popovers. If you do not have popover pans (like small, squat dariole molds), use muffin pans. Grease the inside of each pan generously with extra butter. Preheat the oven to 450°F. Have all the ingredients at room temperature before you start work.

Sift the flour and salt together into a large mixing bowl; gradually stir in the milk, butter, and lightly beaten eggs to make a pouring batter. Fill each small pan half-full with batter and place in the oven. After 15 minutes, turn down the oven temperature to 350°F and bake for a another 20 minutes or until browned and firm. Prick each popover with a fork or tip of a sharp knife a few minutes before removing from the oven to allow steam to escape.

The popovers can be kept waiting for up to 30 minutes if the oven is turned off and the door left slightly ajar but they will be crisper and drier than if you serve them immediately. Take them out of the pans and serve while they are very hot, with butter, honey or maple syrup, jam or marmalade.

———•———

POMEGRANATE FIZZ

1–2 bottles champagne, chilled

1–2 ripe pomegranates

Take all the fleshy seeds out of the pomegranates, place 1 heaping tbsp of them in a chilled stem glass, pour in the champagne, and serve immediately.

Pumpkin and amaretti cake

ENGLISH AFTERNOON TEA

HAM AND WATERCRESS
SANDWICHES

—— • ——

VEGETABLE SAMOSAS

—— • ——

RHUBARB CAKE

—— • ——

CARROT CAKE

—— • ——

CHESTNUT CAKE

—— • ——

ITALIAN EASTER PIE

—— • ——

Spiced Indian Tea

Orange Pekoe, Earl Grey,
Lapsang Souchong

*U*ntil fairly recently, afternoon tea had been out of fashion but now it is starting to stage a revival. Whether it is tea on the lawn between 4 and 5 o'clock in the middle of a lazy summer Sunday afternoon or high tea at about 6 o'clock, the evening equivalent of brunch with everyone sitting round the table provides the perfect occasion for a get-together. Everything can be prepared well in advance, leaving only the tea to be brewed at the last minute. You can entertain as many or as few people as you like, invite just two or three friends, or plan a large tea party, or even a garden tea party.

Included in the menu will be a choice of teas: Orange Pekoe, Earl Grey or Lapsang Souchong, and traditional items such as cucumber sandwiches. The rhubarb cake will be a popular choice, agreeably moist and not at all sickly sweet, while the chestnut cake will be an unfamiliar and unusual treat for most of your guests. The children might be persuaded to try a delicious homemade milk shake with some ham and cress sandwiches and carrot cake. For a high tea, or for those who are very hungry and enjoy a savory dish at any time of day, Easter Pie, accompanied by a cup of Queen Mary's tea with its delicate muscatel flavor, will be a great treat. For those who enjoy a taste of India, samosas, crisp packages of pastry enclosing spicy peas and potato, or sandwiches with a filling of mildly curried chopped egg mayonnaise can be enjoyed with a glass of spiced tea to add an exotic touch.

| 1 | 2 | 3 | 4 | 5 |

CARROT CAKE	p.199
ITALIAN EASTER PIE	p.50
HAM AND WATERCRESS SANDWICHES	p.49
VEGETABLE SAMOSAS	p.210
RHUBARB CAKE	p.96

CHESTNUT CAKE

Preparation: 15 minutes

Cooking: 30 minutes

Easy

2¼ cups chestnut flour

2¼ cups milk

½ cup granulated sugar

2 tbsp pine nuts

1 tbsp fresh rosemary leaves

½ cup light olive oil

pinch salt

Preheat the oven to 400°F. Brush the inside of 2 shallow cake pans about 10in in diameter lightly with olive oil. Sift the chestnut flour (available from Italian delicatessens and other specialist stores) into a large mixing bowl and gradually stir in the milk, mixed with an equal volume of cold water. Add a large pinch of salt and the sugar, stir well, and then beat in the olive oil gradually with a balloon whisk. Spread out half the mixture in each prepared cake pan, smoothing it level with a spatula. Sprinkle the pine nuts, rosemary leaves, and a very little olive oil evenly over the surface.

Bake in the oven for about 30 minutes or until the surface has turned golden brown and a skewer or knife pushed into the cake comes out clean and dry. Leave to cool for about 10 minutes, then loosen by running the spatula round the inside of the pans and turn out onto a rack to cool. Yields 16 portions.

—— • ——

SPICED INDIAN TEA

Preparation: 5 minutes + 15 minutes' simmering

10 cardamom pods

2 cinnamon sticks

10 cloves

3½ tbsp sugar

1¼ cups milk

¼ cup Ceylon or Assam tea

Make a small slit in the tough skin of the cardamom pods so that they will release their flavor but not the seeds enclosed inside. Place the cardamoms, cinnamon sticks, and cloves in a saucepan, add 6¼ cups cold water, bring to a boil, cover, and simmer gently for 10 minutes. Add the sugar and milk and return to a boil; remove from the heat and immediately add to the tea in a warmed teapot. Leave to stand for 2 minutes and then pour through a fine strainer. Serve at once. Serves 6.

Chestnut cake

MARDI-GRAS BUFFET SUPPER

SUN-DRIED TOMATOES IN OIL
WITH BREAD AND CHEESE

———•———

CRUDITÉS WITH SESAME DIP

———•———

MUSHROOM, RICE, AND
TRUFFLE SALAD WITH SHERRY
MAYONNAISE

———•———

CHICKEN AND ENDIVE
GALETTE

———•———

PASTA CAPRI

———•———

MUSHROOM AND SPINACH
STRUDEL WITH RED WINE
SAUCE

———•———

GLOBE ARTICHOKES ROMAN
STYLE

———•———

STRAWBERRY BAVAROIS WITH
KIWI FRUIT SAUCE

———•———

SOFT NOUGAT COOKIES

B*efore the days of efficient food storage, long before canning or deep-freezing techniques had been developed, food supplies would be monotonous or scarce by the end of the winter. In order to cheer people up when they knew that the enforced abstinence and fasting of Lent was about to begin, a carnival with feasting and dancing became traditional on Shrove Tuesday (Mardi Gras). What better excuse to invite friends for a buffet supper and provide them with an enjoyable reminder that Easter and spring are only forty days away.*

Nearly all the preparations can be completed the day before or at least several hours ahead, with just a few finishing touches left until the last minute. The cauliflower timbale or mold can be made a few hours in advance and reheated in a bain-marie. *The galette can be made the day before, then heated and browned just before serving, while the mushroom and spinach strudel with ham sauce can be made well in advance and frozen, then thawed on the day, and reheated. To budget for twenty-four people, for example, the quantities given in the recipes for the crudités and for the artichokes can be multipled by five or six, while quantities given in the recipes for the three main course dishes can be multiplied by three or four, the dessert recipes on the opposite page are for twenty-four.*

After serving your own choice of wines or non-alcoholic drinks with the main course, you may choose to serve liqueurs or sweet dessert wines with the strawberry molded cream mousse and kiwi fruit sauce, and the nougat cookies. Although deep-frozen stawberries and raspberries are widely available, you can always adapt the cream mousse recipe for less expensive fruits that are available in winter and use dried apricots that have been soaked, cooked until tender, then drained and puréed.

1 2   3   4   4   4 5 6   7 8

CHICKEN AND CHICORY GALETTE — p.72

MUSHROOM, RICE, AND TRUFFLE SALAD WITH SHERRY MAYONNAISE — p.250

PASTA CAPRI — p.123

SUN-DRIED TOMATOES IN OIL WITH BREAD AND CHEESE — p.105

CRUDITÉS WITH SESAME DIP — p.162

CAULIFLOWER TIMBALE — p.83

MUSHROOM AND SPINACH STRUDEL WITH RED WINE SAUCE — p.258

GLOBE ARTICHOKES ROMAN STYLE — p.128

STRAWBERRY BAVAROIS MOUSSE WITH KIWI FRUIT SAUCE

Dessert

2¼–2½ lb hulled strawberries or alternative fruit

4½ cups confectioner's sugar

5 envelopes gelatin powder

6 tbsp citron or orange flower water

5½ cups whipping cream

For the kiwi fruit sauce:

1¾ cups ripe kiwi flesh

1 cup superfine sugar

½ lemon

Thaw the frozen fruit slowly, put it in the food processor or blender, and reduce to a purée. Pour into a large mixing bowl and stir in the confectioner's sugar. Pour the citron or orange flower water and an equal volume of hot water into a bowl set over gently simmering water; sprinkle in the gelatin powder and stir until it has completely dissolved. Remove from the heat and stir into the purée. When this mixture

Strawberry Bavarois with Kiwi Fruit Sauce

is just beginning to thicken but is not at all set, beat the chilled cream until stiff and fold in.

Rinse three 4½-cup turban molds with cold water, but do not dry. Fill with the fruit cream, smooth the surface level and chill for at least 3 hours. Blend the kiwi flesh with the sugar and lemon juice in a food processor. Unmold and serve with the kiwi sauce. Serves 24.

·

SOFT NOUGAT COOKIES

Dessert

1¾ cups unsalted butter

⅓ cup clear, runny honey

1½ cups granulated sugar

½ cup candied mixed fruit, chopped

3⅔ cups flaked almonds

½ cup light cream, warmed

few drops vanilla extract

½ lb frozen puff pastry, thawed

Preheat the oven to 400°F. Place the butter, honey, and sugar in a heavy-bottomed saucepan and cook over moderately low heat for 10 minutes, by which time the sugar should have dissolved and the mixture darkened to a pale golden brown. Choose moist candied fruit, preferably a mixture of orange and citron peel, cherries, and angelica and chop fairly finely. Stir in the almonds and fruit. Continue cooking for another 2–3 minutes. Stir the vanilla into the cream and warm it in a small saucepan, bring to a boil quickly, then add to the fruit and honey mixture. Stir well and remove from the heat.

Roll out the puff pastry into a thin sheet and use it to line a shallow cookie tray measuring about 16 × 12 in. Prick with the tines of a fork and bake "blind" for 10 minutes, then take out of the oven and fill with the nougat mixture. Return to the oven for another 10 minutes' baking. Take out of the oven and cool slightly before cutting into diamond shapes. Serves 20.

VEGETARIAN SUPPER FOR TWO

SESAME STRAWS

•

AIDA SALAD

•

CURRIED CREAM OF
CAULIFLOWER SOUP

•

EGGPLANT TIMBALES WITH
FRESH MINT SAUCE

•

MACEDOINE OF VEGETABLES
WITH THYME

•

Chris Evert

Stinger

hat better than a light, delicately-flavored vegetarian supper for a romantic meal? Start with a heavenly blue cocktail, named after tennis star Chris Evert, of vodka and Curaçao, served with sesame puff pastry straws. Then the supper itself begins with a first course of Aida salad, an appetizing prelude to a creamy cauliflower soup with a hint of curry. The main course consists of eggplant timbales with a contrasting sharp, fresh mint sauce and a very special macedoine of vegetables with thyme. Then comes the frozen Curaçao mousse, a soft, light and creamy confection that melts in the mouth, served with a delicious vanilla sauce. Finally, after-supper cocktails, Stingers made with brandy and crème de menthe, conclude the meal.

1   2   3   4

CURRIED CREAM OF
CAULIFLOWER SOUP   p.62

EGGPLANT TIMBALES
WITH FRESH MINT
SAUCE   p.134

MACEDOINE OF
VEGETABLES WITH
THYME   p.196

AIDA SALAD   p.53

FROZEN CURAÇAO MOUSSE

Preparation: 15 minutes + 2 hours' freezing time

Dessert

1¾ cups chilled whipping cream

½ cup granulated sugar

4 very fresh egg yolks

½ cup blue Curaçao

Serve with:

vanilla sauce (see pages 268–269)

Beat the cream until stiff. To make the basic syrup, place the sugar in a heavy-bottomed saucepan with 1/2 cup water, bring to a boil, and simmer for 2 minutes. Remove from the heat and gradually add the lightly beaten egg yolks, beating continuously with a whisk as you do so. Keep beating until the mixture is completely cold. You may prefer to do this in the bowl of a food processor fitted with the whisk or egg-beater attachment. Stir the Curaçao into the cream and then fold into the syrup, mixing gently but thoroughly. Fill small timbale molds, ramekin dishes or your chosen ice cream molds with this mixture and place in the freezer for at least 2 hours. Just before serving, dip the molds in hot water for a couple of seconds and turn the frozen mousses out onto small, chilled plates. Hand round the vanilla sauce separately at the table. Serves 6.

—— • ——

SESAME STRAWS

Preparation: 15 minutes
Cooking: 8 minutes
Cocktail eats

¼ lb frozen puff pastry, thawed
1 egg
3–4 tbsp sesame seeds
salt and freshly ground pepper

Preheat the oven to 400°F. Roll out the pastry on a lightly floured pastry board with a floured rolling pin to a thickness of about ⅛ in. Beat the egg lightly in a small bowl with a pinch of salt and some freshly ground pepper and coat the entire surface of the pastry thinly with it, using a pastry brush. Sprinkle with the sesame seeds. Cut the pastry into long strips about ¾ in wide, then cut these into 2-in lengths. Arrange them carefully on a greased cookie sheet, egg glaze and sesame seeds uppermost. Bake for 8–10 minutes or until they have puffed up and are lightly browned.

CHRIS EVERT

Cocktail

5 tbsp vodka
2 tbsp blue Curaçao
2 tbsp fresh lemon juice
Decorate with
maraschino cherries and small mint leaves

Mix the vodka, Curaçao, and lemon juice in a cocktail shaker or any tightly sealed container and pour into highball glasses containing ice cubes. Place a short straw in each glass and decorate with maraschino cherries and sprigs of mint. Serves one.

—— • ——

STINGER

After-dinner cocktail

2 parts brandy
1 part crème de menthe

Allow 3 tbsp brandy and 2 tbsp crème de menthe for each cocktail. Mix in the shaker and pour into liqueur glasses.

Frozen Curaçao Mousse

INDIAN DINNER

DEEP-FRIED OKRA

•

INDIAN SPICED CABBAGE AND POTATOES

•

EGGPLANT AND MINT RAITA

•

PEAS AND FRESH CURD CHEESE INDIAN STYLE

•

RED LENTILS WITH RICE INDIAN STYLE

•

INDIAN PEA SOUP

•

SAFFRON MILK JELLY WITH ROSE PETALS

•

Bombay

Mango Fizz

Darjeeling Tea

There is a seemingly inexhaustible choice of Indian vegetarian dishes; this menu combines a selection that can be prepared in advance. Ideally the okra fritters should be fried at the last minute but they can also be successfully reheated. Serve with the Bombay cocktail, or with a delicious Mango Fizz.

We begin with spiced cabbage with potatoes. A raita *made with yoghurt, eggplant, and mint acts as a cooling side salad with the hotter dishes. Indian fresh curd cheese* (paneer) *and peas, and the delicious lentils with cumin and coriander provide the main protein content of the meal, accompanied by chapatis. Any of the many Indian recipes in this book can be substituted for the dishes on this menu, or added to them. Usually a choice of two drinks will be enough: a light beer or* lassi *a refreshing yoghurt drink that goes well with curries. A dish of fluffy steamed Basmati rice, grown in the north of the Indian subcontinent, is served as an accompaniment to the main curry dishes and conveniently soaks up liquid from the moister recipes. The delicately spiced green pea soup can be served after the main dishes, refreshing the palate and rounding off the main part of the meal. Then comes a dessert worthy of the Moghul emperors, saffron jelly, decorated with pistachio nuts and rose petals. Serve halva as well (see recipe on page 199) if you wish to offer a choice of desserts.*

Although some of your guests may prefer coffee, tea is really the better choice after Indian food and Darjeeling tea, picked on the hills below the Himalayas, will make the perfect end to the meal.

PEAS AND FRESH CURD CHEESE INDIAN STYLE	p.222
DEEP-FRIED OKRA	p.115
INDIAN SPICED CABBAGE AND POTATOES	p.81
RED LENTILS WITH RICE INDIAN STYLE	p.230
INDIAN PEA SOUP	p.213
EGGPLANT AND MINT RAITA	p.147

SAFFRON JELLY WITH ROSE PETALS
Kesari barfi

Preparation: 10 minutes + 2 hours for chilling and setting
Cooking: 5 minutes
Dessert

½ tsp cardamom seeds

generous pinch saffron threads

½ cup granulated sugar

3½ cups milk

2⅔ envelopes gelatin powder

3 tbsp flaked or slivered almonds

3 tbsp flaked or slivered pistachio nuts

Decorate with:

yellow rose petals

Pound the cardamom seeds to a powder. Mix the saffron threads with 1½ tbsp hot water in a small bowl, stirring and crushing them until they have dissolved. Bring the sugar and milk slowly to a boil; remove from heat and sprinkle in the gelatin and the ground cardamom. Wait until the gelatin has completely dissolved, give the milk a good stir, then pour through a fine sieve into a mixing bowl and stir in the saffron water. Pour this mixture into individual glass dishes and when cool, refrigerate for at least 2 hours or until set. Sprinkle each serving with slivered or flaked almonds and pistachio nuts (the almonds can be bought ready prepared but you will need to slice the pistachios with a strong sharp knife or chop them very coarsely in the food processor). Decorate with fresh rose petals, rinsed and gently dried at the last moment.

·

BOMBAY

Cocktail

3 tbsp brandy

2 tbsp dry vermouth

2 tbsp sweet red vermouth

¼ tsp Pernod

½ tsp Curaçao

Place some small ice cubes in a cocktail shaker, add all the ingredients, and mix. Serve in a well chilled cocktail glass. Serves one.

MANGO FIZZ

4 tbsp mango juice

4 tbsp orange juice

1 tsp lime juice

chilled champagne, sparkling dry white wine or soda water

Decorate with:

thin lime slices, small sprigs of mint, and maraschino cherries

Have all the ingredients well chilled. Pour the mango, orange, and lime juice into a tall glass; fill up with champagne or dry sparkling white wine. For an alcohol-free drink, substitute soda water for the wine. Serves one.

Saffron Milk Jelly with Rose Petals

PROVENÇAL GARDEN LUNCH

SALADE GOURMANDE
WITH SHERRY VINAIGRETTE

—•—

CHILLED TOMATO SOUP WITH
SCAMPI

—•—

PROVENÇAL STUFFED
VEGETABLES

—•—

APPLE SORBET WITH
CALVADOS

—•—

CHOCOLATE DATE COOKIES

*T*he long-awaited warm weather of
early summer has finally arrived and
you feel like celebrating with lunch in
the garden. You may be thousands of miles away
from the South of France, there may be no
ancient olive tree to cast its dappled shade over
the table and chairs, no scent of jasmine or
sound of cicadas or crickets with their pulsating
call, nevertheless, you can still create your own
Provençal summer idyll, even if the scene is a
roof terrace high above the busy city streets.
Having started with Salade Gourmande, a rich
and luxurious salad of crisp tender green beans
with truffled foie gras and a subtle sherry
vinegar dressing, which is very simple and quick
to prepare, you will want to refresh the palate.
What better than a chilled soup made with ripe,
flavorsome tomatoes and garnished with
scampi? Then onto the enchanting petits farcis,
delicious mouthfuls of small, fresh summer
vegetables stuffed and flavored with Provençal
herbs, baked in the oven until just tender.
This sybaritic meal ends with an apple sorbet
flavored with Calvados, followed by coffee and
chocolate date cookies.

CHILLED TOMATO SOUP WITH SCAMPI	p.117
SALADE GOURMANDE WITH SHERRY VINAIGRETTE	p.52
PROVENÇAL STUFFED VEGETABLES	p.136

APPLE SORBET WITH CALVADOS

Preparation: 40 minutes + 2 hours for freezing
Fairly easy
Dessert
2¼–2½ lb dessert apples
2¼ cups granulated cane sugar
1¼ cups corn syrup
Calvados or good-quality apple brandy
Decorate with:
small mint leaves and maraschino cherries

Try to use dessert apples with a pronounced aromatic flavor or substitute a good, dry cider for half or
less of the water content. Mix the sugar and corn
syrup in a saucepan with 4½ cups cold water and

heat slowly to a temperature of 155°F, stirring with a wooden spoon. Quarter, core, and peel the apples; cut them into small pieces and process to a smooth purée in the food processor. Push this purée through a fine sieve into a large mixing bowl and gradually stir in the sugar syrup. Pour into an ice-cream maker and process for 15 minutes or until very thick and frozen. If you do not have an ice-cream maker, follow the method for Melon Sorbet on page 309.

Transfer to a suitable freezer container with a tight-fitting lid and store in the freezer until 10 minutes before serving. Leave the sorbet at room temperature for these 10 minutes to soften slightly.

To serve, use a small ice cream scoop to shape into balls and place 2 or 3 of these in glass or crystal coupes or small dishes. Sprinkle 1-1/2–3 tbsp chilled Calvados over each serving and decorate with mint leaves and maraschino cherries.

*Apple Sorbet
with Calvados*

CHOCOLATE DATE COOKIES

Preparation: 30 minutes
Cooking: 35 minutes
Easy
Dessert or sweet snack
3 tbsp all-purpose flour
½ cup superfine sugar
1 tsp baking powder
2 eggs
½ lb dates, pitted
½ cup walnuts
½ cup finely grated bitter (unsweetened) chocolate
½ tsp ground cinnamon
3 tbsp demerara sugar or light brown cane sugar
butter

Preheat the oven to 320°F. Grease a shallow cookie tray or jelly roll pan about 9 in wide with butter and dust with flour, tipping out excess. Sift the flour with the sugar and baking powder into a large mixing bowl. Beat the eggs briefly with a whisk. Chop the dates and walnuts together coarsely and mix with the eggs. Stir in the grated chocolate.

Mix the eggs, dates, and nuts with the flour and sugar briefly but thoroughly. Spread the mixture out in the prepared cookie tray, keeping it as level as possible. Mix the cinnamon with the brown sugar and sprinkle evenly over the surface. Bake for about 35 minutes, or until the surface is golden brown. Take out of the oven and leave to stand for 10 minutes before turning out onto a pastry board or chopping board; use a large, strong knife to cut into bars measuring about 4 × 1 in. Serve warm with after-dinner coffee or as a nourishing sweet snack any-time.

MIDDLE EASTERN FEAST

TURKISH STUFFED VINE
LEAVES

—•—

ZUCCHINI FRITTERS

—•—

TSATSIKI

—•—

CARROT AND ONION SALAD
TURKISH STYLE

—•—

TURKISH EGGPLANT DIP

—•—

TURKISH STUFFED PEPPERS

—•—

ALMOND CREAM WITH
POMEGRANATE SEEDS

—•—

Raki

Iced vodka

Southern Anatolian Coffee

*T*his menu can be served as a buffet lunch or supper; the choice of menu is particularly practical if your guests are likely to arrive and leave at different times as none of the dishes will spoil if kept waiting for a while. Once all the preparations are completed, you can relax and talk to all the guests without worrying whether the next course will be overdone or not ready on time. Although some of the dishes are known throughout the Middle East, this is essentially a Turkish meal. It is heavily based on vegetables but you can always add your own favorite meat, poultry or fish recipes from the famous Ottoman culinary repertoire.

Starting with a well stocked bar, you can offer your guests a choice of raki, iced vodka or chilled champagne, and for those who prefer non-alcoholic beverages, homemade lemon or rose sherbet drinks, fruit juice or iced mint tea. The Turkish appetizers included in the menu are equally suitable as accompaniments to drinks or as courses in a meal: vine leaves stuffed with rice and pine nuts, zucchini fritters with tsatsiki. Pieces of hot pita bread are served to scoop up small mouthfuls of the delicately flavorsome eggplant dip; more hot pita bread can be served with the grated carrot salad and the main dishes. Turkey's ancient and sophisticated cuisine is typified by the mouthwatering stuffed peppers, rich in subtle spices and aromas. Other vegetarian dishes can be added to the feast, such as Imam bayildi *a particularly delicious recipe for stuffed eggplant (see page 107) and a good example of the elegant simplicity and inspired combination of natural flavors which are the hallmark of traditional Turkish cooking.*

The almond cream dessert is a suitable end to such a meal, satisfying a longing for something sweet after all the savory delights, and attractive, with deep, delicate pink pomegranate seeds against the ivory color of the almond cream.

1 2 3   4   5 6

TURKISH EGGPLANT DIP	p.108
CARROT AND ONION SALAD TURKISH STYLE	p.162
TSATSIKI	p.149
TURKISH STUFFED PEPPERS	p.124
ZUCCHINI FRITTERS	p.143
TURKISH STUFFED VINE LEAVES	p.53

ALMOND CREAM WITH POMEGRANATE SEEDS

Preparation: 20 minutes + 30 minutes' resting time + 1 hour for chilling

Cooking: 20 minutes

Dessert

9 cups milk

9 cups light cream

3 cups granulated cane sugar

6 cups ground almonds

5⅓ envelopes gelatin powder

2 tsp almond extract

Decorate with:

1 cup unsalted, shelled pistachio nuts

seeds from 6 ripe pomegranates

Place the milk, cream, sugar, and ground almonds in a saucepan and heat very slowly to boiling. Remove from the heat, cover, and leave to stand for 30 minutes. Place a large piece of cheesecloth in a large sieve and pour the mixture through it into a very large mixing bowl, scraping all the almond paste out of the saucepan into the cheesecloth. Gather up the ends of the cheesecloth and twist them round to squeeze out all the moisture into the bowl. Return all the strained liquid to the saucepan; heat to boiling once more, remove from the heat, and sprinkle in the gelatin. Stir as the gelatin melts into the very hot liquid and make sure that it has completely dissolved. If the gelatin does not dissolve completely, reheat the mixture very gently while stirring but do not allow to come anywhere near boiling point. Stir in the almond extract. Strain through a very fine sieve and ladle into coupes or very small bowls. Chill for 2 hours or until set and decorate with pomegranate seeds and the pistachio nuts, cut into thin slivers. Serves 15.

Almond Cream with Pomegranate Seeds

SOUTHERN ANATOLIAN COFFEE

1¼ cups very finely ground dark roast coffee

6 tbsp granulated cane sugar

12 cardamom pods

This recipe serves 12; if you wish to make this type of coffee for fewer or more people, allow 2 tsp of coffee, 1 tsp of sugar, and 1 cardamom per person.

Pour 3½ cups water into a long-handled Turkish coffee pot if you have one, or an enameled saucepan; add the coffee, sugar, and the cardamom pods, having first pierced their seed cases or skins with the tip of a sharp knife. Stir well and heat slowly to boiling point over a gentle heat. As soon as the liquid boils, take the saucepan off the heat and leave to stand for 30 seconds, until the foam on the top subsides. Return to the heat and remove again as soon as the liquid returns to a boil; leave to stand again and then return to the heat once more to return to a boil. Leave to stand for 3 minutes before serving.

GREEK SUMMER LUNCH

GREEK SALAD

•

CHICKPEA CROQUETTES

•

DEEP-FRIED EGGPLANT
TURNOVERS

•

GREEK SPINACH AND FETA
CHEESE PIE

•

STUFFED EGGPLANTS GREEK
STYLE

•

TSATSIKI

•

LEMON AND BASIL SORBET

•

RISÓGALO

•

Ouzo

*T*his is a simple, easy menu that can be served for lunch or supper. Relying on readily available, wholesome ingredients, this healthy and delicious Greek meal shows how much can be done with a limited range of basic foodstuffs. First of all, you may choose to serve ouzo, *the Greek version of anisette, with the usual accompanying glasses of iced water, and* melitzano burakakia, *crisply fried eggplant turnovers with a feta cheese filling – a recipe imported into Greece from Turkey many generations ago.*

As a refreshing start to the main meal, prepare a simple Greek salad, followed by crunchy chick pea fritters, a speciality from the island of Rhodes, which reveals the influence of India via North Africa in the use of cumin and mint, and the yoghurt and cucumber dip, tsatsiki. *The delicately flavored and melting puff pastry pie* spanakopita *with its spinach, dill, and cheese filling completes the selection of appetizers. A substantial main dish of eggplants stuffed with a meat filling and covered with a velvety sauce will leave most people feeling that they have eaten extremely well. Still they will be able to enjoy a fresh lemon and basil sorbet to refresh and cleanse the palate. The dessert, a very digestible rice pudding made with goat's milk and flavored with lemon and cinnamon, is typical of eastern Mediterranean cooking. Retsina can be served throughout the meal, although for some people this resin-flavored wine is very much an acquired taste. Otherwise any young, light, dry wine is suitable.*

STUFFED EGGPLANTS GREEK STYLE	p.133
DEEP-FRIED EGGPLANT TURNOVERS	p.109
GREEK SALAD	p.110
GREEK SPINACH AND FETA CHEESE PIE	p.80
CHICK PEA CROQUETTES	p.229
TSATSIKI	p.149

LEMON AND BASIL SORBET

Preparation: 25 minutes + 2½ hours' freezing

Dessert

1¼ cups superfine sugar

20 medium-sized fresh basil leaves

1¼ cups fresh lemon juice

2 egg whites

Pour 3 cups water into a saucepan, add the sugar, and heat slowly to boiling point; simmer for 5 minutes by which time the sugar should have completely dissolved. Remove this syrup from the heat. Chop the basil leaves coarsely and process in the blender with 3 tbsp cold water to reduce to a paste. Add the lemon juice and the syrup and process very briefly. Pour through a fine sieve into an ice-cream maker and process until frozen. Alternatively, strain into a freezer-proof bowl with a tight-fitting lid. Place the container in the freezer for 1½ hours to freeze and solidify. Take out of the freezer and break up the frozen mixture; as it becomes a little softer, beat vigorously to break up all the large ice crystals and make the mixture smooth, ready for adding the egg white. If you have used an ice-cream maker, there will be no large ice crystals. Fold in the stiffly beaten egg whites, combining thoroughly without crushing too much air out of the mixture. Cover and return to the freezer for 2 hours. Take out of the freezer shortly before serving to soften.

———•———

RISÓGALO

Preparation: 5 minutes

Cooking: 35 minutes

Dessert

¾ cup short-grain rice

6 cups goat's or whole milk

½ cup sugar

1 tsp cornstarch or potato flour (optional)

2 egg yolks

grated rind of 1 lemon

ground cinnamon

Place the rice in a heavy-bottomed saucepan or flameproof casserole dish, add all but ½ cup of the milk, and heat slowly until the milk comes to a boil. Turn down the heat to very low and simmer gently for 25–30 minutes, stirring occasionally. The rice should be tender. Mix the cornstarch or potato flour with the sugar, stir in 6 tbsp of the reserved cold milk very thoroughly, and add to the rice and milk while stirring continuously. Keep stirring as you allow the mixture to come to a gentle boil and then simmer for 2 minutes over very low heat. Add the lemon rind. Beat the egg yolks lightly with the remaining milk and then stir this into the rice pudding, adding a little at a time. Continue stirring without interruption as you cook the rice pudding over the lowest possible heat for about 30 seconds; but do not allow to boil. Spoon into small bowls or glass coupes; cool and serve at room temperature. Sprinkle with cinnamon before serving. Serves 6.

Lemon and Basil Sorbet

JAPANESE DINNER PARTY

CHILI PEPPER AND DAIKON
ROOT SALAD WITH SALMON
ROE

———•———

MUSHROOM AND NOODLE
SALAD WITH SESAME SAUCE

———•———

TEMPURA WITH TENTSUYU
SAUCE

———•———

MISO AND TOFU SOUP

———•———

BEAN SPROUTS WITH PORK
AND STEAMED RICE

———•———

MELON SORBET

———•———

Mikado

Saké

Green Tea

Japanese plum brandy

*N*ow that Japanese food is becoming increasingly popular, many Western cooks are venturing into this intricate, artistic, and very healthful cuisine. Japanese cooking techniques lend themselves most readily to the painstaking cook who likes to devote a lot of time and attention to food preparation for the result can be very rewarding, delicious, and beautiful. This meal typifies the simpler, home-cooked version of Japanese cooking; the more elaborate dishes can be sampled in Japanese restaurants in large cities all over the world.

Early fall, when the chrysanthemums are in bloom, might well be a fitting time to invite friends over for a Japanese dinner party. When they arrive, welcome them with a Mikado cocktail. The appetizer of daikon root and salmon roe will look very appealing. Japanese rice wine or saké, *gently warmed and served in tiny cups, can be served with this, switching to chilled beer or green tea to accompany the rest of the meal.*

The next course, a salad of mushroom and soy bean flour noodles, is a relatively bland prelude to the tempura – fresh tender morsels of vegetables dipped in featherlight batter, deep-fried, and served with tentsuyu sauce. You can leave your guests to chat and drink more saké, beer or green tea while you fry these. If you prefer not to set about cooking while your guests wait, you

could omit the tempura and go straight onto the soy bean sprouts with pork and steamed rice. The main part of the meal will end with Miso and Tofu Soup. After a pause, it is time for the dessert course: a refreshing melon sorbet. The meal ends with small diamond or triangle shaped slices of adzuki jelly accompanied by Japanese plum brandy served in tiny glasses or porcelain cups.

TEMPURA WITH
TENTSUYU SAUCE   p.141

BEAN SPROUTS WITH
PORK AND STEAMED
RICE   p.126

MISO AND TOFU SOUP   p.217

CHILI PEPPER AND
DAIKON ROOT SALAD
WITH SALMON ROE   p.163

MUSHROOM AND
NOODLE SALAD WITH
SESAME SAUCE   p.251

MELON SORBET
Misore no moto

Preparation: 15 minutes + 3 hours' freezing time

Easy

Dessert

1½ cups melon flesh

1 cup granulated cane sugar

Decorate with:

small mint leaves

Choose melons that are ripe and full of flavor, with a full, heady scent. Measure out the peeled flesh and any juice when you have removed the seeds and any fibers attached to them. Process in the blender or food processor with 1½ cups cold water and the sugar for 30 seconds at high speed, until smooth and well blended. If you have an electric ice-cream maker, pour the melon purée into the container and process for 15 minutes, after which the sorbet will be frozen; transfer the container to the freezer. If you do not have an ice-cream maker, pour the purée into a freezerproof container with a tight-fitting lid and place in the freezer in the fast-freeze compartment if you have one; take the bowl out after about 15 minutes and beat for about 1 minute with a rotary whisk or hand-held electric beater. Return to the freezer for another 15 minutes and then repeat the operation. By now the sorbet mixture should be thick. Repeat twice more. This is to prevent large ice crystals from forming which would make the texture grainy. Serve decorated with mint leaves.

MIKADO

Cocktail

½ cup brandy

4 dashes lemon barley water

4 dashes Crème de Noyau

2 dashes Angostura bitters

4 dashes Curaçao

Place in a cocktail shaker with ice (or mix in a jug) and pour into chilled cocktail glasses. Serves 2.

SOUTH AMERICAN THÉ DANSANT

GUACAMOLE

———•———

ICED PUMPKIN SOUP

———•———

MEXICAN PEPPERS WITH
POMEGRANATE SAUCE

———•———

BEANS CREOLE WITH RICE
AND FRIED BANANAS

———•———

FROZEN PEACH MOUSSE WITH
CARAMEL SAUCE

———•———

Cuba Libre

Caipirinha

In the late fall or early winter months, when Christmas seems a long way away, a party with a South-American theme is calculated to cheer everyone up. It could also be an imaginative way of celebrating Halloween or Thanksgiving. If none of your rooms is quite large enough to have dancers stepping out in an accomplished tango, you can always take out your best salsa music and play it as background music to help everyone get into a carefree mood. The menu and recipes on these two pages are intended for sixteen people, while the recipes that come from other sections can simply have their ingredients multipled by four. With the exception of the rice, all the dishes can be prepared well in advance and if you choose to have a buffet, the easiest solution, you can serve all the savory dishes at once. Your guests will probably want to start with the guacamole dip and tortilla chips; then comes the iced pumpkin soup which can be served in a hollowed out half pumpkin to make it even more attractive. Stuffed sweet peppers with a topping of sour cream, coriander, and pomegranate seeds come next, followed by a substantial and filling Creole dish of beans with rice and fried bananas.

The Caipirinha cocktail should make everyone suitably uninhibited to dance or at least sway to the most energetic or sensuous Latin American rhythms while those who prefer a long, thirst-quenching drink can have cuba libre. Chilled beer, a Mexican lager perhaps, or some good Chilean wine would be other options.

MEXICAN PEPPERS WITH POMEGRANATE SAUCE	p.138
BEANS CREOLE WITH RICE AND FRIED BANANAS	p.220
ARGENTINE ICED PUMPKIN SOUP	p.118
GUACAMOLE	p.113

FROZEN PEACH MOUSSE WITH CARAMEL SAUCE

Preparation: 25 minutes + 3 hours for freezing
Cooking: 20 minutes
Easy
Dessert
5 cups superfine sugar
5 cups ripe peach flesh
9 cups whipping cream
1¼ cups crumbled amaretti or almond cookies
For the almond and caramel sauce:

1¼ cups granulated sugar

almond extract

Place the sugar in a heavy-bottomed saucepan with 2½ cups cold water and bring slowly to a boil, stirring occasionally. Simmer for 1 minute, then take off the heat and leave to cool. Pour out exactly 4½ cups of this sugar syrup into a very large mixing bowl. Process the peach flesh to a smooth purée in a processor. Put this purée through a fine sieve and measure out 4½ cups of it. When the syrup is completely cold, stir in the peach purée. Beat the cream until stiff and fold into the peach and syrup mixture together with the finely ground amaretti. Pour into rectangular containers. Smooth the surface level with a knife, cover with waxed paper, and freeze for at least 3 hours before turning out onto a large, chilled serving platter and cutting into slices.

While the mousse is freezing, make the sauce: place the sugar in a heavy-bottomed saucepan with 1 cup water; bring slowly to a boil, stirring now and then with a wooden spoon. Simmer over low heat, stirring frequently, for about 20 minutes or until the syrup has started to caramelize and has turned a light golden brown; draw aside from the heat; add 1 cup cold water, taking care to keep your hands and face out of the way of the sudden burst of hot steam; add 4 drops almond extract. Leave to cool. Serves 24.

CAIPIRINHA

Cocktail

¼ cup light rum

1 small lime

2 tsp cane sugar

Cut the lime in quarters and pound these in a mortar with the sugar using a pestle. Discard the skin, reserve the juice and flavored sugar, and place in a glass with ice cubes and the rum. Mix briefly and serve. Serves one.

CUBA LIBRE

Cocktail

¼ cup light rum

juice of ½ lime

well chilled Coca Cola

Pour the rum into a glass, add the lime juice, and fill up the glass with Coca Cola. Serves one.

Frozen Peach Mousse with Caramel Sauce

Adzuki bean
  Japanese sweet red bean paste 239
  jelly 240
Aida salad 53
Almond croquette potatoes 198
Apulian pizza pie 71
Artichoke, globe
  and egg pie 129
  Arugula, artichoke, and egg salad
    57
  bottoms in cheese sauce 144
  bottoms Venetian style 145
  Braised artichokes 144
  Braised peas, artichokes, and
    lettuce 225
  Casserole of peas and beans with
    artichokes 224
  Castilian cream of artichoke soup
    116
  Crispy globe artichokes 145
  fricassée 129
  Globe artichokes in oil 105
  Globe artichokes Roman style 128
  Potato and artichoke pie 187
  risotto 121
  sauce 77
  Spinach, artichoke, and fennel pie
    77
Arugula
  artichoke and egg salad 57
  Green gnocchi 69
  Pasta and beans with arugula 219
  Ribbon noodles with arugula,
    tomato, and cheese 68
Asparagus
  Bamboo shoot and asparagus
    soup 118
  Cream of asparagus soup 115
  Honey mushrooms with
    asparagus tips and basil sauce
    252
  Japanese asparagus salad 107
  Morel mushrooms with asparagus
    and mousseline sauce 264
  molds 128
  Mushroom and asparagus pie with
    fresh tomato sauce 258
  Pea and asparagus soup 214
  Steamed asparagus with
    mousseline sauce 107
  White asparagus with spicy sauce
    127
  with béarnaise sauce 106
  with eggs and parmesan 127
Asparagus chicory and beans
  Apulian style 87

Avocado
  and chicken salad with blue
    cheese dressing 141
  and shrimp in tuna sauce 111
  Boiled or steamed vegetables with
    Guasacaca sauce 191
  Chicken, tomato, and avocado
    salad with tuna sauce 125
  cream dessert 150
  Guacamole 113
  ice cream 151
  Palm heart, shrimp, and avocado
    salad 125
  soup 119
  tomato, and grapefruit salad with
    citronette dressing 111

Baby turnips sautéed with garlic and
  parsley 194
Baked eggplants 146
Baked potatoes with caviar and sour
  cream 166
Baked sweet potatoes with sour
  cream and coriander 165
Baked tomatoes with fava bean
  mousse 211
Baked tomatoes with mint 146
Baked vegetable omelet 130
Bamboo shoot and asparagus soup
  118
Beans and eggs Mexican style 226
Bean and peanut pilaf 235
Bean and pumpkin risotto 218
Beans Creole with rice and fried
  bananas 220
Bean sprout
  cucumber, and crab salad 105
  with pork and steamed rice 126
Beets
  and celery salad 164
  Iced beet soup 171
  Russian beet and cabbage soup
    171
  Russian beet pâté 164
  tomato, and cucumber salad 195
Boiled or steamed vegetables with
  Guasacaca sauce 191
Borlotti bean
  and endive salad with horseradish
    dressing 212
  bean purée with cream sauce 231
  Pasta and beans with rocket salad
    219
Braised artichokes 144
Braised endive 87
Braised lentils 239

Braised lettuce 85
Braised onions and rice 191
Braised red cabbage with apples 93
Broccoli
  and shrimp terrine 84
  Pasta with broccoli in hot sauce 67
  purée 93
  Steamed broccoli with sauce
    maltaise 84
Brussels sprouts
  with fresh coriander and lime juice
    57
  with hollandaise sauce 85
  with roasted sesame seeds 93
Buttered spinach Indian style 88

Cabbage
  and ham rolls 82
  Braised red cabbage with apples
    93
  Cream of cabbage soup 63
  Indian bread with cabbage
    stuffing 49
  Indian spiced cabbage and
    potatoes 81
  Japanese daikon root and cabbage
    salad 168
  Pickled cabbage soup 62
  Rice with cabbage 94
  Russian beet and cabbage soup
    171
  Swiss chard and cabbage soup 60
Caesar salad with quail's eggs 52
Candied pumpkin 151
Cannellini beans with ham and
  tomato sauce 228
Cantonese stir-fried rice and peas
  233
Caponata 135
Cardoons au gratin 94
Carnival fritters 200
Carrot
  and onion salad Indian style 167
  bread 161
  cake 199
  celery, and fennel salad with
    scampi 163
  Grated carrot Turkish style 162
  Grated carrot salad with Indian
    mustard dressing 167
  Indian carrot dessert 199
  Russian carrot and green apple
    salad 167
  Russian carrot mold 184
  with cream and herbs 192
Casserole of peas and beans with

artichokes 224
Castilian cream of artichoke soup 116
Catalan lobster with peppers 140
Cauliflower
    and macaroni pie 65
    and potatoes Indian style 91
    and rice au gratin 64
    Cream of cauliflower soup 61
    Curried cream of cauliflower soup 62
    Fu-yung 82
    Polonais 83
    salad 92
    timbale 83
Celery
    Beet and celery salad 164
    and green apple juice 49
    au gratin 90
    canapés with Roquefort stuffing 52
    Carrot, celery, and fennel salad with scampi 163
    Japanese style 90
Celery root
    Cream of celery root soup 169
    salad 164
Chapatis 81
Chicken and endive galette 72
Chicken and truffle salad with anchovy dressing 252
Chicken, tomato, and avocado salad with tuna sauce 125
Chickpea
    and spinach soup 217
    Greek chickpea croquettes 229
    Hummus 212
    Tuscan chickpea soup 221
Chicory
    Fried chicory Roman style 86
Chilled leek and potato soup 169
Chilled pumpkin soup 118
Chilled tomato soup with scampi 117
Chinese chicken and mushroom soup 253
Chinese cucumber, pork, and chicken soup 119
    Pe t'sai (celery cabbage)
    Sweet-sour celery cabbage Peking style 92
    Stuffed pe t'sai rolls 51
Chinese watercress soup 58
Country lentil soup 221
Cream of asparagus soup 115
Cream of cabbage soup 63
Cream of cauliflower soup 61
Cream of celery root soup 169

Cream of leek and zucchini soup 172
Cream of parsnip soup with bacon and garlic croutons 170
Cream of potato and leek soup 172
Cream of potato soup with crispy leeks 173
Cream of tomato soup 117
Cream of watercress soup 59
Crêpes or pancakes 36
Crispy globe artichokes 145
Crudités with sesame dip 162
Cucumber
    Bean sprout, cucumber, and crab salad 105
    Beet, tomato, and cucumber salad 195
    Chinese cucumber, pork, and chicken soup 119
    Japanese cucumber and ham salad 112
    Tsatsiki 149
    with salmon roe and hard-boiled eggs 112
Curried cream of cauliflower soup 62
Curried fennel 91

Daikon root
    Chili pepper and daikon root salad with salmon roe 163
    Japanese daikon root and cabbage salad 168
Dandelion and scampi soup 58
Deep-fried eggplant and cheese sandwiches 130
Deep-fried eggplant with miso sauce 147
Deep-fried eggplant with yoghurt 110
Deep-fried eggplant turnovers 109
Deep-fried mushrooms with tartare sauce 261
Deep-fried okra 115
Deep-fried potatoes
Deep-fried black salsify 193
Desserts
    Almond cream with pomegranate seeds 301
    Apple sorbet with calvados 296
    Chestnut cake 281
    Chocolate date cookies 297
    Dame Blanche 273
    Floating islands 268
    Frozen Curaçao mousse 288
    Frozen peach mousse with caramel sauce 312
    Lemon and basil sorbet 305

Melon sorbet 285
    Nougat pastries 285
    Risógalo 305
    Saffron milk jelly with rose petals 292
    Strawberry bavarois with kiwi fruit sauce 285
Drinks
    Bombay 293
    Black Russian 273
    Brandy coffee 269
    Caipirinha 313
    Chris Evert 289
    Cuba libre 313
    Mango fizz 293
    Mikado 309
    Pomegranate fizz 277
    Southern Anatolian coffee 301
    Spiced Indian tea 281
    Stinger 289

Eggplant
    and mint raita 147
    Baked eggplants 146
    Caponata 135
    cutlets with tomato relish 133
    Deep-fried eggplant and cheese sandwiches 130
    Deep-fried eggplant with miso sauce 147
    deep-fried eggplant with yoghurt 110
    Deep-fried eggplant turnovers 109
    Fried eggplants with garlic and basil 148
    Greek eggplant dip 110
    Japanese style 109
    mock pizzas 108
    mousseline 223
    Neapolitan eggplant salad 146
    Neapolitan stuffed eggplants 132
    pie 134
    Spaghetti with eggplant 122
    Stuffed eggplants 131
    Stuffed eggplants Greek style 133
    timbales with fresh mint sauce 134
    tomato, and cheese bake 131
    Turkish eggplant dip 108
    Turkish style 107
Endive
    and scampi salad 53
    Belgian endive and cheese mold with tomato sauce 74
    Braised endive 87
    Chicken and endive galette 72
    fennel and pear salad 54

Broiled endive with cheese and chive dumplings 75
salad with port and raisins 55
Risotto with endive 64
with lemon Flemish style 86
with raspberry vinegar and cream 196
Escarole
Apulian pizza pie 71
Green ravioli with walnut sauce 70
Lettuce and black olive pizza 70
with beans and olive oil 75
with olives 87

Fava bean
asparagus chicory and beans Apulian style 87
Baked tomatoes with fava bean mousse 211
Casserole of peas and beans with artichokes 224
with cream 232
Fennel
and lettuce au gratin 76
au gratin 90
Carrot, celery, and fennel salad with scampi 163
Curried fennel 91
Endive, fennel, and pear salad 54
orange and walnut salad 55
Spinach, artichoke, and fennel pie 77
spinach, and beef salad 78
Friar's beard chicory
with garlic and chili 89
with lemon dressing 89
Fried breaded mushrooms 262
Fried chicory Roman style 86
Fried eggplants with garlic and basil 148
Fried rice with mushrooms 254
Fried stuffed puris 209

Garlic
Provençal garlic soup 174
Spaghetti with tomatoes, garlic and chili 122
Gazpacho 120
Genoese seafood and vegetable salad platter 177
Globe artichokes in oil 105
Globe artichokes Roman style 128
Grated carrot salad with Indian mustard dressing 167
Greek eggplant dip 110
Greek chickpea croquettes 229

Greek potato fricadelles 189
Greek salad 110
Greek spinach and feta cheese pie 80
Green beans with tomato sauce 238
Green gnocchi 69
Green ravioli with walnut sauce 70
Green salad with Roquefort dressing 54
Grilled chicory with cheese and chive dumplings 75
Guacamole 113

Ham and cress sandwiches 49
Haricot bean
soup 62
Tuscan bean soup 215
Warm Tuscan bean salad 226
Herb and rice soup with pesto 61
Honey mushrooms with asparagus tips and basil dressing 252
Hot mushroom molds with fresh tomato sauce 260
Hot mushroom terrine with celery root and tomato sauce 249
Hummus 212
Hungarian zucchini with dill sauce 142
Hungarian potato salad 166

Iced beet soup 171
Iced Hawaiian potato soup 175
Indian bread with cabbage stuffing 49
Indian carrot dessert 199
Indian pea soup 213
Indian spiced cabbage and potatoes 81
Italian Easter pie 50

Japanese asparagus salad 107
Japanese cucumber and ham salad 112
Japanese daikon root and cabbage salad 168
Japanese mixed salad with *sambai zu* dressing 168
Japanese spinach and egg soup 60
Japanese stir-fried rice with green peas and mixed vegetables 232
Japanese sweet red bean paste 239
Jerusalem artichoke
Jerusalem artichoke salad 166
Jerusalem artichoke vol-au-vents Russian style 179
Jerusalem artichokes with cream sauce 198

Mushroom and Jerusalem artichoke salad 250

Leek
and cheese pie 183
Chilled leek and potato soup 169
Cream of leek and zucchini soup 172
Cream of potato and leek soup 172
Turkish rice with leeks 192
Lentil
Braised lentils 239
Country lentil soup 221
Red lentils with rice Indian style 230
Spiced lentils and spinach 238
Lettuce
and black olive pizza 70
and pea soup 58
Braised lettuce 85
Braised green peas, artichokes, and lettuce 225
Stir-fried lettuce with oyster sauce 86
Stuffed lettuce 73
Stuffed lettuce rolls 68
Lima bean
and pumpkin soup 216
Bean and pumpkin risotto 88

Macedoine of vegetables with thyme 196
Mexican beans with garlic and coriander 231
Mexican peppers with pomegranate sauce 138
Milanese eggs and peas 223
Minestrone 216
Miso and tofu soup 217
Mung dhal fritters 229
Mushroom
and asparagus pie with fresh tomato sauce 258
and cheese puff pastry pie 259
and herb omelet 261
and Jerusalem artichoke salad 250
and noodle salad with sesame sauce 251
and Parmesan salad 249
and spinach strudel with red wine sauce 258
and truffle salad 249
Bordeaux style 264
Chinese chicken and mushroom soup 253
Deep-fried mushrooms with

tartare sauce 261
Fried breaded mushrooms 262
Fried rice with mushrooms 254
Honey mushrooms with
    asparagus tips and basil
    dressing 252
Hot mushroom molds with fresh
    tomato sauce 260
Hot mushroom terrine with celery
    root and tomato sauce 249
Morel mushrooms with asparagus
    and mousseline sauce 264
Mushroom and potato bake 257
Pasta with mushrooms 255
Pasta with mushrooms and pine
    nuts 256
Polenta with mushroom topping
    262
Potato gnocchi with mushroom
    sauce 257
rice, and truffle salad with sherry
    mayonnaise 250
rissoles 263
risotto 254
salad with Swiss cheese and
    celery 249
soup 253

Neapolitan potato cake 186
Neapolitan stuffed eggplants 132
Nettle risotto 64
Nettle soup 61
North African vegetable casserole
    228

Okra
    Deep-fried 115
Onion
    and bacon pancakes 180
    Braised onions and rice 191
    Carrot and onion salad Indian
        style 167
    Pissaladière 176
    salad with cold meat 161
    soup 173
    soup au gratin 174
    Spanish omelet 179
    Sweet-sour baby onions 192
    with sharp dressing 161

Pakoras 209
Palm heart, shrimp, and avocado
    salad 125
Pan-fried potatoes, peppers, and beef
    186
Parsnip

Cream of parsnip soup with bacon
    and garlic croutons 170
Pasta and beans with arugula 219
Pasta and potatoes 176
Pasta Capri 123
Pasta with broccoli in hot sauce 67
Pasta with mushrooms 255
Pasta with mushrooms and pine nuts
    256
Pasta with peas and Parma ham 219
Pasta with spinach and anchovies 66
Pasta with turnip tops 89
Pea
    and asparagus soup 214
    and fresh curd cheese Indian style
        222
    and wild rice timbales 234
    Braised green peas, artichokes,
        and lettuce 225
    Cantonese stir-fried rice and peas
        233
    Casserole of peas and beans with
        artichokes 224
    Indian pea soup 213
    Milanese eggs and peas 223
    green pea molds with eggplant
        mousseline 223
    Pasta with peas and Parma ham
        219
    purée 233
    Venetian rice and pea soup 214
    with bacon and basil 234
Pepper
    au gratin 148
    Lobster Catalan 140
    Mexican peppers with
        pomegranate sauce 138
    Pan-fried potatoes, peppers, and
        beef 186
    Peperonata 137
    Stir-fried peppers in sharp sauce
        149
    Stuffed baked pepper and anchovy
        rolls 111
    Stuffed peppers 138
    Turkish stuffed peppers 124
    zucchini, and tomato fricassée 139
Persian rhubarb frappé 95
Persian spinach with yoghurt
    dressing 56
Persian stuffed zucchini 143
Persian vegetable and egg timbale
    182
Philadelphia pepper-pot 170
Pickled cabbage soup 62
Pili-pili tomato salad 113

Pissaladière 176
Pizza dough 70–71
Poached turnip tops 89
Polenta with mushroom topping 262
Popovers 277
Potato
    Almond croquette potatoes 198
    and artichoke pie 187
    baked with caviar and sour cream
        166
    Carnival fritters 200
    Cauliflower and potatoes Indian
        style 91
    Chilled leek and potato soup 169
    Cream of potato and leek soup 172
    Cream of potato soup with crispy
        leeks 173
    Deep-fried potatoes 197
    dumplings with basil sauce 177
    gnocchi with mushroom sauce
        257
    gratin 197
    Greek potato fricadelles 189
    Hungarian potato salad 166
    Iced Hawaiian potato soup 175
    Indian spiced cabbage and
        potatoes 81
    mousseline 197
    Mushroom and potato bake 257
    Neapolitan potato cake 186
    Pan-fried potatoes, peppers, and
        beef 186
    Pasta and potatoes 176
    Rösti potatoes 195
    Rösti potatoes with shallots 196
    salad with scallions and chives
        169
    soufflé with artichoke sauce 187
    Spanish omelet 179
    String bean and potato pie 227
    String bean and potato salad with
        mint 237
    Turkish potato salad 165
    Vegetable and cheese bake 187
    Vegetables with potato stuffing
        189
Provençal garlic soup 174
Provençal pancakes 180
Provençal stuffed vegetables 136
Pumpkin
    and Amaretti cake 152
    Argentine iced pumpkin soup 118
    Bean and pumpkin risotto 218
    Butter bean and pumpkin soup
        216
    Candied pumpkin 151

soup with Amaretti 118
Turkish pumpkin dessert 152

Radish
    and cress salad in cream dressing
        163
    Radishes with cream, coriander,
        and pomegranate sauce 194
Ratatouille 137
Red kidney bean
    Beans and eggs Mexican style
        226
    Mexican beans with garlic and
        coriander 231
Red lentils with rice Indian style 230
Rhubarb
    cake 96
    Persian rhubarb frappé 95
Ribbon noodles with arugula,
    tomato, and cheese 68
Rice with cabbage 94
Risotto with endive 64
Risotto with zucchini flowers 120
Rösti potatoes 195
Rösti potatoes with shallots 196
Russian beet and cabbage soup 171
Russian carrot and green apple salad
    167
Russian carrot mold 184
Russian spinach and sorrel soup
    with smetana 59

Salads
    Aida salad, 53
    Arugula, artichoke, and egg salad
        57
    Avocado and chicken salad with
        blue cheese dressing 141
    Avocado, tomato, and grapefruit
        salad with citronette dressing
        111
    Bean sprout, cucumber, and crab
        salad 105
    Beet and celery salad 164
    Beet, tomato, and cucumber salad
        195
    Borlotti bean and endive salad
        with horseradish dressing 212
    Caesar salad with quail's eggs 52
    Carrot and onion salad Indian
        style 167
    Carrot, celery, and fennel salad
        with scampi 163
    Cauliflower salad 92
    Celery root salad 164
    Chicken, tomato, and avocado

salad with tuna sauce 125
Chili pepper and daikon root salad
    with salmon roe 163
Eggplant salad 146
Endive and scampi salad 53
Endive, fennel, and pear salad 54
Endive salad with port and raisins
    55
Fennel, orange, and walnut salad
    55
Fennel, spinach, and beef salad 78
Genoese seafood and vegetable
    salad platter 177
Grated carrot salad with Indian
    mustard dressing 167
Greek salad 110
Green salad with Roquefort
    dressing 54
Hungarian potato salad 166
Japanese asparagus salad 107
Japanese cucumber and ham salad
    112
Japanese daikon root and cabbage
    salad 168
Japanese mixed salad 168
Jerusalem artichoke salad 166
Mushroom and Jerusalem
    artichoke salad 250
Mushroom and noodle salad with
    sesame sauce 251
Mushroom and Parmesan salad
    249
Mushroom and truffle salad 249
Mushroom, rice, and truffle salad
    with sherry mayonnaise 250
Mushroom salad with Swiss
    cheese and celery 249
Onion salad with cold meat 161
Palm heart, shrimp, and avocado
    salad 125
Pili-pili tomato salad 113
Potato salad with scallions and
    chives 169
Radish and cress salad in cream
    dressing 163
Russian carrot and green apple
    salad 167
Salad appetizer with pomegranate
    seeds and foie gras 50
Salade Gourmande with sherry
    vinaigrette 52
Spinach salad with raisins and
    pine nuts 55
String bean and potato salad with
    mint 237
Turkish potato salad 165

Warm Tuscan bean salad 226
Salsify, white (oyster plant)
    Polish style 185
Salsify, black (scorzonera)
    Black salsify and fish soup 174
Savory bread stuffing 114
    Deep-fried 193
    Greek style 185
Sauces, dressings, and stocks
    Anchovy dressing 252
    Anchovy-flavored citronette
        dressing 52
    Artichoke sauce 77
    Basil dressing 252
    Béarnaise sauce 106
    Béchamel sauce 36
    Beef stock 37
    Blanc de cuisine 165
    Chicken stock 37
    Citronette 35
    Cheese sauce 144
    Clarified butter (ghee) 36
    Coriander chutney 211
    Coconut chutney 209
    Court-bouillon 37
    Cream dressing 163
    Creamy velouté sauce 72
    Dill sauce 142
    Eggplant mousseline 223
    Fish stock (fumet) 38
    Fresh mint sauce 134
    Fresh tomato sauce 258
    Guasacaca sauce 191
    Hollandaise sauce 85
    Horseradish and apple sauce 188
    Japanese dressing 107
    Kimi-zu dressing 168
    Mayonnaise 35
    Mediterranean sauce 122
    Miso sauce 147
    Mousseline sauce 107
    Mustard-flavored citronette
        dressing 78
    Oyster sauce 86
    Pili-pili hot relish 113
    Pesto 177
    Pomegranate sauce 139
    Red wine sauce 259
    Roquefort dressing 54
    Sambai zu dressing 106
    Sauce maltaise 84
    Sesame dip 162
    Sesame sauce 127
    Shellfish stock 38
    Sherry mayonnaise 251
    Sherry vinaigrette 52

Smetana 59
Spicy ketchup 206
Spicy sauce 127
Sweet-sour sauce 92
Tartare sauce 261
Tentsuyu sauce 141
Tomato relish 133
Tomato sauce 74
Tuna sauce 125
Vegetable and herb sauce 69
Vegetable stock 38
Velouté sauce 35
Vinaigrette 35
Walnut citronette dressing 53
Walnut sauce 66
Scallion
   Scrambled eggs with scallions Indian style 182
   Spaghetti with scallion, tomato, and herb sauce 175
Scrambled eggs with scallions Indian style 182
Sesame straws 289
Shallot
   Rösti potatoes with shallots 196
   Snow peas with shallots and cream 238
   Spinach with shallots 88
Snow peas
   Snow peas and shrimp Cantonese 225
   in cream and basil sauce 237
   stir-fried 237
   with shallots and bacon 238
Spaghetti with eggplant 122
Spaghetti with zucchini flowers, cream, and saffron 123
Spaghetti with Mediterranean sauce 122
Spaghetti with scallion, tomato, and herb sauce 175
Spaghetti with tomatoes, garlic, and chili 122
Spanish omelet 179
Spiced lentils and spinach 238
Spiced vegetables in creamy sauce 181
Spinach
   and cheese dumplings with sage butter 67
   and rice Indian style 88
   artichoke, and fennel pie 77
   Buttered spinach Indian style 88
   Chickpea and spinach soup 217
   Fennel, spinach, and beef salad 78
   Greek spinach and feta cheese pie 80
   Italian Easter pie 50
   Japanese spinach and egg soup 60
   Japanese style 56
   molds 56
   Mushroom and spinach strudel with red wine sauce 258
   noodles with walnut sauce 66
   Pasta with spinach and anchovies 66
   Persian spinach salad with yoghurt dressing 56
   Russian spinach and sorrel soup with smetana 59
   salad with raisins and pine nuts 55
   soufflé 78
   Spiced lentils and spinach 238
   stir-fry with pork, eggs, and ginger 79
   Sweet spinach pie 95
   with shallots 88
Split green pea soup 213
Springtime risotto 218
Spring vegetable soup 115
Steamed asparagus with mousseline sauce 107
Steamed broccoli with sauce maltaise 84
Steamed rice 126
Steamed vegetables with curry sauce 183
Stir-fried zucchini with parsley, mint, and garlic 149
Stir-fried lettuce with oyster sauce 86
Stir-fried peppers in sharp sauce 149
Stir-fried snow peas 237
String bean
   and potato pie 227
   and potato salad with mint 237
   maître d'hotel 235
   with crispy bacon 236
   with tomatoes and sage 236
Stuffed eggplants 131
Stuffed eggplants Greek style 133
Stuffed baked pepper and anchovy rolls 111
Stuffed Chinese leaf rolls 51
Stuffed lettuce 73
Stuffed lettuce rolls 68
Stuffed peppers 138
Stuffed tomatoes au gratin 114
Stuffed vegetable selection 190
Sun-dried tomatoes in oil with bread and cheese 105
Sweet potato
   Baked sweet potatoes with sour cream and coriander 165
   cream Indian style 200
   pancakes with horseradish and apple sauce 188
Sweet-sour baby onions 192
Sweet spinach pie 95
Swiss chard
   and cabbage soup 60
   au gratin 76
   Italian Easter pie 50
Tagliatelle with truffle 255
Tempura with Tentsuyu sauce 141
Tomato
   au gratin 146
   Baked tomatoes with fava bean mousse 211
   Baked tomatoes with mint 146
   Chicken, tomato, and avocado salad with tuna sauce 125
   Chilled tomato soup with scampi 117
   Cream of tomato soup 117
   Eggplant, tomato, and cheese bake 131
   Fresh tomato sauce 258
   Gazpacho 120
   Pepper, zucchini, and tomato fricassée 139
   Pili-pili tomato salad 113
   Ribbon noodles with arugula, tomato, and cheese sauce 68
   sauce 74
   Spaghetti with scallion, tomato, and herb sauce 175
   Spaghetti with tomatoes, and chili 122
   Stuffed tomatoes au gratin 114
   Sun-dried tomatoes in oil with bread and cheese 105
Tsatsiki 149
Turkish eggplant dip 108
Turkish potato salad 165
Turkish pumpkin dessert 152
Turkish rice with leeks 192
Turkish stuffed peppers 124
Turkish stuffed vine leaves 53
Turnip
   and rice purée 193
   Baby turnips sautéed with garlic and parsley 194
   Pasta with turnip tops 65
   Poached turnip tops 89
Tuscan bean soup 62
Tuscan chickpea soup 221
Truffle

Baked truffled eggs with polenta 263
Chicken and truffle salad with anchovy dressing 252
Mushroom and truffle salad 249
Mushroom, rice, and truffle salad with sherry mayonnaise 250
risotto 255
Tagliatelle with truffle 255

Vegetable and cheese bake 187
Vegetable samosas 210
Vegetables with potato stuffing 189
Venetian rice and pea soup 214
Vine leaf

Turkish stuffed vine leaves 53

Watercress
Chinese watercress soup 58
Cream of watercress soup 59
Ham and cress sandwiches 49
Radish and cress salad in cream dressing 163
Warm Tuscan bean salad 226
White asparagus with spicy sauce 127

Zucchini
Cream of leek and zucchini soup 172

Escabeche 114
fritters 143
Pepper, zucchini, and tomato fricassée 139
Persian stuffed zucchini 143
purée 150
Risotto with zucchini flowers 120
Russian style 150
soup with basil and lemon 120
Spaghetti with zucchini flowers, cream, and saffron 123
Stir-fried zucchini with parsley, mint, and garlic 149
with dill sauce 142